30
Topics in Heterocyclic Chemistry

Series Editor: Bert U.W. Maes

Topics in Heterocyclic Chemistry

Series Editor: Bert U.W. Maes

Recently Published and Forthcoming Volumes

β-Lactams: Unique Structures of Distinction for Novel Molecules
Volume Editor: Bimal K. Banik
Volume 30, 2013

Metalation of Azoles and Related Five-Membered Ring Heterocycles
Volume Editor: Gordon W. Gribble
Volume 29, 2012

Click Triazoles
Volume Editor: J. Košmrlj
Volume 28, 2012

Halogenated Heterocycles
Volume Editor: J. Iskra
Volume 27, 2012

Heterocyclic Scaffolds II: Reactions and Applications of Indoles
Volume Editor: G.W. Gribble
Volume 26, 2010

Synthesis of Heterocycles via Multicomponent Reactions II
Volume Editors: R.V.A. Orru, E. Ruijter
Volume 25, 2010

Anion Recognition in Supramolecular Chemistry
Volume Editors: P.A. Gale, W. Dehaen
Volume 24, 2010

Synthesis of Heterocycles via Multicomponent Reactions I
Volume Editors: R.V.A. Orru, E. Ruijter
Volume 23, 2010

Heterocyclic Scaffolds I: β-Lactams
Volume Editor: B. Banik
Volume 22, 2010

Phosphorous Heterocycles II
Volume Editor: R.K. Bansal
Volume 21, 2009

Phosphorous Heterocycles I
Volume Editor: R.K. Bansal
Volume 20, 2009

Aromaticity in Heterocyclic Compounds
Volume Editors: T. Krygowski, M. Cyrański
Volume 19, 2009

Heterocyclic Supramolecules I
Volume Editor: K. Matsumoto
Volume 17, 2008

Bioactive Heterocycles VI
Flavonoids and Anthocyanins in Plants, and Latest Bioactive Heterocycles I
Volume Editor: N. Motohashi
Volume 15, 2008

Heterocyclic Polymethine Dyes
Synthesis, Properties and Applications
Volume Editor: L. Strekowski
Volume 14, 2008

Synthesis of Heterocycles via Cycloadditions II
Volume Editor: A. Hassner
Volume 13, 2008

Synthesis of Heterocycles via Cycloadditions I
Volume Editor: A. Hassner
Volume 12, 2008

Bioactive Heterocycles V
Volume Editor: M.T.H. Khan
Volume 11, 2007

Bioactive Heterocycles IV
Volume Editor: M.T.H. Khan
Volume 10, 2007

Bioactive Heterocycles III
Volume Editor: M.T.H. Khan
Volume 9, 2007

Bioactive Heterocycles II
Volume Editor: S. Eguchi
Volume 8, 2007

β-Lactams: Unique Structures of Distinction for Novel Molecules

Volume Editor:
Bimal K. Banik

With contributions by

B.K. Banik • I. Banik • B. Basu • I. Fernández • B. Mandal • I. Ojima • J.D. Seitz • M.A. Sierra • T.T. Tidwell • E. Turos • E.S. Zuniga

Springer

The series *Topics in Heterocyclic Chemistry* presents critical reviews on "Heterocyclic Compounds" within topic-related volumes dealing with all aspects such as synthesis, reaction mechanisms, structure complexity, properties, reactivity, stability, fundamental and theoretical studies, biology, biomedical studies, pharmacological aspects, applications in material sciences, etc. Metabolism will also be included which will provide information useful in designing pharmacologically active agents. Pathways involving destruction of heterocyclic rings will also be dealt with so that synthesis of specifically functionalized non-heterocyclic molecules can be designed.
The overall scope is to cover topics dealing with most of the areas of current trends in heterocyclic chemistry which will suit to a larger heterocyclic community.
As a rule, contributions are specially commissioned. The editors and publishers will, however, always be pleased to receive suggestions and supplementary information. Papers are accepted for *Topics in Heterocyclic Chemistry* in English.
In references, *Topics in Heterocyclic Chemistry* is abbreviated *Top Heterocycl Chem* and is cited as a journal.

Springer www home page: springer.com
Visit the THC content at springerlink.com

Topics in Heterocyclic Chemistry ISSN 1861-9282
ISBN 978-3-642-33187-9 ISBN 978-3-642-33188-6 (eBook)
DOI 10.1007/978-3-642-33188-6
Springer Heidelberg New York Dordrecht London

Library of Congress Control Number: 2012948393

Printed on acid-free paper

Springer is part of Springer Science+Business Media (www.springer.com)

Topics in Heterocyclic Chemistry Also Available Electronically

Topics in Heterocyclic Chemistry is included in Springer's eBook package *Chemistry and Materials Science*. If a library does not opt for the whole package the book series may be bought on a subscription basis. Also, all back volumes are available electronically.

For all customers who have a standing order to the print version of *Topics in Heterocyclic Chemistry*, we offer free access to the electronic volumes of the Series published in the current year via SpringerLink.

If you do not have access, you can still view the table of contents of each volume and the abstract of each article by going to the SpringerLink homepage, clicking on "Chemistry and Materials Science," under Subject Collection, then "Book Series," under Content Type and finally by selecting *Topics in Heterocyclic Chemistry*.

You will find information about the

- Editorial Board
- Aims and Scope
- Instructions for Authors
- Sample Contribution

at springer.com using the search function by typing in *Topics in Heterocyclic Chemistry*.

Aims and Scope

The series *Topics in Heterocyclic Chemistry* presents critical reviews on present and future trends in the research of heterocyclic compounds. Overall the scope is to cover topics dealing with all areas within heterocyclic chemistry, both experimental and theoretical, of interest to the general heterocyclic chemistry community. The series consists of topic related volumes edited by renowned editors with contributions of experts in the field.

Chapters in volumes are cited like journal articles.

Example:
Top Heterocycl Chem (2012) 27: 1–32
DOI 10.1007/7081_2011_63
© Springer-Verlag Berlin Heidelberg 2011
Published online: 23 September 2011

Contents

Top Heterocycl Chem (2013) 30: 1–64
DOI: 10.1007/7081_2012_86

Published online: 7 August 2012

Advances in the Use of Enantiopure β-Lactams for the Synthesis of Biologically Active Compounds of Medicinal Interests

Iwao Ojima, Edison S. Zuniga, and Joshua D. Seitz

Abstract β-Lactams play a significant role in organic synthesis in addition to its importance as the core structure of β-lactam antibiotics. The "β-lactam synthon method" introduced in late 1970s has greatly advanced the use of β-lactams as key intermediates for the synthesis of biologically active compounds such as nonprotein amino acids, peptides, peptidomimetics, and complex natural products and congeners. This chapter describes the advances in the synthesis of β-lactams with excellent enantiopurity, useful patterns of β-lactam ring cleavage, ring-opening coupling, and applications of the β-lactam synthon method to the synthesis of biologically active compounds of medicinal interests. In addition, novel β-lactams, exhibiting potent activities, not as antibacterials but as anticancer and cholesterol-controlling agents, are described.

Keywords β-lactam · β-lactam synthon method · Asymmetric synthesis · Cycloaddition · Cyclocondensation · Ring-opening · Amino acid · Peptide · Petidomimetics · Paclitaxel · Taxoid · Cryptophycin · Lankacidin · Combretastatin · Ezetimibe

Contents

I. Ojima (✉), E.S. Zuniga and J.D. Seitz
Institute of Chemical Biology & Drug Discovery, Stony Brook University, Stony Brook, NY, USA

Department of Chemistry, Stony Brook University, Stony Brook, NY, USA
e-mail: iojima@notes.cc.sunysb.edu

1 Introduction

The β-lactam skeleton has attracted significant interest among synthetic and medicinal chemists over the years mainly because it is the core structure of natural and synthetic β-lactam antibiotics [1]. However, β-lactams have been playing an important role in organic synthesis as well. The use of β-lactams as synthons for compounds with biological and medicinal interests has been greatly advanced since the introduction of the "β-lactam synthon method" in late 1970s [2–12]. Through the β-lactam synthon method, various β-lactams have been used as precursors to nonprotein amino acids and amino alcohols, which are essential building blocks for the synthesis of peptides and peptidomimetics. Furthermore, the β-lactam synthon method has been applied to the synthesis of a variety of medicinally important compounds such as indolizidine and isoquinoline alkaloids, taxoids, and other complex natural products.

The synthesis of these biologically important and stereochemically defined compounds can be facilitated by chemical transformations of enantiopure β-lactams (Fig. 1). β-Lactams with high enantiopurity can be obtained through asymmetric ester enolate–imine cyclocondensation, diastereoselective Staudinger ketene–imine cycloaddition reaction, followed by enzymatic resolution of enantiomers or asymmetric Staudinger [2+2] ketene–imine reaction. The chirality of these chiral β-lactams can be transferred directly to an array of synthetic intermediates, which are useful for the synthesis of enantiopure compounds of biological and medicinal significance.

Once β-lactams with high enantiopurity are obtained, any of the C–C and N–C bonds of the four-membered ring can be cleaved to yield useful chiral synthetic intermediates. The ring strain of the β-lactam ring promotes bond cleavage. The most widely exploited bond cleavage is that of the N1–C2 amide bond. This bond is susceptible to nucleophilic agents. For example, the nucleophilic cleavage of the N1–C2 amide bond of *N*-acyl-3-hydroxy-β-lactam derivatives by alcohols,

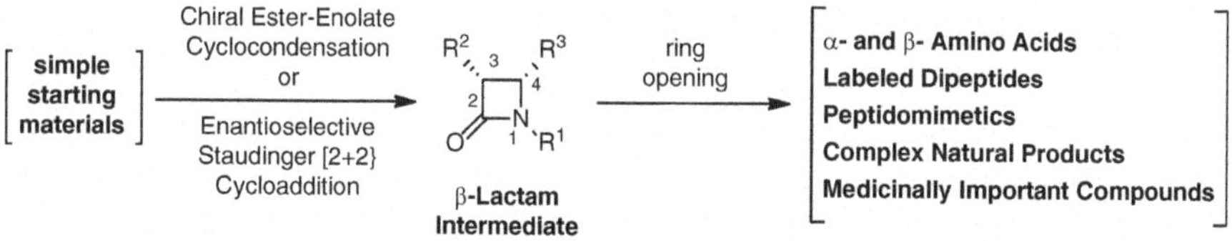

Fig. 1 β-Lactam Synthon Method

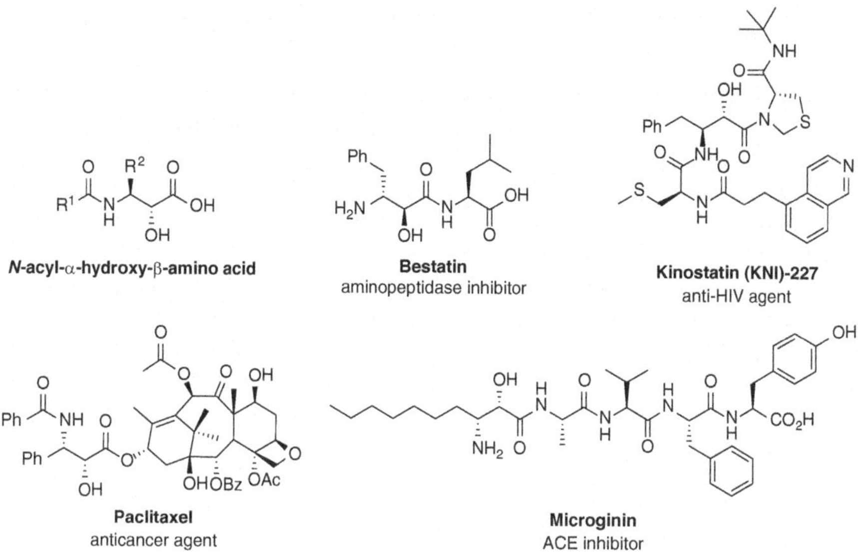

Fig. 2 Examples of biologically active compounds of medicinal interests, containing α-hydroxy-β-amino acid components

amino acids, and thiazoles provides the corresponding *N*-acyl-α-hydroxy-β-amino acid esters, amides, and dipeptides, including the highly efficient introduction of (3*R*,4*S*)-*N*-benzoylphenylisoserine moiety to a baccatin III derivative, forming paclitaxel, one of the most important anticancer drugs, after simple deprotection (Fig. 2).

The β-lactam synthon method has been successfully applied to (1) the synthesis of biologically active oligopeptides, (2) the synthesis and stereoselective labeling of dipeptides, (3) the asymmetric synthesis of peptidomimetics, (4) the synthesis of antimicrobial agents and complex natural products, and (5) the highly efficient synthesis of paclitaxel and new taxoids that are highly potent anticancer agents (Fig. 2). This chapter describes the advances in the asymmetric synthesis of β-lactams and highlights the applications of the β-lactam synthon method to the synthesis of biologically active compounds of medicinal interests. In addition, novel β-lactams, exhibiting potent activities, not as antibacterials but as anticancer and cholesterol-controlling agents, are described.

2 Asymmetric Synthesis of β-Lactams

Selected methods for the asymmetric synthesis of β-lactams are discussed in this section. β-Lactams with high enantiomeric excess can be obtained by means of two major methods, i.e., (1) Staudinger [2+2] ketene–imine cycloaddition and (2) ester enolate–imine cyclocondensation.

2.1 Staudinger [2+2] Ketene–Imine Cycloaddition Reaction

The first synthesis of β-lactams was reported by Staudinger in 1907, in which a Schiff base derived from aniline and benzaldehyde underwent thermal cycloaddition with diphenylketene to form 1,3,3,4-tetraphenylazetidin-2-one [13]. Since its introduction, the Staudinger ketene–imine cycloaddition has been regarded as the most direct and widely used method for the synthesis of β-lactams. Although the Staudinger reaction has been documented for over a century, uncertainties still remain, even now, regarding the details of its [2+2] cycloaddition mechanism. Exhaustive discussions on the factors influencing the mechanism of the Staudinger reaction are beyond the scope of this chapter. However, two recent reviews on this subject summarize the current standing [14, 15]. The most widely accepted mechanism of the Staudinger ketene–imine cycloaddition involves a two-step process through a nucleophilic attack of the imine nitrogen to the electrophilic central carbon of a ketene (generated in situ from an acid chloride and a base), forming a zwitterionic intermediate, which undergoes conrotatory ring closure to give the four-membered cycloadduct, as shown in Scheme 1 (pathway I). When a monosubstituted ketene and an aldimine are used, the reaction produces two stereocenters in the β-lactam ring, hence *cis*- and/or *trans*-β-lactam products can be obtained exclusively or as a mixture in a variable ratio, depending on the substrate structures and reaction conditions [14, 15]. If there is no *Z* to *E* isomerization in the zwitterionic intermediate, the reaction gives *cis*-β-lactam as the exclusive product. Although Scheme 1 only shows *E*-imine, the stereochemistry of the imine surely influences the stereochemistry of the resulting β-lactam, i.e., *E*-imines preferentially lead to the formation of *cis*-β-lactams, while *Z*-imines predominately give *trans*-β-lactams [16–19]. The studies conducted to establish the origin of *cis*/*trans* stereoselection revealed that the relative transition state energy in the rate-determining step is dictated by electronic torquoselectivity [19–21].

Two other pathways have also been proposed for the reaction mixing an acid chloride, a base and an imine all at once (pathways II and III). This reaction involves acylation of the imine nitrogen first, followed by deprotonation at the α-carbon by a base to generate the zwitterionic intermediate same as that in the pathway I (pathway II) or another process wherein the chloride counter ion, generated in the first step, adds to the iminium bond to form the corresponding *N*-α-chloroalkylamide anion, which undergoes cyclization through an intramolecular S_N2 reaction (pathway III), which can accommodate the formation of *trans*-β-lactam as the major product under certain conditions [14, 15].

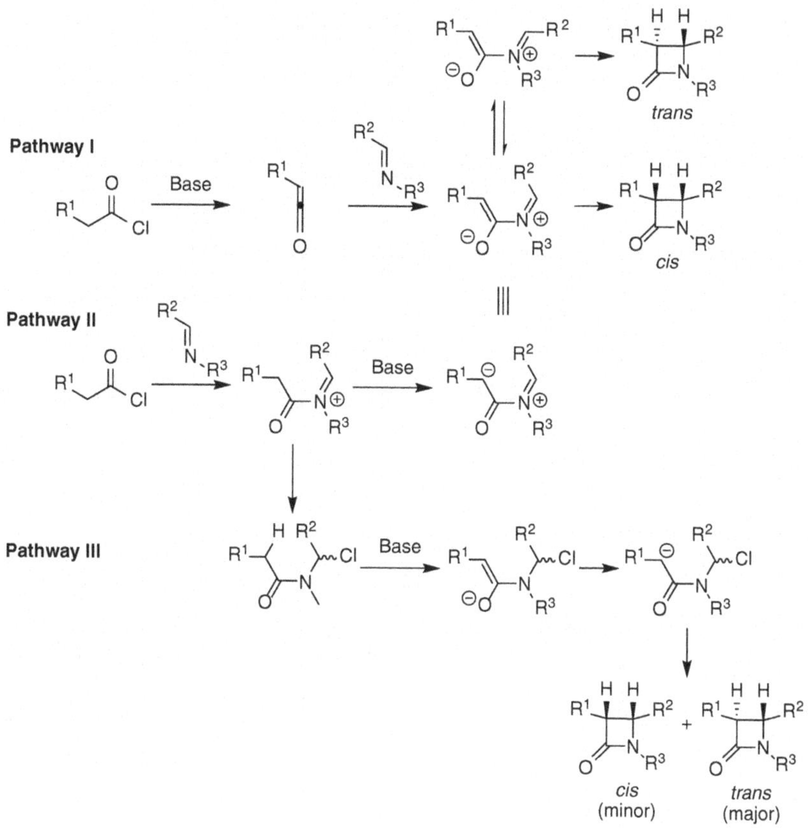

Scheme 1 Proposed mechanistic pathways for β-lactam formation via Staudinger ketene-imine cycloaddition

It has been reported that reaction variables such as reaction temperature, solvent, base, additives, and the order of addition of reactants influence the stereochemistry of the products. Nonpolar solvents are favorable for the formation of *cis*-β-lactams, whereas polar solvents are better for the formation of *trans*-β-lactams [22]. Presumably this effect is due to the stabilization of the zwitterionic intermediate in polar solvents, allowing isomerization of the double bond prior to ring closure. The stereoselectivity is also dependent on the order of addition of reactants, i.e., an acid chloride and an imine, which can be explained by the three pathways shown in Scheme 1.

2.1.1 Through [2+2] Cycloaddition of Achiral Ketenes to Chiral Imines

Asymmetric synthesis of β-lactams can be realized through the [2+2] cycloaddition of an achiral ketene to a chiral imine. Chiral imines can be derived from chiral aldehydes and achiral amines or from chiral amines and achiral aldehydes. Modest to good diastereoselectivities, in most cases, are observed for the reactions using

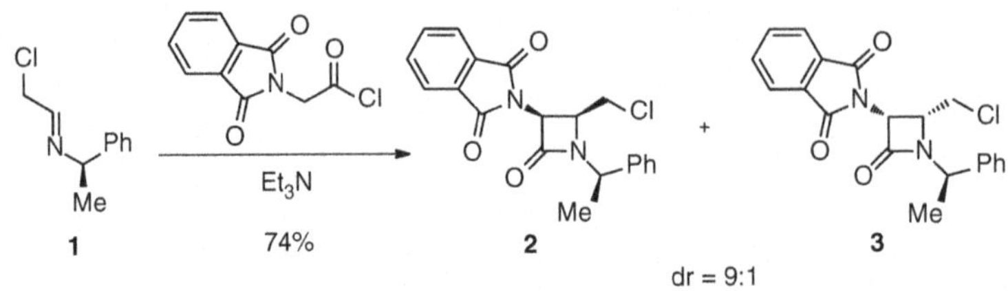

Scheme 2 Diastereoselective chiral imine-phthalimidoketene cycloaddition

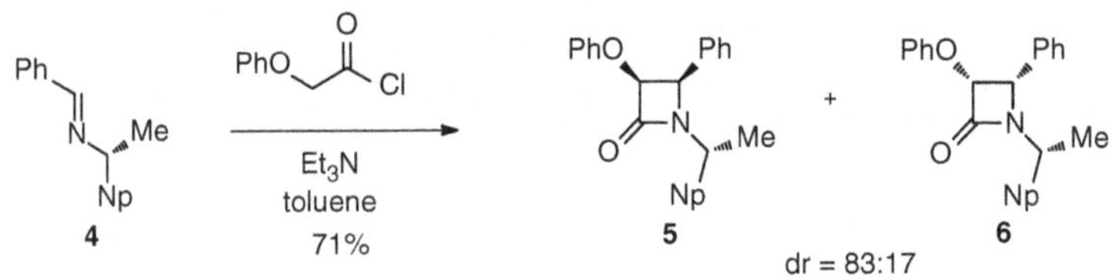

Scheme 3 Diastereoselective chiral imine-phenoxyketene cycloaddition

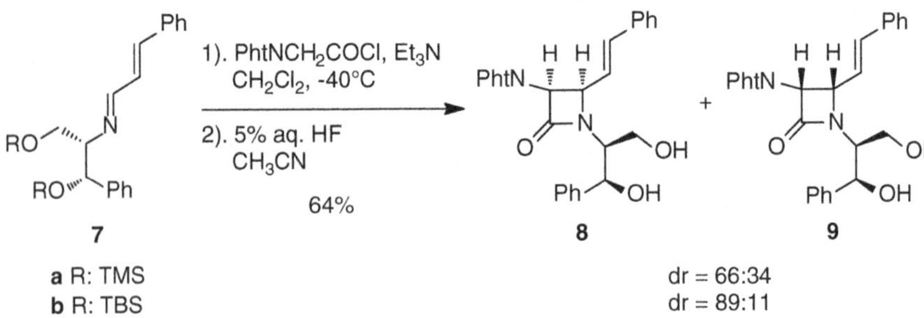

Scheme 4 Diastereoselective chiral phenylethenylimine - ketene cycloaddition

imines derived from simple chiral amines such as (*R*)-1-phenylethylamine or (*R*)-1-naphthylethylamine [8, 23, 24]. For example, 3-phalimido-β-lactams **2** and **3**, were obtained in 74% yield and good diastereomer ratio (dr = 9:1), using imines **1** derived from (*R*)-1-phenylethylamine (Scheme 2) [23]. Imines **4** derived from (*R*)-1-naphthylethylamine reacted with phenoxyketene to give 3-phenoxy-β-lactams **5** and **6** in 71% yield with modest diasteroselectivity (dr = 83:17) (Scheme 3) [24]. It is worthy of note that the two diastereomers were easily separated by silica gel chromatography providing a facile route to the enantiopure β-lactams. The reaction of imine **7a** (R = TMS) with phthalylketene gave β-lactams, **8** and **9**, in a low diastereomer ratio (dr = 2:1), while the reaction with imine **7b** bearing a bulkier *O*-protecting group (R = TBS) gave β-lactam **8** with much higher diastereoselectivity (dr = 8:1) (Scheme 4) [25]. The reaction of the imine **10a** (R = TPS) derived from D-threonine with azidoacetyl chloride in the presence of a base gave β-lactams **11** and **12** with excellent diastereoselectivity (dr = 19:1) (Scheme 5) [26]. The [2+2] cycloaddition of azidoketene and

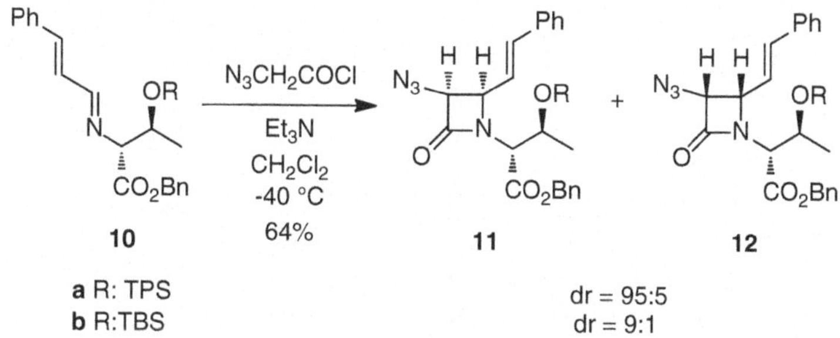

Scheme 5 Efficient asymmetric cycloaddition of azidoketene and chiral imine derived from threonine

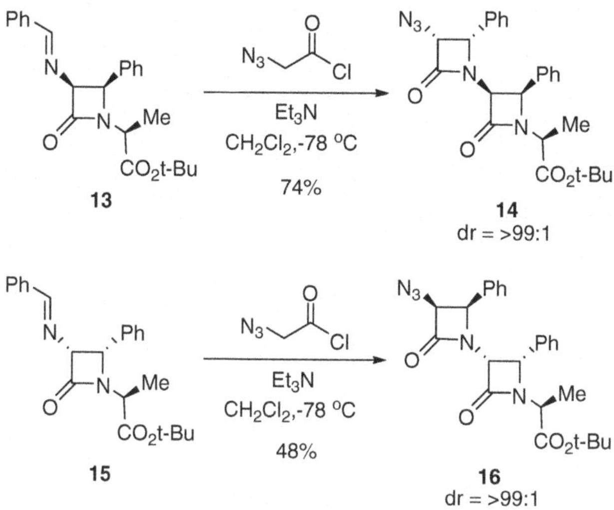

Scheme 6 Highly stereoselective bis-β-lactam formation via azidoketene - chiral imino-β-lactam cycloaddition

benzylideneamines with chiral β-lactam backbones **13** and **15** proceeded in excellent diastereoselectivity (dr = >99:1), resulting in the formation of enantiopure bis-β-lactams **14** and **16**, respectively, in modest to good yields (Scheme 6) [27].

Better stereoselectivities, in general, have been achieved when imines derived from chiral aldehydes are used, e.g., imines derived from α-oxyaldehydes [28], sugar-derived aldehydes [28–30], α,β-epoxyimines [31], and α-amino aldehydes [32]. In most of these cases, *cis*-β-lactams are produced exclusively or with dr greater than 9:1 [28–30]. For example, the reactions of chiral α,β-epoxyimines **17** with a variety of ketenes afforded the corresponding 3-amino-β-lactam derivatives **18** and **19** in 9:1–93:7 diastereomer ratio (Scheme 7) [31]. In many other cases, the formation of a single diastereomer of a β-lactam was observed, as represented by the reactions of two sugar-derived chiral imines **20** and **22** with benzyloxyketene, which gave β-lactams **21** and **23** in 63% and 87% yields, respectively (Schemes 8

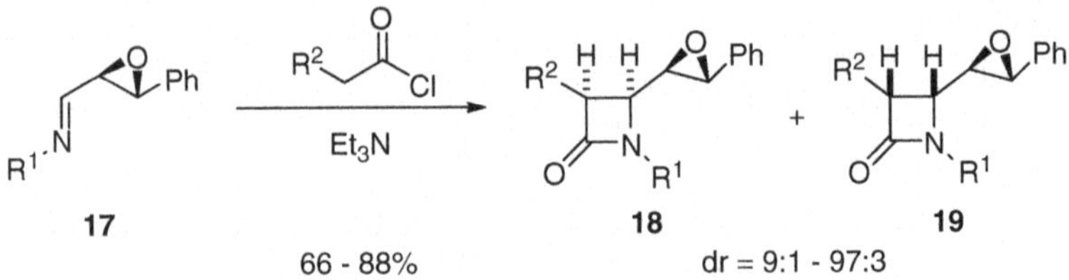

R_1 = Bz, 2,4-DMB, PMP, CH_2CO_2t-Bu, CH_2(Me)C=CH_2
R^2 = PhthN, CbzNH, OxN (4,5-Ph_2oxazolidinonyl)

Scheme 7 Stereoselective ketene - chiral α,β-epoxyimine cycloaddition

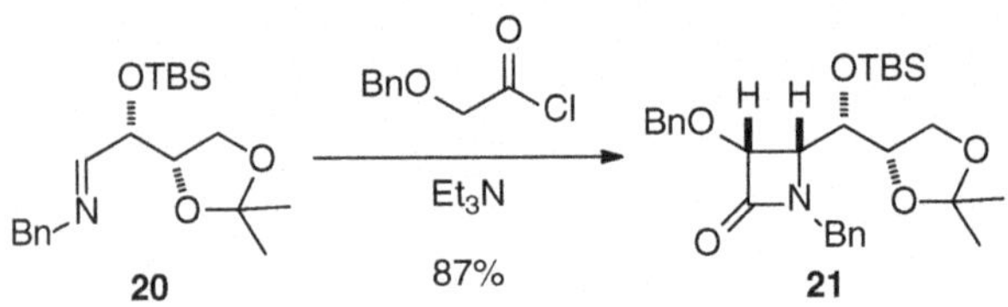

Scheme 8 Stereoselective cycloaddition of ketene and chiral imine derived from sugar derivative

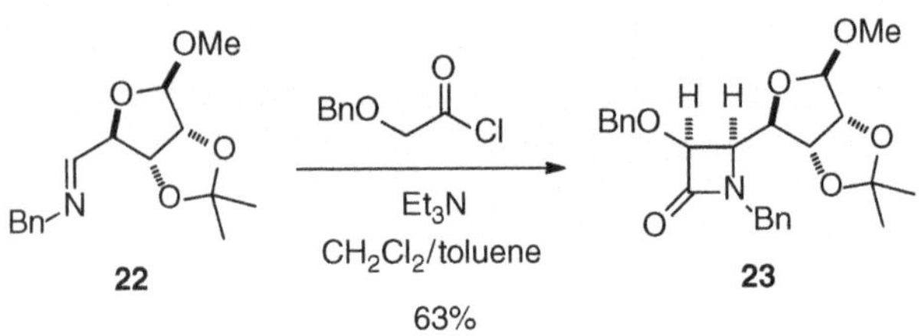

Scheme 9 Stereoselective cycloaddition of ketene and chiral imine derived from sugar

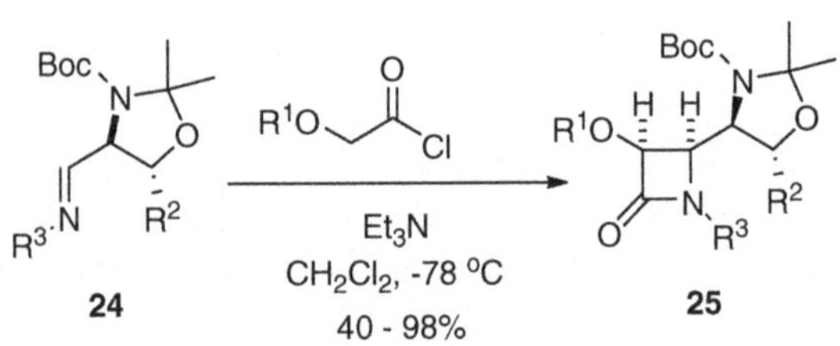

R_1 = Ac, Me, Bn
R_2 = H, Me, Bu, i-Pr, i-Bu
R_3 = Bn, PMP

Scheme 10 Highly stereoselective cycloaddition of ketene and chiral oxazolidinylimine

and 9) [28]. In addition, the reactions of chiral oxazolidinylimines **24** with various hydroxyketene derivatives gave the corresponding β-lactams **25** as single diastereomers in modest to an excellent 98% yields (Scheme 10).

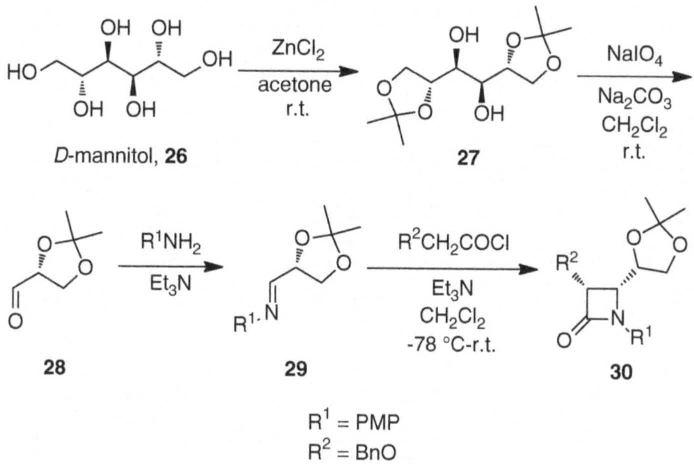

Scheme 11 Highly stereoselective ketene - chiral imine cycloaddition using glyceraldehyde ketal, derived from D-mannitol as chiral source

The reaction of chiral imines **29**, derived from D-glyceraldehyde ketal **28**, with various ketenes in situ generated from acid chloride and an amine base provides an easy access to a large-scale synthesis of the corresponding β-lactams **30** (Scheme 11) [33]. D-Glyceraldehyde ketal **28** can be readily prepared from D-mannitol (**26**), a cheap, enantiopure, and commercially available starting material (Scheme 11) [33]. The terminal glycol moieties of **26** were protected as acetals to form **27**. The C–C bond of the remaining glycol of **27** was then oxidatively cleaved by sodium periodate to give **28**. The [2+2] cycloaddition of the corresponding imines **29** with acyl chloride gave optically active β-lactams **30** as single diastereomers in 44–70% yields [33, 34].

2.1.2 Through [2+2] Cycloaddition of Chiral Ketenes to Achiral Imines

Another approach to the asymmetric ketene–imine [2+2] cycloaddition is to use chiral ketenes. This strategy has been very successful for the asymmetric synthesis of 3-amino-β-lactams. Oxazolidinones derived from (*S*)- and (*R*)-phenylglycine are excellent chiral auxiliaries for the asymmetric ketene–imine [2+2] cycloaddition [35]. For example, (*S*)-4-phenyloxazolidinon-3-ylacetyl chloride (**31**) with *N*-benzylaldimines **32** gave β-lactam **33** with 95:5–97:3 dr in 80–90% yield (Scheme 12) [35]. This methodology has been successfully applied to solid phase synthesis [36, 37]. A chiral ketene generated from **31** was reacted with imines immobilized on a solid phase bead **35** for the combinatorial synthesis of a library of β-lactams after cleavage from the resin-bound β-lactams **36** (Scheme 13) [36, 38]. In another example of ketene-induced chirality, the reaction of phenantridine **38**

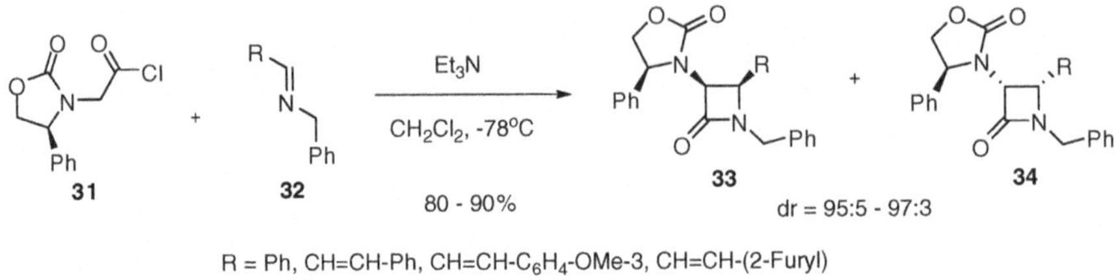

Scheme 12 Highly efficient chiral ketene - imine cycloaddition using 2-phenyloxazolidinone as chiral auxiliary

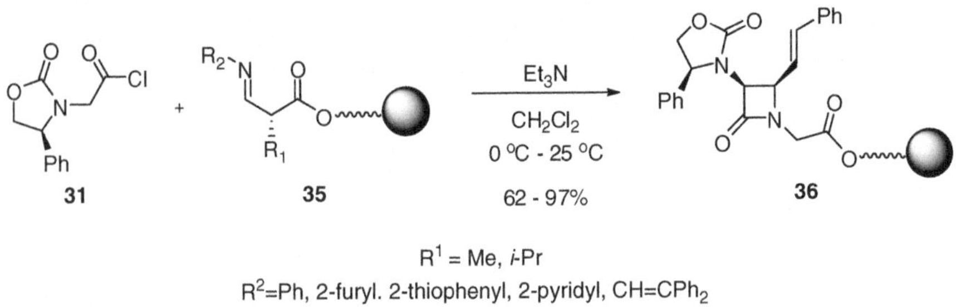

Scheme 13 Stereoselective cycloaddition of chiral ketene and imine anchored on a polymer resin

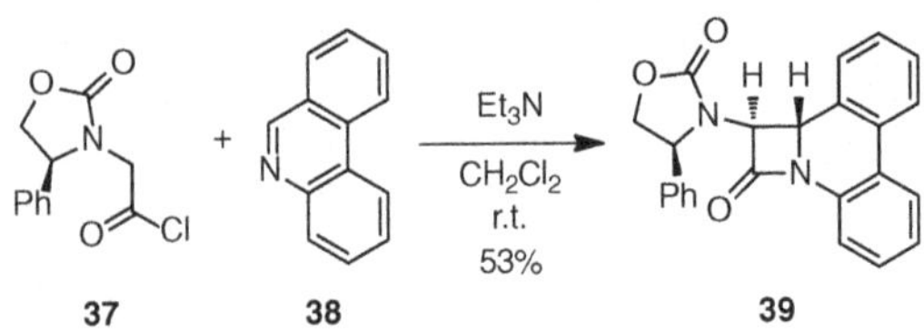

Scheme 14 Asymmetric synthesis of *trans*-β-lactam via chiral ketene - *cis*-imine cycloaddition

with **37** in the presence of triethylamine gave enantiopure polycyclic β-lactam **39** in 53% yield (Scheme 14) [39].

Even when complete diastereoselectivity is not achieved, the resulting diastereomers can be separated by conventional methods, and the chiral auxiliary is cleaved to produce two enantiopure β-lactams. For example, the reaction of ketene **40**, bearing an α-glycoside moiety, with imine **41** gave two *trans* diastereomers in a 45:55 ratio (Scheme 15) [40]. After the separation of the two diastereomers by column chromatography and the subsequent acid-catalyzed hydrolysis, the two enantiomers were obtained in the enantiopure form [40]. The reaction of ketene **45** with imine **46** derived from a glycinate afforded β-lactam **47**, which was obtained enantiopure after recrystallization [41]. β-Lactam **47** served as the key intermediate for the optimization of β-lactam-based vasopressin V1a receptor agonists (Scheme 16) [41]. The use of chiral ketenes in [2+2] cycloadditions has also been extended to the synthesis of *spiro*-β-lactams [42–44].

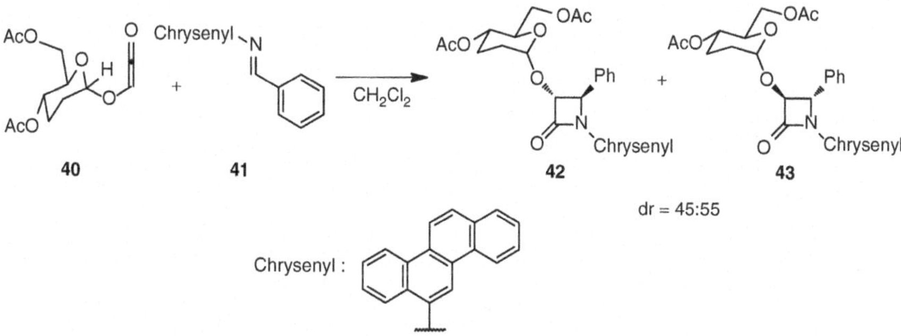

40 41 42 43

dr = 45:55

Scheme 15 Formation of *trans*-β-lactams via chiral ketene - bulky imine cycloaddition using a sugar derived chiral auxiliary

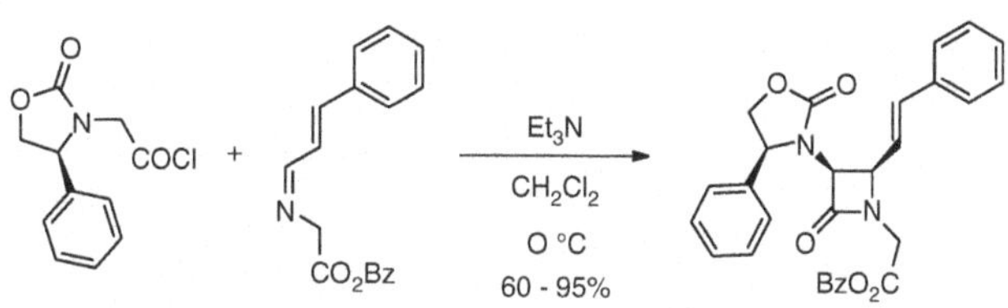

45 46 47

Scheme 16 Asymmetric synthesis of the key intermediate for β-lactam-based vasopressin V1a receptor agonists

2.1.3 Enantiopure β-Lactams Through Enzymatic Kinetic Resolution

Enzymatic kinetic optical resolution of racemic β-lactams obtained by the Staudinger ketene–imine cycloaddition provides another route to enantiopure β-lactams. This protocol is especially useful for the synthesis of enantiopure 3-hydroxy-β-lactams [11, 45–50]. Although at least a half of the racemic β-lactam cannot be converted to the desired enantiomer in this process, the starting materials are inexpensive. Thus, the enzymatic resolution process is economically attractive albeit the "atom economy" is not good. Also, the other enantiomer can be used as a versatile synthetic building block and may not be wasted.

Among various hydrolytic enzymes examined, amano lipases and pig liver acetone powder (PLAP) have been found to provide the best results in the kinetic optical resolution of racemic *cis*-3-acetoxy-4-phenyl-β-lactams, yielding the corresponding (3*R*,4*S*)- and (3*S*,4*R*)-β-lactams with high enantiopurities [45, 48]. The "PS-Amano" lipase has been successfully applied to the kinetic resolution of racemic *cis*-3-acetoxy-4-(2-methylbut-2-enyl)azetidin-2-one [11, 49, 50]. The "PS-Amano" lipase preferentially hydrolyzes the acetate moiety at the C-3 position of this β-lactam. Therefore, (3*R*,4*S*)-β-lactam is obtained with extremely high enantiopurity (>99% ee) when the reaction is stopped over 50% conversion. The actual examples are described together with their synthetic applications later in this chapter.

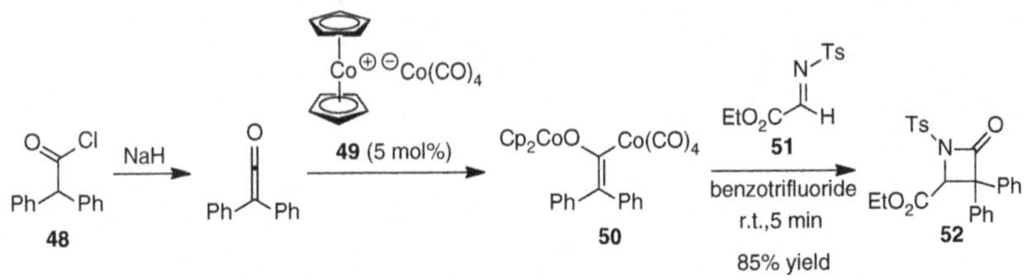

Scheme 17 Cobalt complex - catalyzed β-lactam synthesis

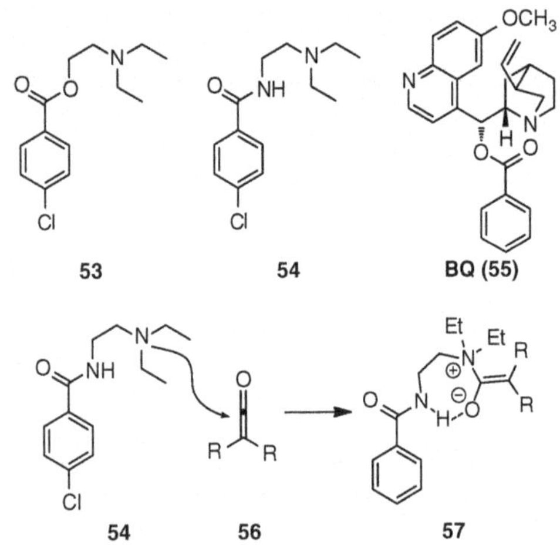

Fig. 3 Organocatalysts for the asymmetric ketene-imine cycloaddition and a proposed key intermediate

2.1.4 Catalytic Asymmetric Synthesis of β-Lactams Through [2+2] Cycloaddition of Zwitterionic Enolates and Imines

A catalytic process for the asymmetric [2+2] ketene–imine cycloaddition to form enantio-enriched β-lactams has been developed [51]. Cobalticenium tetracarbonyl cobaltate (**49**) contains both a nucleophilic anion and Cp_2Co cation that could act in concert with each other. When applied to the [2+2] cycloaddition of diphenylketene, generated from **48** and NaH, to *N*-tosylimino ester **51**, cobalt-complex **49** exhibited a good catalyst activity, giving β-lactam **52** in up to 85% yield after only 5 min at room temperature (Scheme 17) [52].

It was found that the nucleophilic catalyst does not need to be a metal anion or charged. Thus, an organocatalyst was found to effect this process [53]. Moreover, it was hypothesized that an organocatalyst, containing a nucleophilic center associated with an electrophilic center could rigidify the key intermediate **57** (Fig. 3) and potentially generates β-lactams with higher diastereoselectivity. When tertiary amines **53** and **54** were used as catalysts for reaction of diphenylketene with imine **56**, **53** gave β-lactam in a 34:66 *cis*:*trans* ratio, and tertiary amine **54**, capable of hydrogen bonding, outperformed **53** to afford β-lactam in a 3:97 *cis*:*trans* ratio.

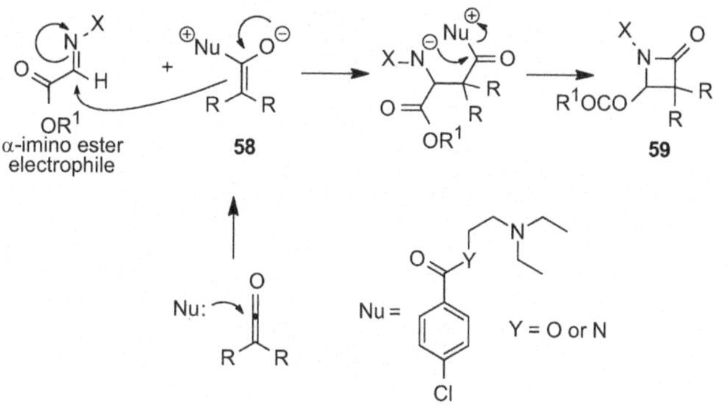

Fig. 4 Proposed mechanism for nucleophilic organocatalysis

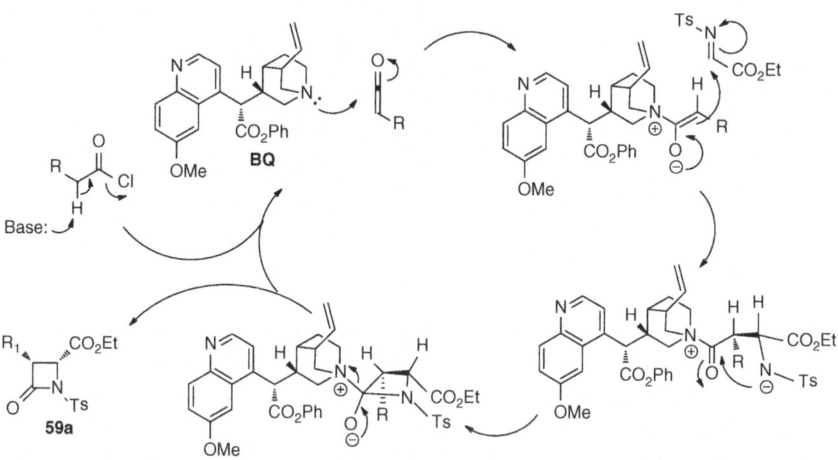

Fig. 5 Enantioselective β-lactam formation through organocatalysis of a cinchona alkaloid

First, a novel catalytic process was studied using achiral nucleophilic catalysts. In general, the imine acts as the electrophile and the ketene component acts as the nucleophile in this catalytic process (Figs. 4 and 5). The imine is made highly electrophilic by incorporating an electron-withdrawing group to the imine nitrogen and a carbalkoxy substituent to the imine carbon. The transformation of the electrophilic ketene component to a nucleophilic reactant was devised with the use of a nucleophilic catalyst (Nu:) that reversibly binds to the ketene to form a zwitterionic enolate **58**. Nucleophilic attack of **58** on the α-carbon of the imine leads to the asymmetric C–C bond formation, followed by cyclization to give β-lactam **59** and regenerates the nucleophilic catalyst (Fig. 4).

Based on these results, optically active cinchona alkaloid derivatives were screened as potential organocatalysts for enantioselective reactions with high diastereoselectivity. Then, it was found that benzoylquinine (BQ) (**55**) catalyzed the reaction of **60** with **61** to give β-lactam **63** with 99% ee, but in low yield

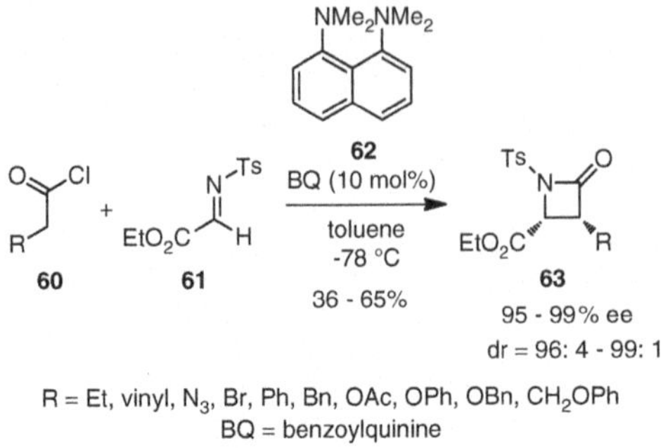

Scheme 18 Enantioselective β-lactam formation through organocatalysis in the presence of a proton sponge

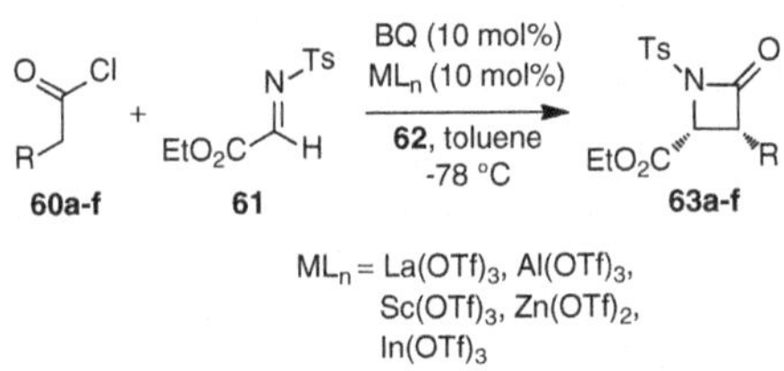

Scheme 19 Enantioselective β-lactam formation through organocatalysis in the presence of Lewis acids

(Scheme 18) [53]. It was also found that BQ promoted the reaction of imine **61** with monosubstituted ketenes, generated in situ from acid chlorides in the presence of a proton sponge **62** at a low temperature, to give β-lactams **63** with very high % ee and dr, but in low to modest yields (Scheme 18) [54]. The low yield in this process can be attributed to the formation of ammonium salts as interfering byproducts, which arise from the use of amine bases as dehydrohalogenating agents. These byproducts leave residual bases that catalyze the formation of a racemic mixture of β-lactam **63** [53].

In order to improve the chemical yield, metal salts were explored as effective cocatalysts to further activate imines [55]. Then, it was found that the use of triflates of Sc(III), Al(III), Zn(II), and In(III) (10 mol %) with BQ gave β-lactams **63** in high yields with excellent enantioselectivity and high diastereoselectivity (Scheme 19). Among the metal salts examined for the reaction of **60a** (R = Ph) with **61**, $In(OTf)_3$ gave the best results (95% yield), followed by $Zn(OTf)_2$ (85% yield), while $Al(OTf)_3$ (78% yield) and $Sc(OTf)_3$ (80% yield) were less effective [55]. Having established that $In(OTf)_3$ was the best overall Lewis acid cocatalyst, it was applied to various substrates to determine the effect on reaction enantio- and diastereoselectivity (Table 1).

Recently, *N*-heterocyclic carbenes (NHCs) have been reported to catalyze the enantioselective formation of β-lactams [56, 57]. For example, NHC **64a** derived from (*S*)-pyroglutamic acid catalyzed the [2+2] ketene–imine cycloaddition of various arylketenes **65** with arylimines **66** to give the corresponding β-lactams **67** in moderate to good yields and with excellent enantioselectivities [56] (Scheme 20).

Table 1 Asymmetric synthesis of β-lactams through [2+2] cycloaddition promoted by the BQ–In(OTf)$_3$ catalyst system

Entry	R	Yield (%)	% ee	dr (*cis*:*trans*)
1	C_6H_5	95	98	60:1
2	$CH_2C_6H_5$	94	98	9:1
3	$CH_2OC_6H_5$	93	96	12:1
4	OC_6H_5	93	97	22:1
5	$OC(O)CH_3$	92	98	34:1
6	$OCH_2C_6H_5$	98	96	11:1

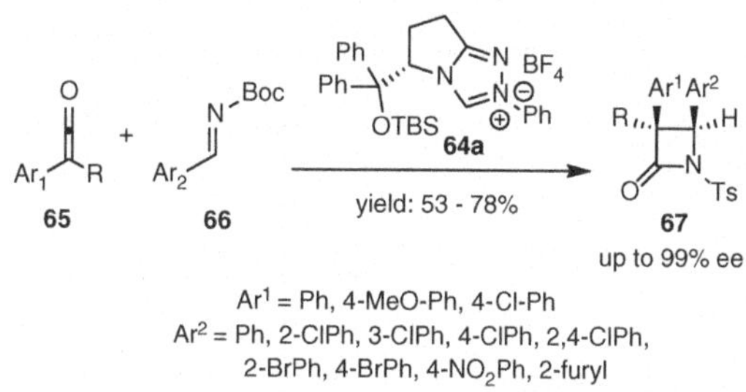

Scheme 20 Asymmetric synthesis of β-lactam catalyzed by *N*-heterocyclic carbene

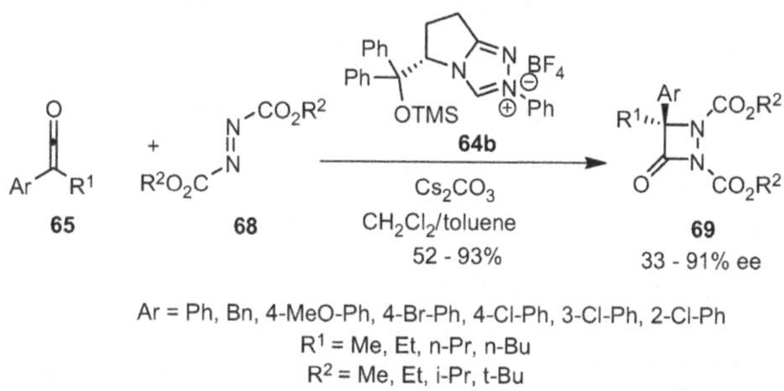

Scheme 21 Asymmetric synthesis of aza-β-lactam catalyzed by *N*-heterocyclic carbene

Another NHC **64b** was used as a catalyst for the reaction of arylketene **65** with azodicarboxylate **68** to afford aza-β-lactams **69** with 33–91% ee [57] (Scheme 21). Moderate to excellent enantioselectivity was also achieved using a C-2 symmetric imidazolinium catalyst **70a** or triazolium catalyst **70b** in the reaction of diphenylketenes **71** with various arylimines **72**, yielding the corresponding β-lactams **73** with 55 ~> 99% ee (Scheme 22) [58].

2.2 *Asymmetric Synthesis of β-Lactams Through Chiral Ester Enolate–Imine Cyclocondensation*

The chiral ester enolate–imine cyclocondensation is an efficient and versatile method for the asymmetric synthesis of β-lactams and has been widely applied

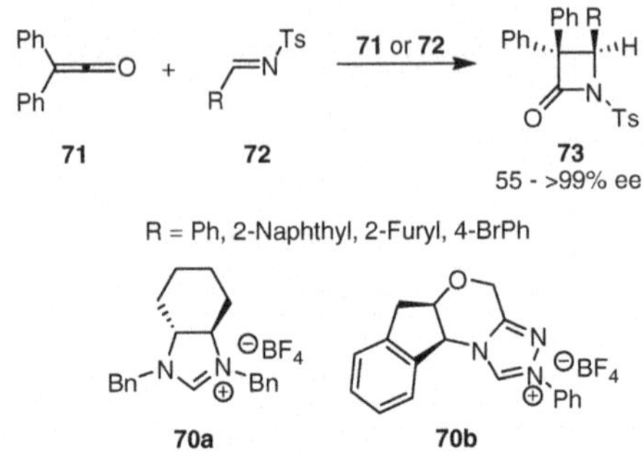

Scheme 22 Enantioselective cycloaddition of diphenylketene and *N*-tosylimine catalyzed by *N*-heterocyclic carbenes

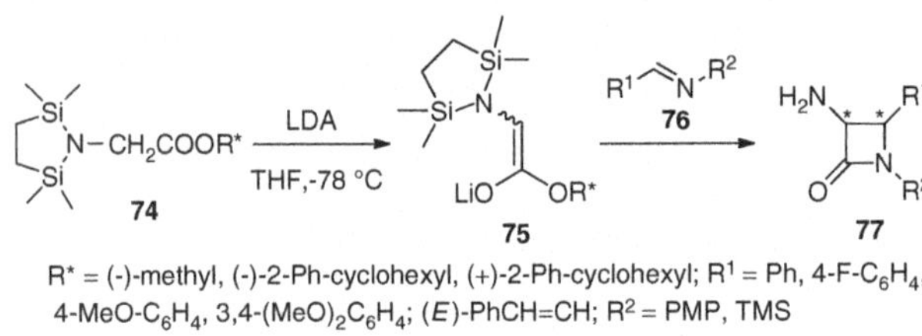

Scheme 23 Asymmetric synthesis of 3-amino-β-lactam via chiral ester enolate - imine cyclocondensation

for the synthesis of 3-amino- and 3-hydroxy-β-lactams with excellent enantiopurities [6, 8, 59–61].

The reaction of *N*-PMP-arylaldimines **76** (PMP = *p*-methoxyphenyl) with chiral ester enolates **75** generated from *N,N*-bis(silyl)glycinates **74** bearing enantiopure ester moieties gives mostly *trans*-3-amino-β-lactams **77** with excellent enantiopurity in fairly good yields (Scheme 23) [59]. For example, the reaction of (−)-menthyl glycinate **75a** with benzaldimine **76a** in the presence of LDA in THF at −78°C for 4 h gave (3*R*,4*R*)-β-lactam **77a** with >99% ee in 65% yield. However, when (+)-neomenthyl glycinate was used under the same conditions, a 26:74 *trans:cis* mixture of **77a** with 54% ee and 21% ee, respectively, were formed [59]. The use of (−)-bornyl ester resulted in the formation of 37:63 *trans:cis* mixture of **77a** with very low enantiopurity [59]. Thus, (−)-menthyl and (−)-2-phenylcyclohexyl glycinates were found to be the chiral auxiliaries of choice in this reaction. When *N*-TMS-cinnamaldimine **76f** was used, the reaction gave *cis*-β-lactam **77f** exclusively in modest yield with low to good enantioselectivity. Representative results are summarized in Table 2 [59].

The most plausible mechanism of this process based on those results is illustrated in Scheme 24 [59]. The addition of (*E*)- or (*Z*)-enolates **75** to imine **76** is likely to proceed through chair transition states and (*Z*)-enolate **Z-75** yields *trans*-**78**, while (*E*)-enolate *E*-**75** affords *cis*-**78**. Because of the bulkiness of the *N,N*-bissilylamino moiety, the **TS**-*cis* transition state for arylaldimines **76** (R^1 = aryl) appears unfavorable due to the steric conflict between the bissilylamine and R^1 as well as

Table 2 Asymmetric synthesis of 3-amino-β-lactams through chiral ester enolate–imine cyclocondensation

Entry	R*	R^1	R^2	Yield (%)	% *trans*-**77** (% ee)	% *cis*-**77** (% ee)
1	(−)-Menthyl	Ph	PMP	68	100 (>99)	
2	(−)-2-Ph-cyclohexyl	Ph	PMP	58	100 (>99)	
3	(−)-Menthyl	4-F-C_6H_4	PMP	55	100 (>99)	
4	(−)-Menthyl	4-CF_3-C_6H_4	PMP	59	100 (>99)	
5	(−)-Menthyl	4-MeO-C_6H_4	PMP	70	89 (>99)	11 (38)
6	(−)-Menthyl	3,4-$(MeO)_2C_6H_4$	PMP	70	91 (>99)	9 (38)
7	(−)-2-Ph-cyclohexyl	(*E*)-PhCH=CH_2	TMS	46		100 (78)

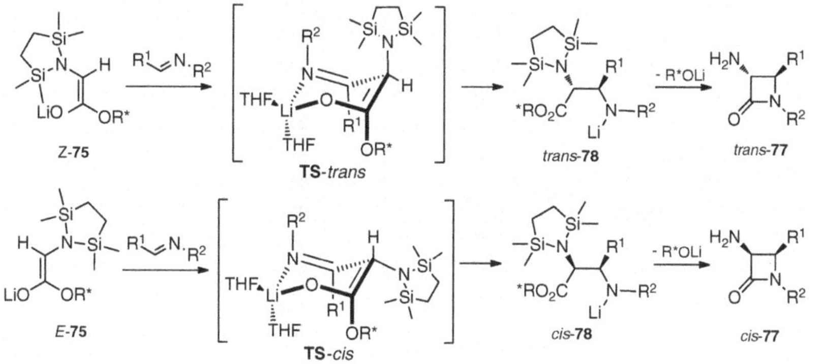

Scheme 24 Proposed mechanisms for the formation of *trans*- and *cis*-β-lactams via chiral ester enolate - imine cyclocondensation

R* moieties. Thus, if *E*-**75** and *Z*-**75** are in equilibrium under the reaction conditions, the reaction should proceed through the **TS**-*trans* transition state to give *trans*-**77** via lithium amide *trans*-**78**. However, in the reaction with cinnamaldimine **76f**, the steric hindrance between the bissilylamine and styryl moieties in the **TS**-*cis* transition state is substantially decreased. Thus, *cis*-**77** is the exclusive product in this reaction.

The chiral ester enolate–imine cyclocondensation has also proven to be a highly efficient method for the asymmetric synthesis of 3-hydroxy-β-lactams with excellent enantiopurity [7, 62]. In contrast to the rigid bissilylamino moiety in glycinates **75**, the *O*-protected hydroxyl group is very flexible in hydroxyacetates **79**. Accordingly, both the chiral auxiliary moieties and *O*-protecting groups were screened to find the optimal chiral ester enolate **80** for this process. Thus, benzyl, *t*-butyldimethylsilyl (TBS) and triisopropylsilyl (TIPS) groups as well as (−)-menthyl, (+)-*N*-methylephedrinyl, (−)- and (+)-*trans*-2-phenylcyclohexyl groups were examined, and the combination of TIPS and (−)- or (+)-2-phenylcyclohexyl groups was found to give the best results. It should be noted that *cis*-β-lactams were formed exclusively in this process.

Chiral enolate **80** was generated from (−)-2-phenylcyclohexyl TIPSO-acetate **79** with LDA in THF at −78°C for 2 h. *N*-TMS-arylaldimines **76** (R^2 = TMS) was added to the enolate solution and reacted at −78°C and then gradually warmed to room temperature to give (3*R*,4*S*)-3-TIPSO-4-aryl-β-lactams **81** with 96–98% ee in

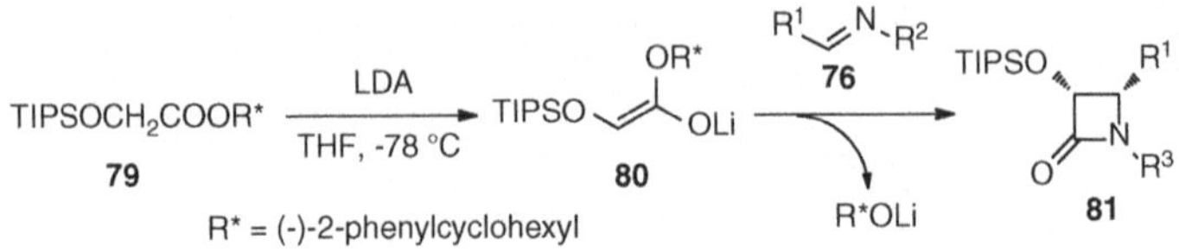

Scheme 25 Enantioselective synthesis of 3-TIPSO-β-lactam via chiral ester enolate - imine cyclocondensation

Table 3 Asymmetric synthesis of 3-TIPSO-β-lactams through chiral ester enolate–imine cyclocondensation

Entry	R^1	R^2	R^3	β-Lactam	Yield (%)	% ee
1	Ph	TMS	H	**81a**	85	96
2	4-MeO-C_6H_4	TMS	H	**81b**	80	96
3	3,4-$(MeO)_2C_6H_3$	TMS	H	**81c**	80	98
4	Ph	PMP	PMP	**81a**	89	98
5	4-F-C_6H_4	PMP	PMP	**81d**	81	98
6	4-CF_3-C_6H_4	PMP	PMP	**81e**	84	99
7	2-Furyl	PMP	PMP	**81f**	78	92
8	(*E*)-PhCH=CH_2	PMP	PMP	**81g**	85	96
9	2-Furyl-CH=CH_2	PMP	PMP	**81h**	72	94
10	c-$C_6H_{11}CH_2$	PMP	PMP	**81i**	85	90
11	Me_2CHCH_2	PMP	PMP	**81j**	85	92
12	Me_2C=CH	PMP	PMP	**81k**	60	94

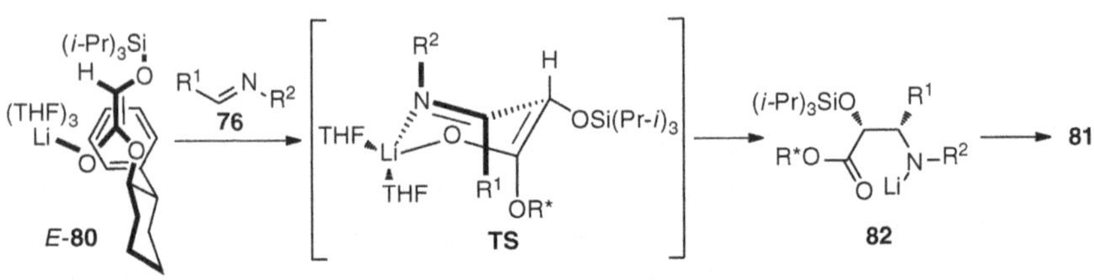

Scheme 26 Proposed mechanism for chiral ester enolate - imine cyclocondensation using 2-phenylcyclohexanol as chiral auxiliary

80–85% yields (Scheme 25) [6, 60]. It is worthy of note that (3*R*,4*S*)-3-TIPSO-4-phenyl-β-lactam **81a** serves as a key intermediate for the highly efficient semi-synthesis of paclitaxel (vide infra) [6].

Since *N*-TMS-imines are limited to arylaldimines due to instability of *N*-TMS-alkylaldimines, *N*-PMP-imines **76** (R^2 = PMP) were employed to expand the scope of this process, and excellent results were obtained (Scheme 25). Representative results, using (−)-2-phenylcyclohexanol as the chiral auxiliary, are summarized in Table 3 together with those for *N*-TMS-imines **76** (R^2 = TMS) [6, 60, 63].

The plausible mechanism of this process includes the kinetically favorable *E*-enolate *E*-**80** formation, the chair transition state **TS**, lithium amide **82** formation, and cyclization as illustrated in Scheme 26 [6]. In the transition state **TS**, the bulky

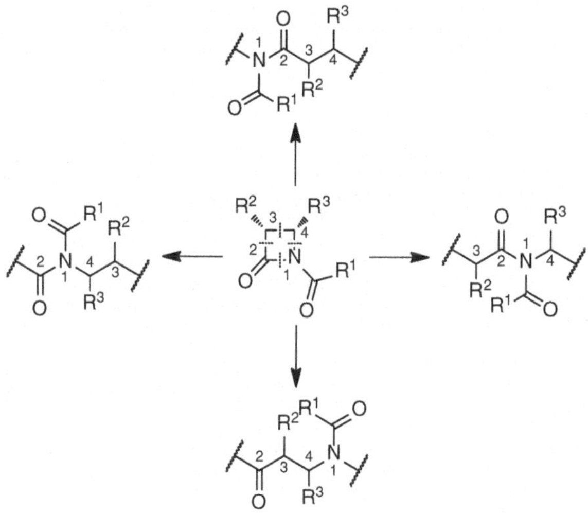

Fig. 6 Four possible ring-opening modes of *N*-acyl-β-lactam

TIPSO group occupies an equatorial position, but the chiral auxiliary moiety as well as R^1 and R^2 of imine **76** take axial positions, which enables highly efficient enantioface selection of imine **76**. It should be noted that Z-enolate *Z*-**80** could also lead to the same stereochemical outcome via boat-like transition state. However, molecular mechanics calculations indicate that ***E*-80**.3THF is more favorable than *Z*-**80**.3THF by 2.5 kcal/mol [6], in addition to the fact that *E*-**80** formation is kinetically favorable, which strongly support the proposed mechanism.

3 Ring-Opening Reactions of β-Lactams

β-Lactams serve as useful and versatile synthetic intermediates, as they are not only readily synthesized with high enantiopurity, but also offer four distinct modes of reactivity (Fig. 6). The strain energy imposed on each of all four bonds, arising from the four-membered ring amide structure, allows for unique reactivity not observed in larger ring systems such as γ- and δ-lactams. Cleavage can occur in any of the four bonds in the β-lactam skeleton, and the selectivity in bond cleavage depends on the substitution pattern and the reaction conditions, as described below.

The most obvious ring-opening reaction is the cleavage between N1 and C2 through nucleophilic acyl substitution at the amide carbonyl moiety. A less obvious, but extensively studied reaction is the N1–C4 bond cleavage either reductively or through catalytic ring expansion or contraction. The C3–C4 bond cleavage is not obvious, but this ring opening is possible through pericyclic reactions and catalytic ring expansions. The least common and not obvious, but interesting reaction is the C2–C3 bond cleavage. Examples of those ring-opening reactions are described below to highlight the unique reactivity of the β-lactam ring system.

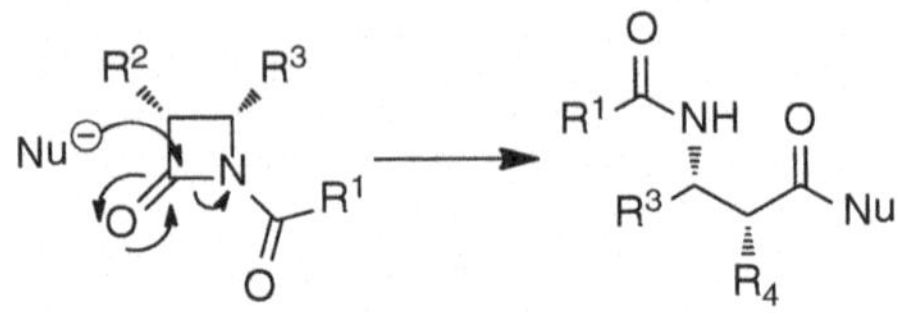

Fig. 7 1,2-Bond cleavage of *N*-acyl-β-lactam by nucleophile

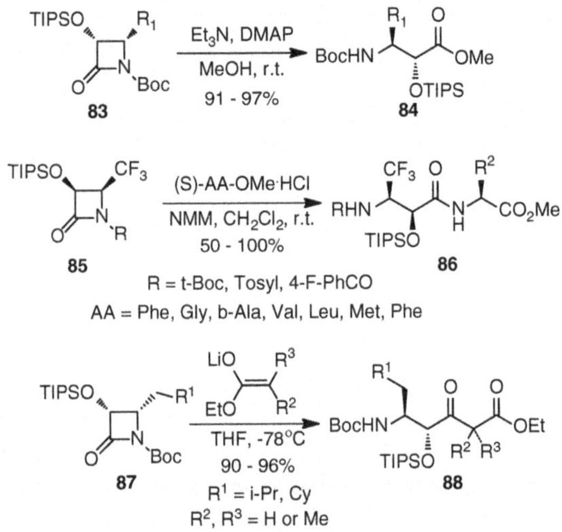

Scheme 27 Examples of 1,2-bond cleavage reactions

3.1 Ring Opening via N1–C2 Bond Cleavage

Ring opening of *N*-acyl-β-lactams through the cleavage of the N1–C2 bond by intermolecular or intramolecular attack of nucleophiles has been extensively studied and applied to numerous systems to give a variety of compounds of biological and medicinal interests (Fig. 7). These reactions constitute the major component of the "β-lactam synthon method," wherein various β-lactams serve as synthetic equivalents to β-amino acid residues [5, 7–12]. For example, the intermolecular reactions with water, alcohols, amines, and amino acid esters afford the corresponding β-amino acids, β-amino esters/amides, and dipeptides, respectively [12]. On the other hand, intramolecular reaction results in ring expansions to highly functionalized cyclic products such as amino-γ-lactones [64], five-membered enaminolactones [65], γ-lactams [66], and macrocycles [67]. Reductive N1–C2 bond cleavage leads to the formation of γ-amino alcohols [68].

Three examples of the N1–C2 bond cleavage by intermolecular reaction with nucleophiles are shown in Scheme 27. Alcohols or alkoxides are good nucleophiles in many N1–C2 ring-opening reactions. The reaction of *N*-Boc-β-lactam **83** with methanol in the presence of Et_3N and 4-dimethylaminopyridine (DMAP) gives

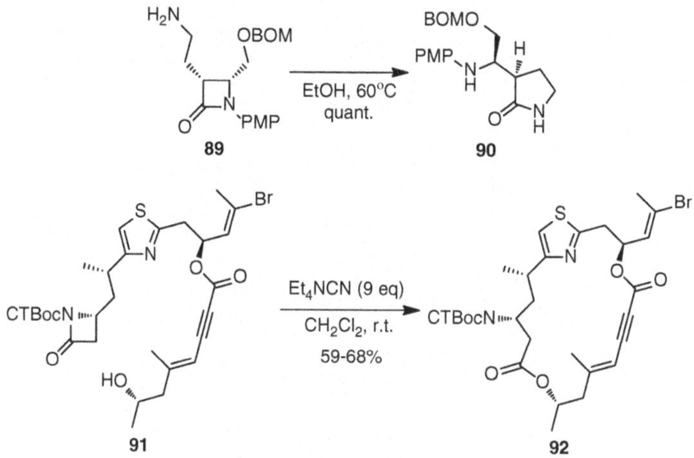

Scheme 28 Examples of intramolecular 1,2-ring-cleavage reaction leading to ring expansion

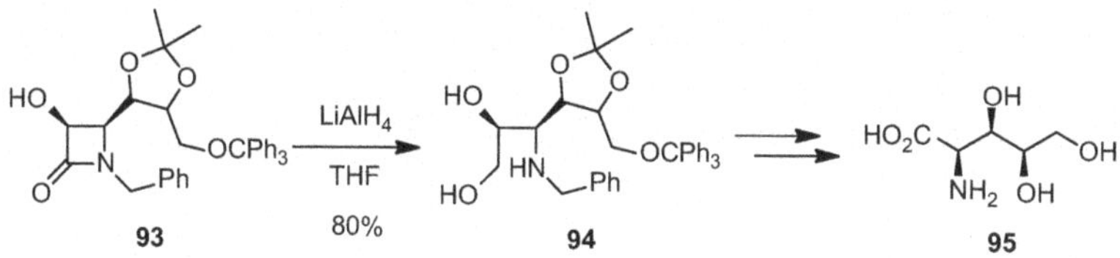

Scheme 29 Reductive 1,2-bond cleavage, leading to the formation of γ-amino alcohol

α-hydroxy-β-amino acid methyl ester **84** in excellent yield [69]. The reaction of *N*-acyl-β-lactam **85** with α-amino acid methyl esters affords dipeptide **86**, bearing α-hydroxy-β-amino acid unit in the N terminus [49]. The use of an ester enolate as the nucleophile for the ring opening of *N*-Boc-β-lactam **87** gives hydroxyl(keto) ethylene dipeptide isostere **88** in excellent yield [70].

Two examples of the N1–C2 bond cleavage through reaction with intramolecular nucleophiles are illustrated in Scheme 28. The intramolecular ring opening of the N1–C2 bond by a primary amine moiety at the C3 position of β-lactam **89** led to the formation of a rearrangement product, functionalized γ-lactam **90** in quantitative yield [66]. In a similar manner, 3-aminoethoxy-β-lactams and 1-aminoethy-β-lactams were transformed to morpholinones and seven-membered azalactams, respectively, in high yields [66]. The macrolactonization of **91** was achieved through intramolecular ring-opening coupling of the alcohol moiety and the β-lactam moiety to give **92** in 59–68% yield, which is an advanced key intermediate for the total synthesis of Pateamine A [67].

Reductive cleavage of the N1–C2 bond β-lactams of by $NaBH_4$ or $LiAlH_4$ provides a viable route to γ-amino alcohols [68, 71]. The reduction of 3-hydroxy-4-trihydroxyalkyl-β-lactam derivative **93** prepared from D-mannitol with $LiAlH_4$ gave polyhydroxy-γ-amino alcohol derivative **94** in 80% yield, which is the key intermediate in the synthesis of (−)-polyoxamic acid (**95**) [71] (Scheme 29). In a

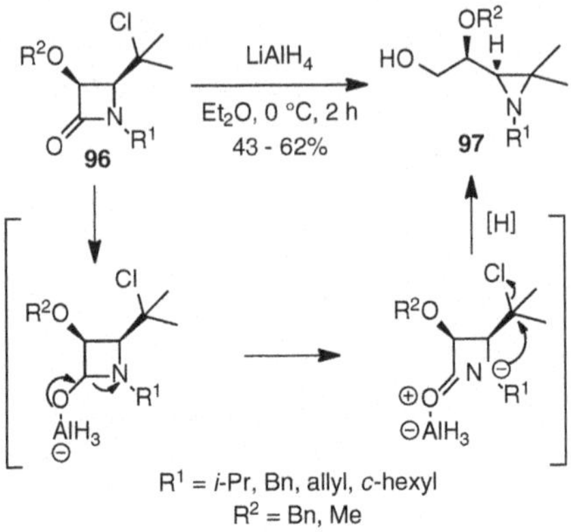

Scheme 30 Formation of aziridine via tandem reductive 1,2-bond cleavage - cyclization process

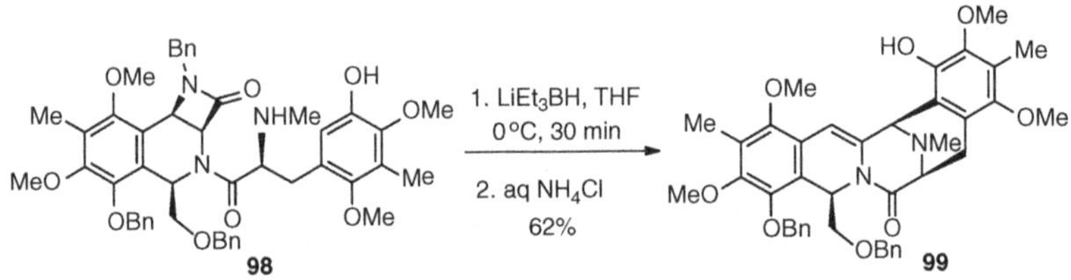

Scheme 31 Tandem 1,2-ring opening - reductive elimination to form a key intermediate for renieramycin H

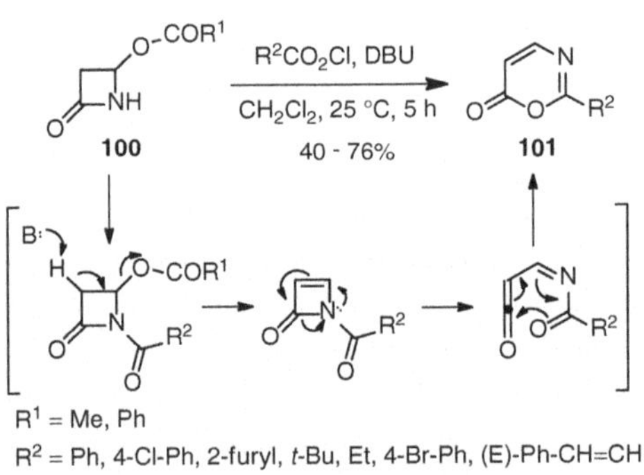

Scheme 32 Formation of 1,3-oxazinone via acylation-elimination-electrocyclic rearrangement

similar manner, 3-hydroxy-4-polyhydroxyalkyl-β-lactam derivatives have been successfully used as the key intermediates for the synthesis of gentosamine, 6-epilincosamine, and γ-hydroxythreomine [71]. The reductive cleavage of the N1–C2 bond of β-lactam **96**, bearing 2-chloroisopropyl group at the C4 position, with $LiAlH_4$ gave β-hydroxyethylaziridine **97** in modest to fairly good yield through a tandem reduction–cyclization process (Scheme 30) [72]. An ingenious ring-opening-reductive elimination reaction of fused β-lactam **98** with $LiBEt_3H$

gave pentacyclic alkaloid **99** in 62% yield, which is an advanced key intermediate for the total synthesis of renieramycin H [73] (Scheme 31).

The reaction of 4-acyloxy-β-lactam **100** with an acid chloride and DBU was found to undergo a rather sophisticated acylation–elimination–electrocyclic rearrangement to afford 1,3-oxazin-6-ones **101** in modest to good yields, as illustrated in Scheme 32 [74].

3.2 Ring Opening via N1–C4 Bond Cleavage

When an aryl substituent is attached to the C-4 position of a β-lactam, a facile N1–C4 bond cleavage takes place exclusively through palladium-catalyzed hydrogenolysis, which is ascribed to the strain energy of the β-lactam skeleton. This reaction made a foundation of the first generation β-lactam synthon method [4, 5, 7, 62], which has been applied to the synthesis of a variety of enantiopure aromatic α-amino acids, α-hydroxy acids, dipeptides, and oligopeptides. Those enantiopure β-lactams can be reduced to the corresponding azetidines, which are further transformed to polyamines, polyamino alcohols, and polyamino ethers [62, 75].

Examples of the application of the reductive N1–C4 bond cleavage of β-lactams are shown in Schemes 33, 34, 35. In general, the hydrogenolysis of stereochemically defined 4-aryl-β-lactam **102** bearing either a benzyloxy or azido group at C3 over Pd-C under mild conditions, followed by acid hydrolysis, gives the corresponding α-hydroxy acid or α-amino acid **103** in excellent yield [5, 7, 62] (Scheme 33). In the same manner, the reaction of enantiopure β-lactam **104**, derived from an imine of a chiral α-amino acid ester, affords the corresponding dipeptide **105** without racemization [5, 7, 62] (Scheme 33).

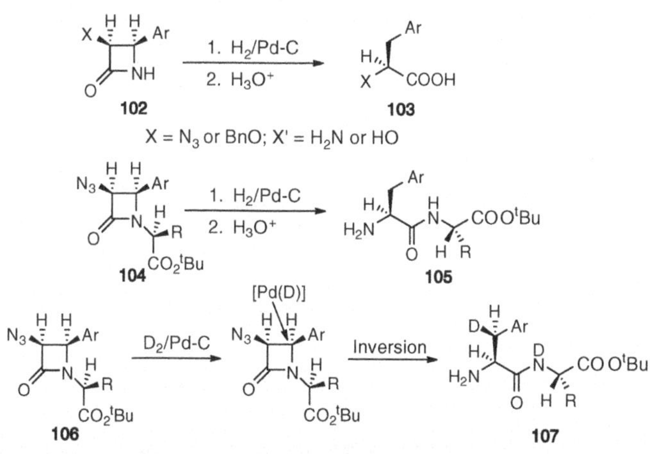

Scheme 33 1,4-Bond cleavage of 4-aryl-β-lactams by stereospecific hydrogenolysis

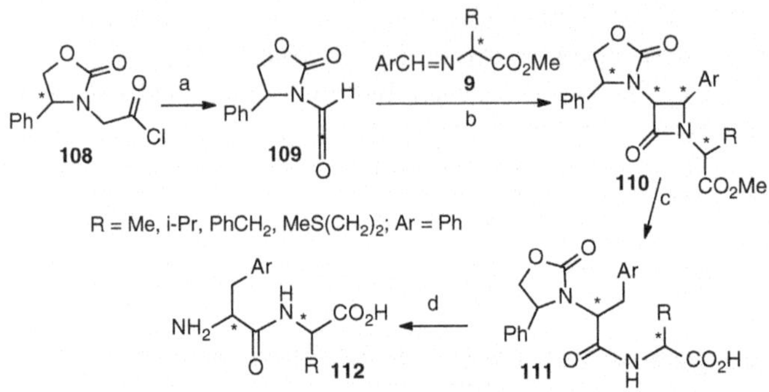

Reagents: a) NEt_3, CH_2Cl_2, -78 °C; b) CH_2Cl_2, -78~0 °C, 2h; c) (i) H_2, Pd / C, MeOH, 50 °C, 5 h, (ii) 1 N NaOH / THF, r.t., 1 h, c) H_3O^+; d) Li / NH_3 / ^tBuOH, -78 °C, 15 min.

Scheme 34 Asymmetric synthesis of dipeptides via stereoselective ketene-imine cycloaddition followed by reductive 1,4-bond cleavage and deprotection

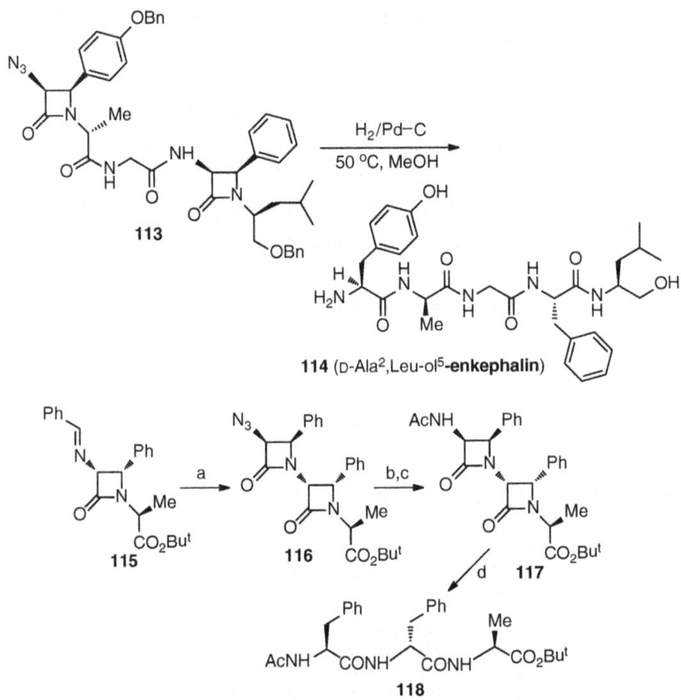

Reagents: (a) N_3CH_2COCl, Et_3N, CH_2Cl_2, -78 °C to room temp. (b) H_2 (1 atm), 5% Pd-C, MeOH, 0-5 °C. (c) Ac_2O, *N*-methylmorpholine, $CHCl_3$. (d) H_2 (1 atm), 10% Pd-C, EtOH, 50 °C.

Scheme 35 Applications of β-lactam synthon method to the synthesis of oligopeptides

It is worthy of note that the stereochemical course of the N1–C4 bond cleavage was found to be complete inversion of configuration at the C4 position, as illustrated in the reaction of **105** to give deuterium-labeled dipeptide **107** [76].

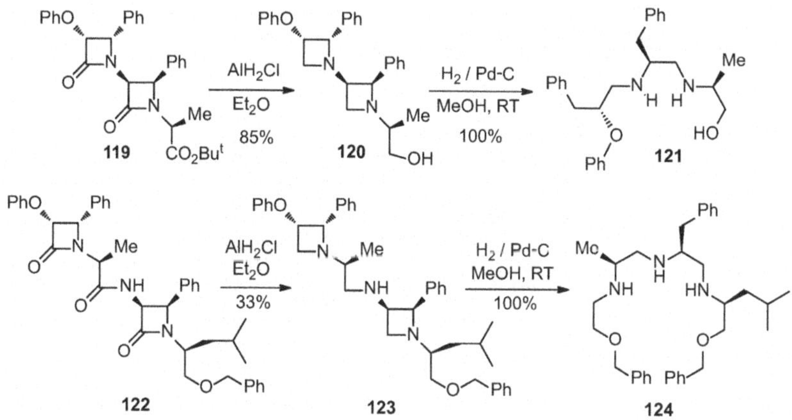

Scheme 36 Transformation of bis-β-lactams to bis-azetidines and then polyamino polyol derivatives

This finding opened a reliable route to the stereospecific isotope labeling of the β-position of aromatic α-amino acid residues in peptides, which are useful in enzyme and metabolism research (Scheme 33).

Enantiopure dipeptide synthon **110** can be obtained through [2+2] cycloaddition of chiral ketene **108**, generated from enantiopure oxazolidin-1-ylacetyl chloride **108** and Et_3N at −78°C, with α-amino ester imine **109**. The hydrogenolysis of **110** over Pd-C, followed by Birch reduction affords dipeptide **112** without epimerization [7, 77, 78] (Scheme 34).

Pentapeptide synthon **113**, prepared through coupling of two enantiopure 4-aryl-β-lactam units, was subjected to hydrogenolysis over Pd-C at 50°C to give [D-Ala^2, Leu^5-ol]enkephalin **114** in high yield [79]. In a similar manner, the reaction of enantiopure bis-β-lactam **117** afforded Ac-Phe-D-Phe-Ala-OMe (**118**) in high yield [27]. Bis-β-lactam **117** was prepared through extremely diastereoselective [2+2] cycloaddition of azidoketene to imino-β-lactam **115**, giving azido-bis-β-lactam **116**, which was selectively reduced to **117** at 0–5°C, followed by acetylation [27] (Scheme 35).

Enantiopure β-lactams can be easily converted to the corresponding azetidines without racemization using monochloroalane. For example, the selective hydroalane reduction of bis-β-lactam **119** afforded bisazetidine **120** in 85% yield and the subsequent N–C cleavage of **120** by hydrogenolysis over Pd-C gave the corresponding polyamino alcohol **121** in quantitative yield (Scheme 36) [75]. In the same manner, tetrapeptide equivalent bis-β-lactam **122** was converted to bisazetidine **123** in moderate yield and then to polyamino ether **124** in quantitative yield. It is worthy of note that the reductive cleavage of the two azetidine moieties of **123** were much faster than debenzylation in this case. Thus, **124** was obtained as dibenzyl ether (Scheme 36) [75].

Reduction of 4-ω-haloalkyl-β-lactam **125** with monochloroalane, followed by reflux in acetonitrile, led to the formation of pyrrolidine ($n = 0$) or piperidine ($n = 1$) **126** through ring expansion (Scheme 37) [80]. The ring expansion reaction

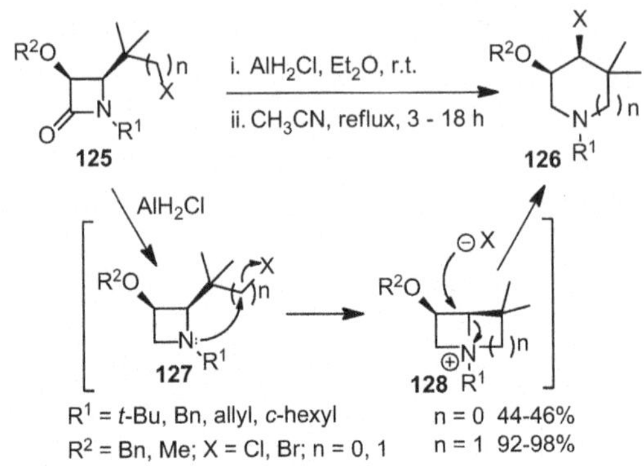

Scheme 37 Unique reductive ring-expansion process via bicyclic ammonium intermediate

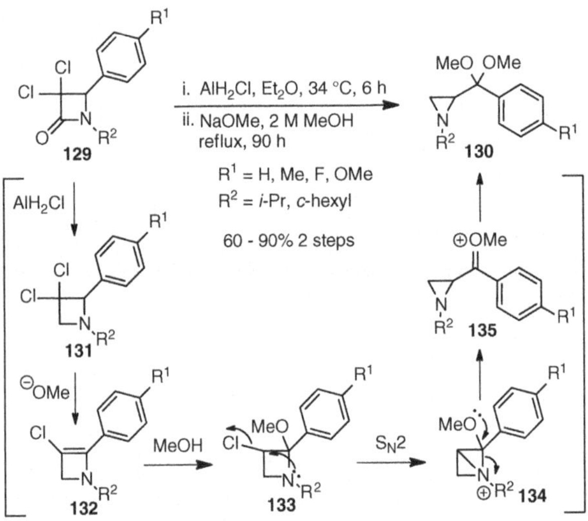

Scheme 38 Formation of aziridines via reductive ring-contraction process

appears to proceed via bicyclic ammonium intermediate **128**, which can accommodate the 3,4-*syn* stereochemistry of **126**. The intermediate **128** was found to react with external nucleophiles, including hydroxide and acetate [81].

When 2-aryl-3,3-dichloroazetidine **131**, derived from β-lactam **129** through monochloroalane reduction, was reacted with sodium methoxide, aziridine ketal **130** was obtained in 60–90% yield for two steps (Scheme 38). The proposed mechanism of this reaction includes a highly strained 2-azetine **132** and azabicyclo [1.1.0]cyclobutane **134** as intermediates. Then, **134** undergoes C–N bond cleavage and ring contraction to form aziridine oxonium salt **135**, which reacts with methoxide ion to give **130** (Scheme 38) [82].

In the course of an attempt to synthesize a fused tricyclic system containing β-lactams, 3,4-difunctionalized β-lactam **136** was treated with DBU to give

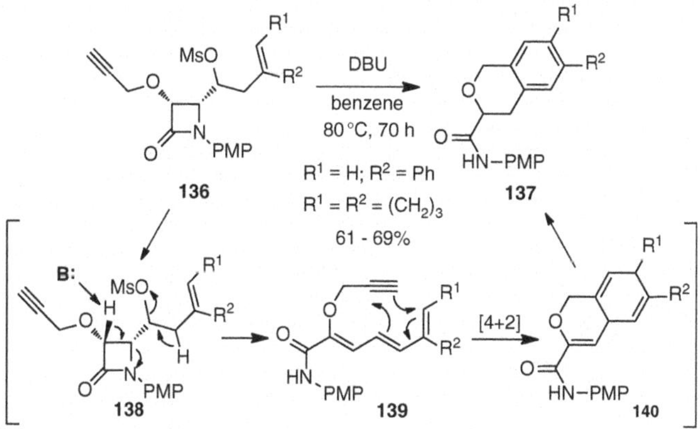

Scheme 39 Ring-expansion via 1,4-bond cleavage followed by Diels-Alder reaction

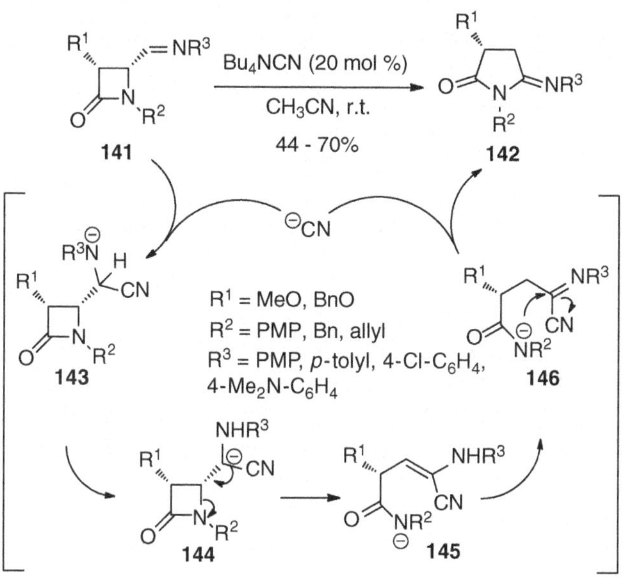

Scheme 40 Cyanide-catalyzed ring-expansion of 4-imino-β-lactam via 1,4-bond cleavage and recyclization

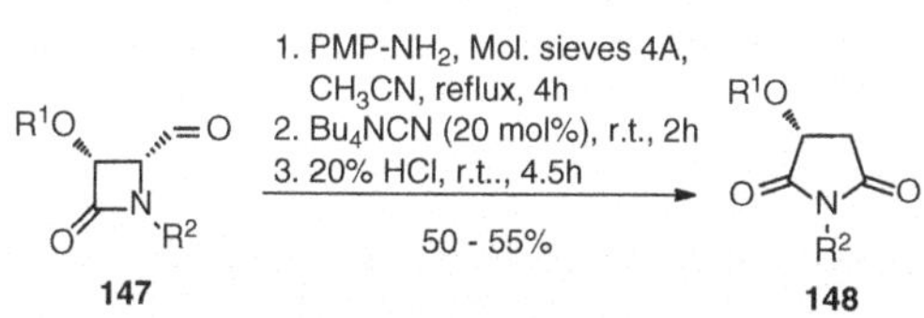

Scheme 41 Ring-expansion of 4-formyl-β-lactams to succinimides

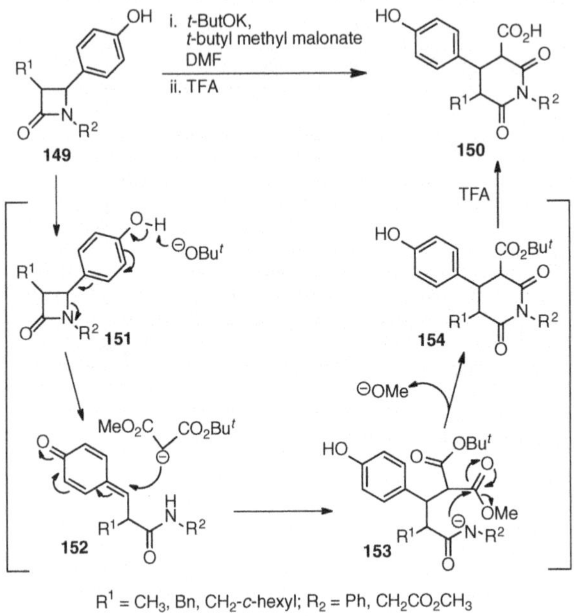

Scheme 42 Ring-expansion reaction to glutarimides

unexpected 2-carbamoylbenzodihydropyran **137** in 60–69% yield (Scheme 39) [83]. A proposed mechanism includes (a) novel N1–C4 bond cleavage initiated by the double deprotonation of two methines to form enynetriene **139**, (b) Diels–Alder cycloaddition to form **140**, and (c) bond isomerization of **140** to form aromatized product **137**, as illustrated in Scheme 39.

The reaction of 4-(arylimino)-β-lactam **141** with a catalytic amount of tetrabutylammonium cyanide (Bu_4NCN) gave chiral 5-aryliminopyrrolidin-2-one **142** via ring expansion through N1–C4 bond breakage (Scheme 40) [84]. A proposed mechanism is illustrated in Scheme 40. This process can be applied to the one-pot transformation of 4-formyl-β-lactam **147** to the corresponding succinimide **148**, including imine formation to **147** and hydrolysis of the imino moiety (Scheme 41).

Ring expansion of β-lactams to six-membered ring imides through N1–C4 bond cleavage has also been reported (Scheme 42). 4-Hydroxyphenyl-β-lactam **149** reacted with *t*-butyl methyl malonate in the presence of potassium *t*-butoxide to give 1,3,4,5-substituted glutarimide **150** through a mechanism illustrated in Scheme 42 [85]. An acid-catalyzed intramolecular variant on this ring expansion reaction was later applied to the synthesis of "tyrosyl" peptidomimetics [86].

3.3 *Ring Opening via C2–C3 Bond Cleavage*

The ring opening of β-lactams through the C2–C3 bond cleavage is less common as compared to the N1–C4 cleavage in the literature. Nevertheless, this process has

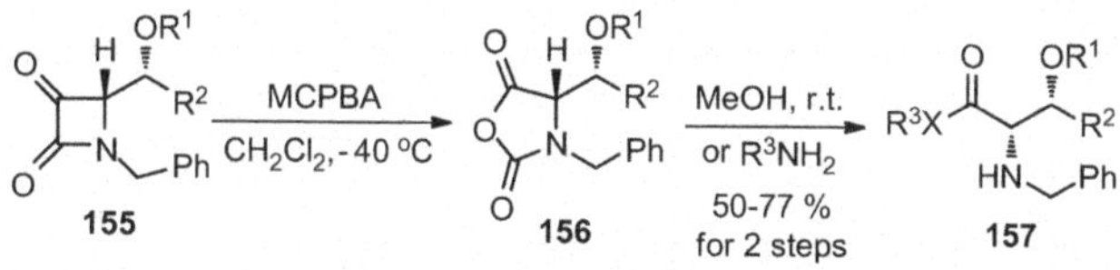

Scheme 43 Formation of *N*-carboxy anhydrides, versatile synthetic intermediates, via oxidative 2,3-bond cleavage of 3-keto-β-lactams

Scheme 44 Formation of *N*-carboxy anhydrides from 3-hydroxy-β-lactams

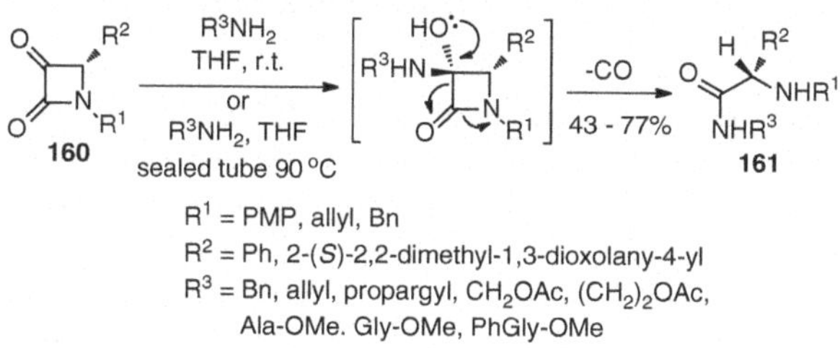

Scheme 45 Formation of α-amino acid amides via 2,3-bond cleavage of 3-keto-β-lactams

been used for the synthesis of various α-amino acids, *N*-carboxy anhydrides (NCAs), and spirobicyclic compounds [87, 88]. Future study on this process may lead to novel routes to biologically active compounds.

NCAs of α-amino acids have found broad applications in peptide synthesis as an NCA is an activated ester with simultaneous protection of the amino group. Direct access to NCA **156** from 3-keto-β-lactam **155** was found to be feasible through Bayer–Villiger oxidation, and **156** was reacted with methanol or amines to give the corresponding serine/threonine ester or amides **157** (Scheme 43) [89]. It was also found that NCA **159** could be obtained though a one-pot double oxidation of 3-hydroxy-β-lactam **158** using TEMPO-NaOCl as shown in Scheme 44 [90]. Decarboxylation of NCAs using TMSCl in methanol was reported to give the methyl esters of the corresponding amino acids in moderate to excellent yields [91].

The ring-opening coupling of 3-keto-β-lactam **160** with amines in THF gave the corresponding α-amino amides **161** via the C2–C3 bond cleavage and decarbonylation in moderate to good yield (Scheme 45) [92]. When an α-amino ester was used as an amine, the corresponding enantiopure dipeptide was formed in fairly good yield. Cyclic amines, e.g., pyrrolidine, piperidine, morpholine, and proline methyl ester, were also successfully employed in this reaction.

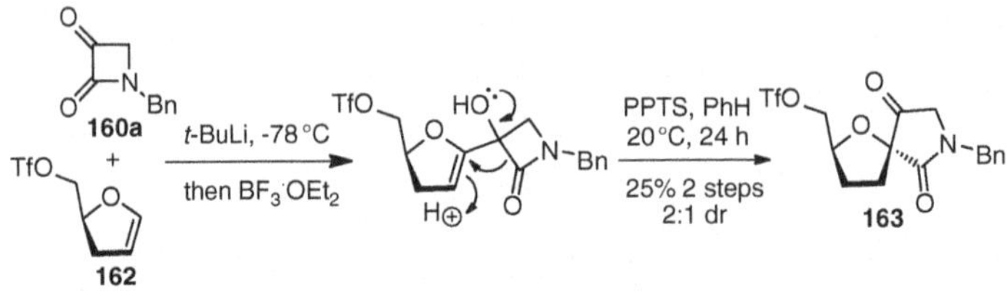

Scheme 46 Spirobicyclization via addition - ring-expansion involving 2,3-bond cleavage

An interesting application of the C2–C3 bond cleavage was reported for the reaction of dihydrofuran **162** with 3-keto-β-lactam **160a**. The reaction of **162** with **160a** in the presence of *t*-butyllithium and then $BF_3{\cdot}OEt_2$, followed by pyridinium *p*-toluenesulfonate (PPTS) gave spirobicyclic product **163** in 25% for two steps (Scheme 46) [88].

3.4 Ring Opening via C3–C4 Bond Cleavage

Ring opening reactions of β-lactams through the C3–C4 bond cleavage are fairly abundant in the literature, which include catalytic ring expansions, electrocyclic reactions, and radical processes. These reactions lead to a myriad of heterocyclic compounds with well-defined stereochemistry.

The acid-catalyzed ring rearrangement of 4-formyl-β-lactam **164** gave pyrrolidinedione **165** in 75% yield (Scheme 47) [93]. This reaction proceeded through the C3–C4 bond cleavage, initiated by the protonated aldehyde species (i.e., oxonium ion **166**) and skeletal rearrangement to pyrrolidinone cation **167**, followed by 1,2-hydride shift to form **165** and regenerate proton.

The reaction of 3-alkoxy-4-formyl-β-lactam **168** with *t*-$BuMe_2SiCN$ (TBSCN) catalyzed by molecular iodine gave 5-cyanopyrrolidin-2-one **169** in moderate to high yield with high *syn* selectivity (Scheme 48) [94]. This skeletal rearrangement is proposed to involve the C3–C4 bond cleavage, acyliminium ion formation, and

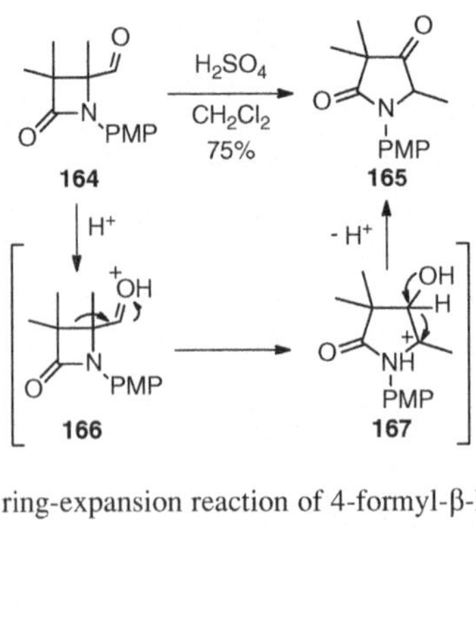

Scheme 47 Acid-catalyzed ring-expansion reaction of 4-formyl-β-lactam

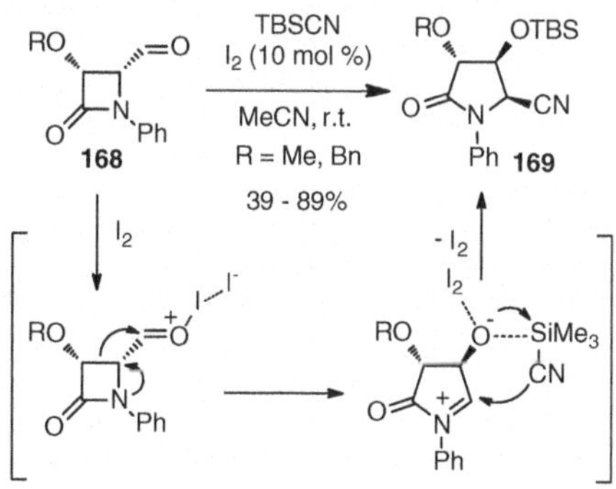

Scheme 48 Cyanide-catalyzed ring-expansion of 4-formyl-β-lactams

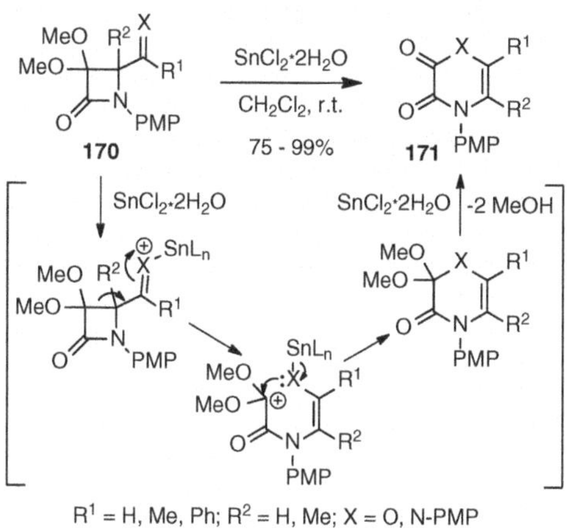

Scheme 49 Lewis acid - catalyzed ring-expansion of 3,3-dimethoxy-4-formyl(or imino)-β-lactams

cyanide addition. The proposed mechanism was supported by the DFT calculation using 1-phenyl-3-methoxy-4-formylazetidinone (**168a**) and TMSCN. This unique process is applicable to other nucleophiles such as allyltrimethylsilanes and propargylsilane [94].

4-Formyl-β-lactam **170a** (X = O) and 4-arylimino-β-lactam **170b** (X = ArN) underwent stannous chloride-mediated ring expansion via the C3–C4 bond cleavage to give 1,4-dihydrooxane **171a** (X = O) and pyrazine-2,3-dione **171b** (X = ArN), respectively, in good to excellent yield (Scheme 49) [93].

The thermal [1,3]-sigmatropic rearrangement of 3,3-bis(2-phenylethenyl)-β-lactam **172** gave 6-(2-phenylethenyl)-5,6-didehydropiperidin-2-one **173** in excellent yield (Scheme 50) [95]. The β-lactam **172** was prepared from a ketimine of bis(benzylidene)acetone with diphenylketene or dimethylketene via [2+2] cycloaddition and used for the rearrangement process in one pot. The piperidinone-diene **173** was used for hetero-Diels–Alder reaction with electrodeficient dienophiles to afford bicyclic and tricyclic alkaloid skeletons in excellent yields.

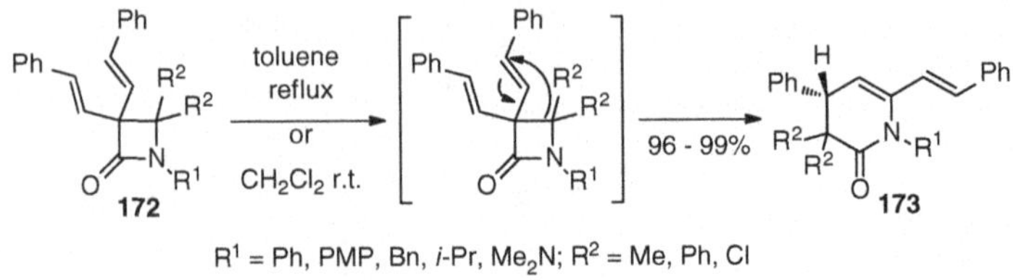

Scheme 50 3,4-Ring-cleavage via [1.3]-sigmatropic rearrangement, leading to ring-expansion

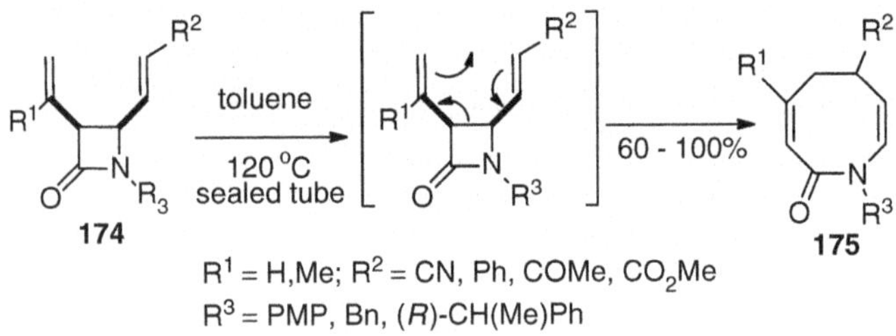

Scheme 51 Ring-expansion via [3,3]-sigmatropic rearrangement involving 3,4-bond cleavage

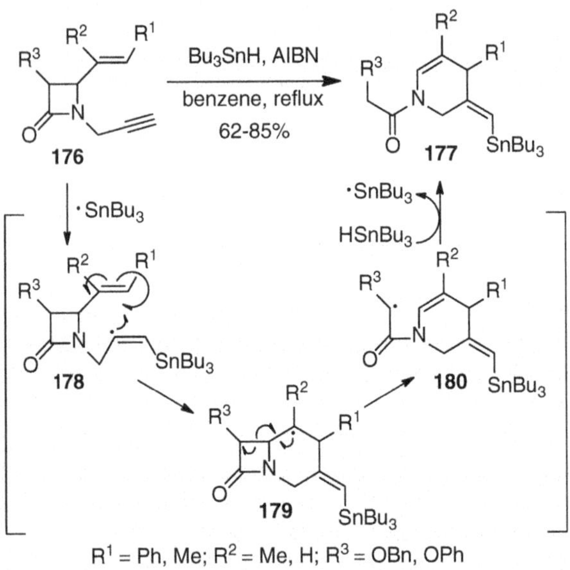

Scheme 52 Ring-expansion via radical cyclization involving 3,4-bond cleavage

The [3.3] sigmatropic (Cope) rearrangement of 3,4-dialkenyl-β-lactam **174**, involving the C3–C4 bond cleavage, took place at 120°C to give the corresponding eight-membered lactam, tetrahydroazocinone **175**, in fairly good to quantitative yield (Scheme 51) [96].

The tin-initiated radical cyclization of 1-propargyl-4-alkenyl-β-lactam **176** gave 1-acyl-5-tributylstannylmethylene-2,3-didehydropiperidine **177** in 62–85% yield, through cyclization to bicyclic radical intermediate **179** and bond migration to form a monocyclic radical intermediate **180** via C3–C4 bond cleavage (Scheme 52) [97].

4 Use of Enantiomerically Enriched β-Lactams as Intermediates for the Synthesis of Biologically Active Compounds of Medicinal Interest

4.1 β-Lactam Peptidomimetics

β-Turn peptidomimetics play an important role in drug design and medicinal chemistry [98]. Turns are common motifs in peptides, where a peptide chain reverses its overall direction. A β-turn is formed when this directional shift occurs over four residues in such a way that the carbonyl oxygen of the first residue (*i*) comes in close proximity to the amide proton of the fourth residue (*i* + 3) (Fig. 8) [99–101]. In the case of types I, II, and III β-turns, this conformation involves the formation of an intramolecular hydrogen bond between the two residues to give a pseudo-10-membered ring (Fig. 8). β-Turns are critical to the

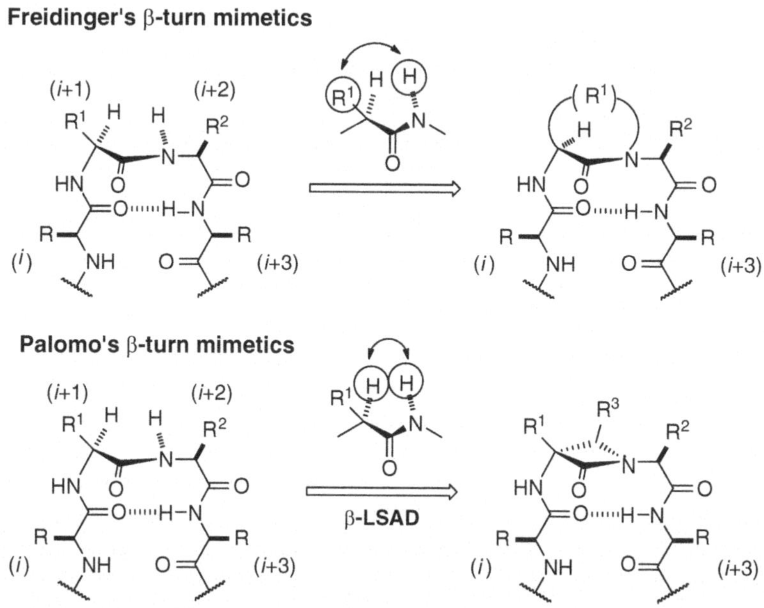

Fig. 8 β-Turn mimetics

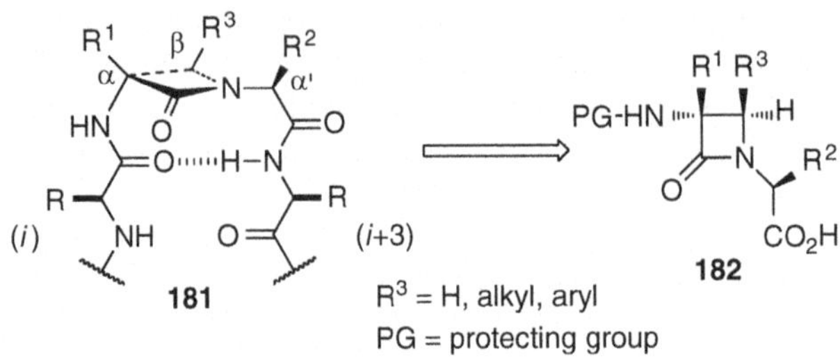

Fig. 9 β-lactam-based β-turn surrogate dipeptide

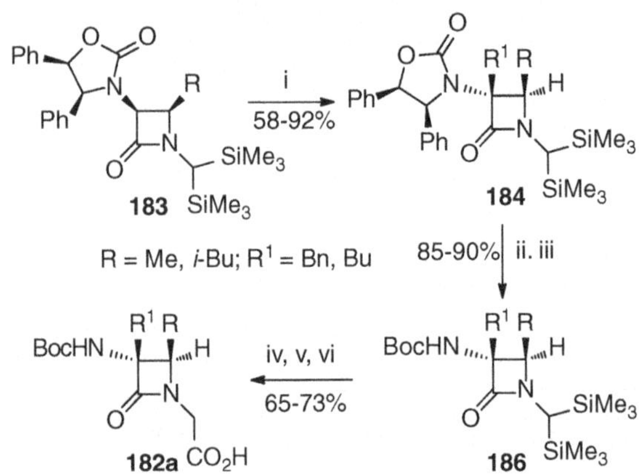

i) R^1X, LDA, THF, -78~20 °C; ii) H_2 (140 psi)/Pd-C, MeOH; iii) $(Boc)_2O$; iv) CAN, then $NaHCO_3$, MeOH; v) $BrCH_2CO_2Me$, $CsCO_3$, CH_3CN, 80 °C; vi) LiOH, THF, H_2O

Scheme 53 Synthesis of β-lactam-based β-turn surrogate dipeptide unit

stability of protein conformation [102], protein–protein interaction [103] and are preferred sites for protein degradation by proteolytic enzymes [104]. Accordingly, the development of an efficient method for the synthesis of enantiopure molecular templates for β-turns has a significant meaning in peptide and medicinal chemistry.

Novel β-turn surrogate dipeptides for structurally defined β-lactam peptides were designed and developed based on the "β-lactam scaffold-assisted design" (β-LSAD) (Fig. 8) [99–101, 105]. For the rational design of an efficient β-turn mimetic, two essential elements need to be incorporated into the formal modification of a native bioactive peptide, i.e., (1) a β-turn-constraining element is needed to force the peptide backbone to overlay the desired conformation and (2) at least one recognition group must be placed in a stereocontrolled manner at the desired position for interaction with the receptor or enzyme active site. Although the previous β-turn mimetics, developed by Freidinger et al., employed a tether connecting the R^1 moiety of the $i + 1$ amino acid residue and the NH hydrogen of the

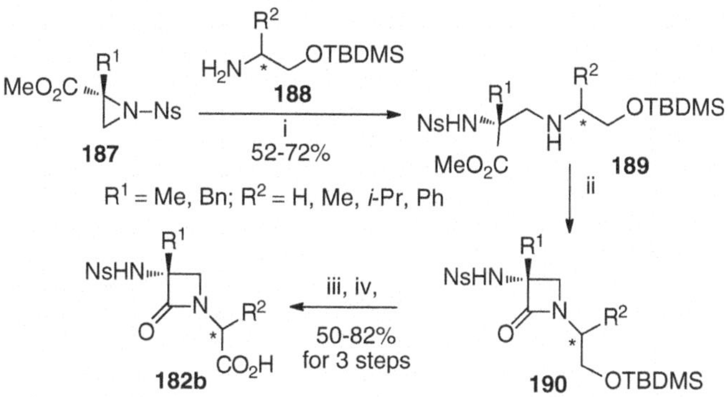

i) CH_3CN,r.t.; ii) LiHMDS,0°C; iii) HF; iv) CrO_3/H_2O or TCCA-TEMPO

Scheme 54 Synthesis of β-lactam-based β-turn surrogate dipeptide units from aziridines

i + 2 residue, forming six-, seven-, and eight-membered rings, the rigidity of those mimetics was often not sufficiently high [106–108]. Therefore, Palomo et al. introduced a highly rigid β-lactam skeleton, which appears to be very beneficial. In this case, the methine moiety of the i + 1 residue and that of the i + 2 residue needed to be connected by a methylene moiety (Fig. 8).

The β-turn mimetic **181** resulting from this approach differs in one substituted or unsubstituted methylene group at the β-position from the native peptides, as highlighted in the 3-alkyl-3-amino-β-lactam ring moiety and a linear disposition of the Cα, N, and Cα′ atoms (Fig. 9) [99]. This analysis led to the design of novel β-lactam-based β-turn surrogate dipeptide **182** (Fig. 9) [99].

Monotopic or ditopic β-lactam scaffolds **182** as the β-turn surrogate dipeptide units were synthesized through *syn*-selective 3-alkylation of enantiopure 3-amino-β-lactam derivatives **183**, giving 3-alkyl-3-amino-β-lactam **184**, followed by deprotection and oxidation (Scheme 53) [100], as well as ring-opening coupling of *N*-nosylaziridine **187** with α-amino silyl ether **188**, giving β-amino ester **189**, followed by cyclization to the corresponding β-lactam **190** and subsequent installation of carboxylic acid terminus via deprotection and oxidation (Scheme 54) [101].

4.2 *Asymmetric Synthesis of Nonprotein Amino Acids*

Nonprotein amino acids are amino acids that do not arise from protein amino acids by posttranslational modification and they do not have a specific transfer-RNA and codon triplet, thus they are not found in the main chain of proteins. Several nonprotein amino acids are important components of compounds of medicinal importance, thus the synthesis of enantiomerically enriched nonprotein amino acids is vital for medicinal chemistry and organic synthesis.

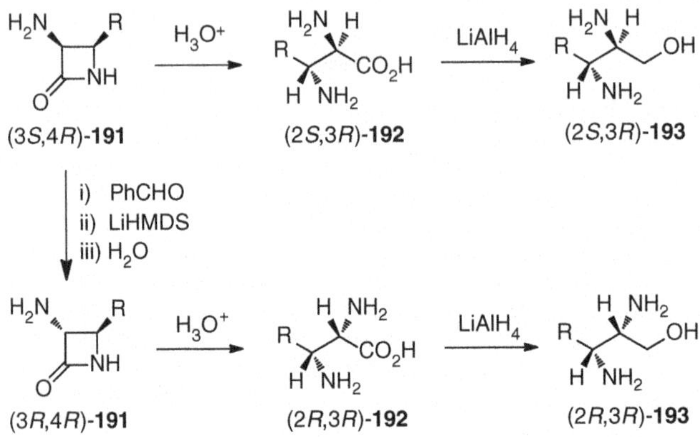

Scheme 55 Transformations of enantiopure 3-amino-β-lactams to diamino acids and diamino alcohols

4.2.1 Asymmetric Synthesis of α,β-Diamino Acids

α,β-Diamino acids are often components of peptidic antibiotics such as lavendomycin or glumamycin [109]. It has been shown that the hydrolysis of enantiopure 3-amino-β-lactams gives the corresponding α,β-diamino acid (Scheme 55) [7, 8]. Acidic hydrolysis of β-lactam (3*S*,4*R*)-**191**, which was prepared through chiral ester–enolate cyclocondensation or asymmetric [2+2] ketene–imine cycloaddition reaction (vide supra), gave α,β-diamino acid (2*S*,3*R*)-**192** as hydrochloride in quantitative yield. *Cis*-β-lactam (3*S*,4*R*)-**191** was epimerized to the corresponding *trans*-β-lactam (3*R*,4*R*)-**191** through imine formation, deprotonation, and protonation. β-Lactam (3*R*,4*R*)-**191** was converted to α,β-diamino acid (2*R*,3*R*)-**192** as hydrochloride through acidic hydrolysis in the same manner as that for (3*S*,4*R*)-**191**. Moreover, (3*R*,4*S*)-**191** can be prepared just by switching the chiral auxiliary to the other enantiomer in the asymmetric ketene–imine [2+2] cycloaddition (vide supra), (3*S*,4*S*)-**191,** (2*R*,2*S*)-**192** and (2*S*,2*S*)-**192** can be obtained in the same manner. Furthermore, α,β-diamino acid **192** was reduced to the corresponding α,β-diamino alcohols **193** in high yields using $LiAlH_4$. Thus, four diastereomers of enantiopure α,β-diamino acid **192** as well as α,β-diamino alcohol **193** can be readily obtained by this protocol.

4.2.2 Asymmetric Synthesis of α-Alkyl-α-Amino Acids and Their Derivatives

A good number of α-alkyl-α-amino acids serves as powerful substrate-based inhibitors of enzymes such as decarboxylases and aminotransferases [109]. Also, α-alkyl-α-amino acid residues can be introduced into physiologically active peptides to impose conformational constraints, which is highly beneficial for 3D

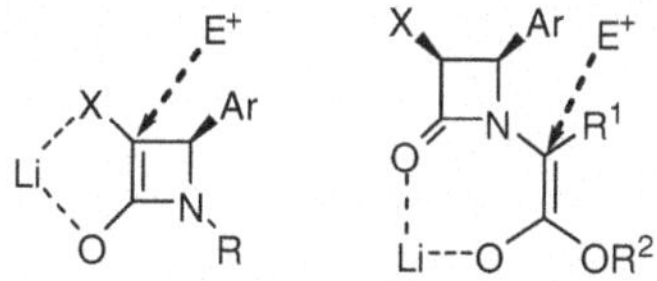

Scheme 56 Type 1 and Type 2 asymmetric alkylation of β-lactams

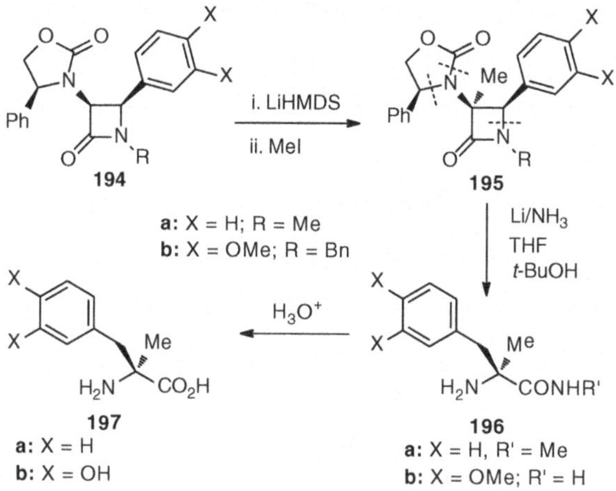

Scheme 57 Synthesis of enantiopure (*S*)-α-Me-Phe-OH and α-Me-DOPA via Type 1 alkylation

structure–activity/function studies [109]. The synthesis of α-alkyl-α-amino acids is challenging because conventional enzymatic resolution cannot be applied effectively. Thus, it is necessary to use asymmetric synthesis to obtain these amino acids. Among the synthetic methods investigated, the β-lactam synthon method provides one of the most efficient routes to α-alkyl-α-amino acids with high enantiopurity through extremely stereoselective alkylation of 3-amino-β-lactams.

Two types of asymmetric alkylations have been developed, i.e., (1) the alkylation of the C-3 position of a β-lactam (Type 1) and (2) the alkylation of the side chain ester enolate (Type 2) (Scheme 56) [7, 8]. In the Type 1 alkylation, an electrophile attacks the β-lactam enolate at the C-3 position from the back side of the C-4 aryl group to avoid steric hindrance. In the Type 2 alkylation, an electrophile attacks the C1′ position of the lithium enolate, which forms a chelate with the β-lactam oxygen, from the opposite side of the C-4 aryl group.

For example, the Type 1 alkylation was successfully applied to the asymmetric synthesis of (*S*)-methylphenylalanine (**197a**: X = H) and (*S*)-*O*,*O*-dimethyl-α-methyl-DOPA (**197b**: X = MeO) with >99.5% ee (Scheme 57) [7, 78, 110]. Methylation of β-lactam **194** with MeI and LiHMDS gave 3-methyl-β-lactam **195** with >99.5% de in excellent yield (Scheme 57). Birch reduction of **195** gave the corresponding α-methyl-α-amino acid amide **196** in good yield. Acidic hydrolysis

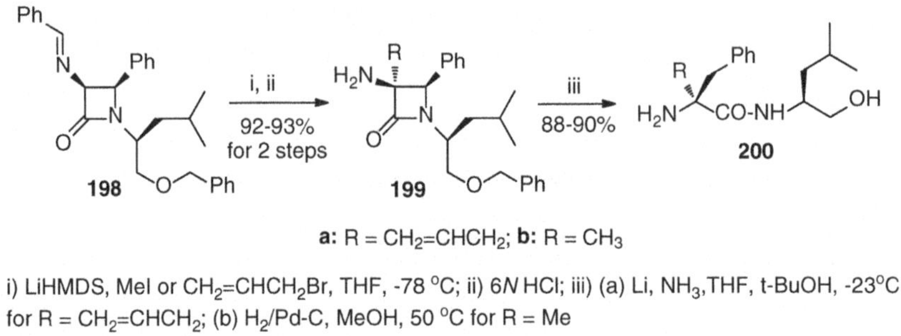

Scheme 58 Synthesis of enantiopure (*S*)-α-alkyl-Phe-(*S*)-Leu-ol via Type 1 alkylation

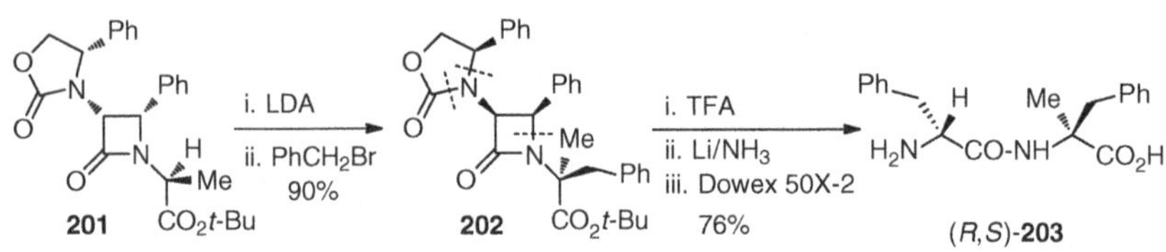

Scheme 59 Synthesis of enantiopure (*R*)-Phe-(*S*)-α-Me-Phe-OH via Type 2 alkylation

of **196** should afford **197**. The Type 1 alkylation was also applied to (*S*)-α-allyl-Phe-(*S*)-Leu-OH (**200a**) and (*S*)-α-Me-Phe-(*S*)-Leu-OH (**200b**) [7, 111] (Scheme 58). The allylation of imino-β-lactam **198** with LiHMDS and allyl bromide, followed by hydrolysis of the imine moiety gave 3-amino-3-allyl-β-lactam **199a** with >99.5% de in 93% yield. Birch reduction of **199a** gave **200a** in 90% yield. In a similar manner, **199b** with >99.5% de was obtained via the hydrogenolysis of **200b** on 10% Pd-C in high overall yield.

The Type 2 alkylation of β-lactam ester **201** with benzyl bromide and LDA gave C1′-dialkyl-β-lactam ester **202** with >99.5% de in excellent yield, and the subsequent Birch reduction afforded (*R*)-Phe-(*S*)-α-Me-Phe-OH (**203**) with >99.5% de in high yield (Scheme 59) [78]. The Type 2 alkylation was successfully applied to the sequential asymmetric double alkylation of chiral β-lactam ester **204**, which was prepared via asymmetric [2+2] cycloaddition of chiral ketene to *tert*-butyl *N*-benzylideneglycinate with >99% enantioselectivity. In this reaction, the double alkylation took place at the C1′ position and the absolute configuration of the newly formed chiral center was controlled just by changing the order of the addition of two alkyl halides (Scheme 60) [8, 110]. For example, the reaction of **204** with MeI and then benzyl bromide gave the C1′-dialkyl-β-lactam (3*R*,4*S*,1′*S*)-**206a** with >99% de in 79% yield, which was subjected to deprotection with trifluoroacetic acid, followed by Birch reduction to give (*S*)-Phe-(*R*)-α-Me-Phe-OH, (*S*,*S*)-**207** with >99.5% de in 76% yield.

The triple alkylation of **204** was also achieved through a combination of the Type 2 and Type 1 alkylations (Scheme 61) [8, 110]. After the Type 2 dialkylation with MeI and allyl bromide, the side chain of C1′-dialkyl-β-lactam ester **206b** had no acidic proton. Thus, Type 1 alkylation with MeI took place at the C3 position of

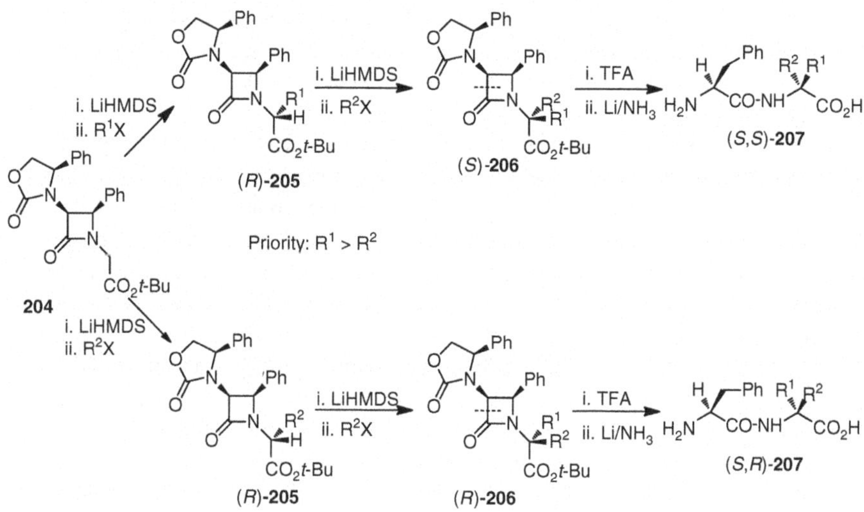

Scheme 60 Synthesis of (*S*,*S*)-and (*S*,*R*)-dipeptides via sequential Type 2 dialkylation

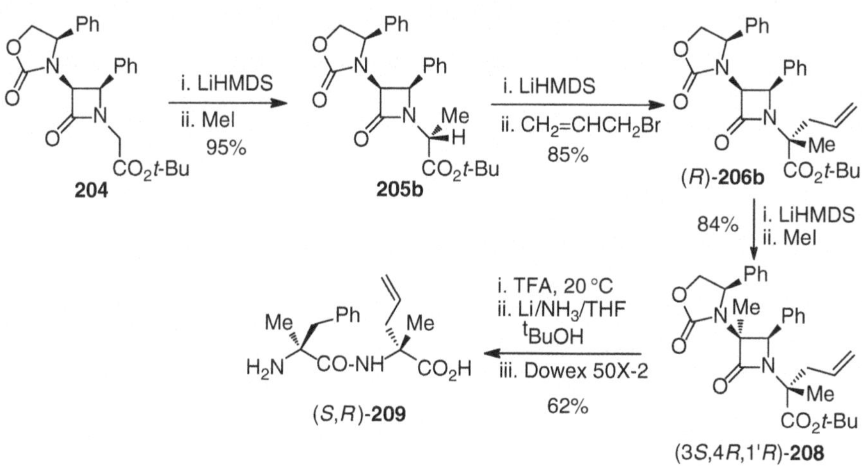

Scheme 61 Synthesis of enantiopure (*S*)-α-Me-Phe-(*S*)-α-allyl-Ala-OH via tandem triple alkylation

206b to give trialkyl-β-lactam (3*S*,4*R*,1′*R*)-**208** in high yield, which was converted to (*S*)-α-Me-Phe-(*R*)-α-allyl-Ala-OH, (*S*,*R*)-**209**, with >99.5% de in 62% yield.

4.2.3 Asymmetric Synthesis of *α*-Hydroxy-β-Amino Acids and Their Derivatives

α-Hydroxy-β-amino acids (isoserines) are key components of a variety of therapeutically important compounds. For instance, (2*R*,3*S*)-3-amino-2-hydroxy-5-

methylhexanoic acid (norstatine) and its analogs have been incorporated, as critical amino acid residues, into peptide-based inhibitors of enzymes such as rennin [112], HIV-I protease [113], and angiotensin converting enzyme (ACE) [114]. *N*-Benzoyl-(2*R*,3*S*)-3-phenylisoserine and *N*-*tert*-butoxycarbonyl-(2*R*,3*S*)-3-phenylisoserine moieties are essential components of paclitaxel and docetaxel, respectively, which are two of the most widely used anticancer drugs in cancer chemotherapy [115, 116]. Accordingly, extensive efforts have been made to develop efficient methods for the syntheses of isoserines with excellent enantiopurity, and the β-lactam synthon method has provided highly efficient solution to this synthetic challenge.

Norstatine and other isoserines **218** are readily accessible through acidic hydrolysis of 3-hydroxy-β-lactam **217** with excellent enantiopurity, which was obtained through chiral ester enolate–imine cyclocondensation or Staudinger ketene–imine cycloaddition followed by enzymatic resolution (vide supra) (Schemes 61, 62, 63, 64) [7, 61].

The N1–C2 bond cleavage of β-lactams is substantially enhanced when a strongly electron-withdrawing group, e.g., acyl, carbalkoxy, carbamoyl, sulfonyl, etc., is introduced to the N1 position. Thus, *N*-acyl-β-lactams can exploit additional

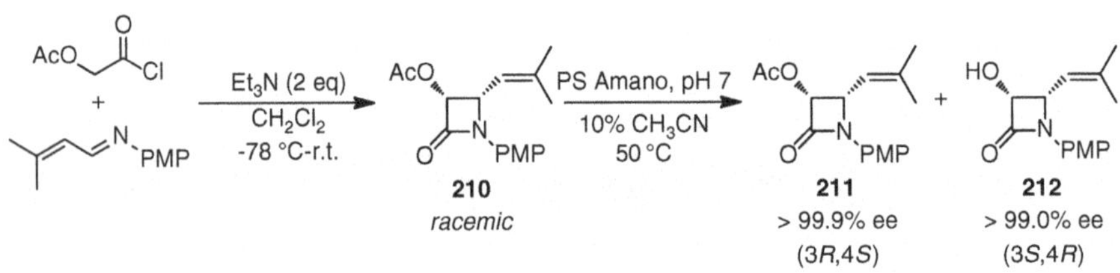

Scheme 62 Synthesis of 3-acetoxy- and 3-hydroxy-β-lactams via ketene-imine cycloaddition followed by enzymatic optical resolution

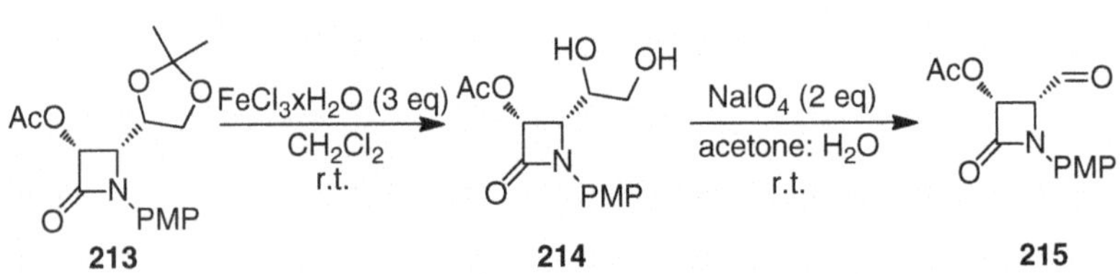

Scheme 63 Synthesis of 3-acetoxy-β-lactams from asymmetric ketene-imine cycloaddition product

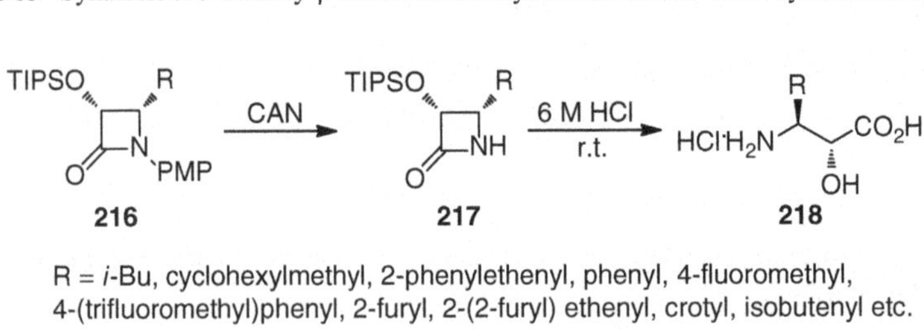

Scheme 64 Synthesis of α-hydroxy-β-amino acids from enantiopure 1-PMP-3-TIPSO-β-lactams

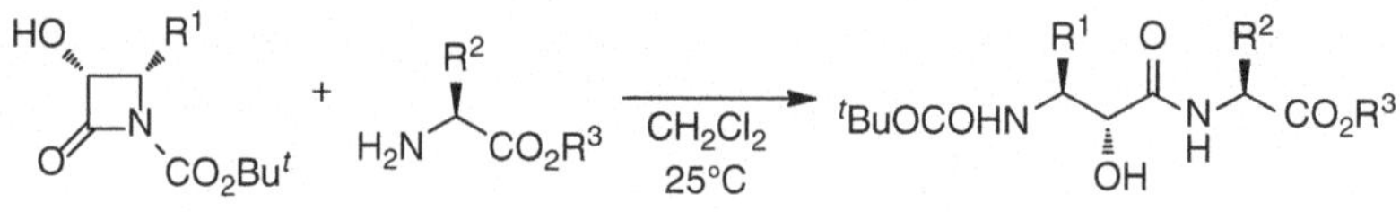

(a) R^1 = *i*-Bu; (b)R^1 = Ph; (c)R^1 = c-$C_6H_{11}CH_2$; (d) R^1 = PhCH=CH; R^2 = $PhCH_2$, *i*-Bu, *i*-Pr, indolylmethyl; R^3 = Me,*t*-Bu, Wang resin

Scheme 65 Synthesis of dipeptides via ring-opening coupling of *N*-*t*-Boc-3-hydroxy-β-lactam with α-amino acid ester

activation by the *N*-acyl group in addition to the inherent ring strain associated with the β-lactam skeleton. For example, *N*-Boc-3-hydroxy-β-lactams **219** undergo ring-opening coupling with various amino acid esters **220** to give the corresponding dipeptides **221** in high yields under mild conditions in the absence of any coupling agent without racemization (Scheme 65). This coupling method is applicable to bulkier proline methyl ester as well as the Wang resin-bound amino acids [7, 117]. This highly efficient and atom-economical ring-opening coupling represents one of the most salient features of the β-lactam synthon method. Furthermore, enantiopure *N*-Boc-3-siloxy-β-lactams have been successfully converted to the corresponding hydroxyethylene isosteres and dihydroxylethylene isosteres, which are critical components of various enzyme inhibitors [118, 119].

The incorporation of fluorine(s) into a biologically active compound often improves the pharmacological properties of the compound, resulting in increased membrane permeability, enhanced hydrophobic binding, and stability against metabolic oxidation among other benefits [49, 120, 121]. In addition, fluorine is not present in living tissue. Thus, the addition of fluorine(s) into biologically active compounds as marker(s) for ^{19}F NMR studies provides a simple and valuable means to monitor protein structures and drug–protein interactions in vitro and in vivo. Thus, the development of efficient methods for the synthesis of various fragments and components to construct those fluorine-containing biologically relevant compounds for pharmaceutical, pharmacological, medicinal, and chemical biology studies is in high demand.

The β-lactam synthon method has been successfully applied for the synthesis of fluorine-containing α-hydroxy-β-amino acid derivatives and congeners. Enantiopure 3-hydroxy-4-Rf-β-lactams (Rf = fluorine-containing substituent) were obtained through (1) ketene–imine [2+2] cycloaddition followed by enzymatic resolution of the resulting racemic 3-acetoxyl-4-Rf-β-lactams with the PS Amano lipase or (2) functional group transformations of enatiopure 1-PMP-3-TIPSO-4-(2-methylprop-1-enyl)azetidin-2-one, (3*R*,4*S*)-**216a** or (3*S*,4*R*)-**216a** [12]. For example, (3*R*,4*S*)-**216a** was converted to the corresponding 4-formyl-β-lactam (3*R*,4*R*)-**222** by ozonolysis, which was further reacted with diethylaminosulfur trifluoride (DAST) and CBr_2F_2/Zn to give 4-difluoromethyl-β-lactam (3*R*,4*R*)-**223** and 4-difluorovinyl-β-lactam (3*R*,4*R*)-**224**, respectively, with >99.9% ee in high yields (Scheme 66) [12, 50, 122]. 4-Trifluoromethyl-β-lactams, (+)-**226** and (−)-**227**, with >99% ee

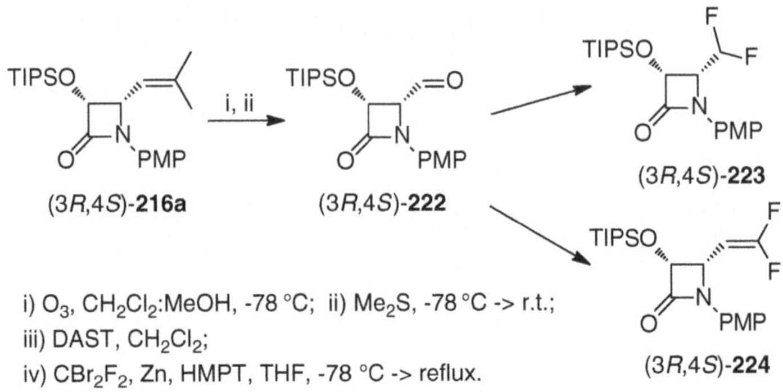

Scheme 66 Synthesis of 4-difluoromethyl- and 4-difluorovinyl-β-lactams via 4-formyl-β-lactam

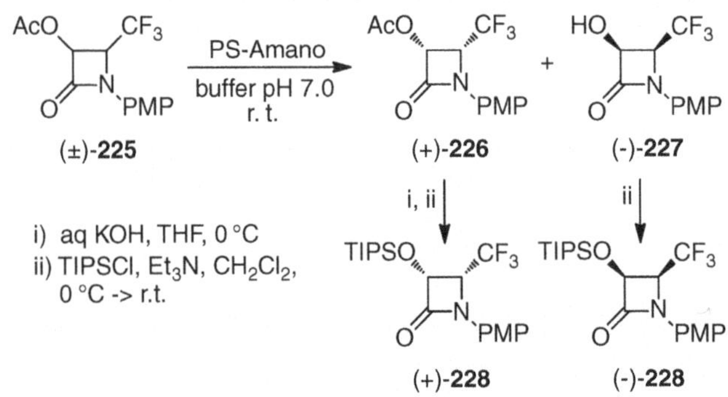

Scheme 67 Synthesis of enantiopure 3-TIPSO-4-trifluoromethyl-β-lactams via enzymatic optical resolution of racemic 3-acetoxy-β-lactam

were obtained through enzymatic resolution of racemic *cis*-3-acetyl-4-CF_3-β-lactam (±)-**225**, and converted to 3-TIPSO-β-lactams, (+)- and (−)-**228** (Scheme 67) [12, 50]. The PMP group of the enantiopure 4-Rf-β-lactams was removed by cerium ammonium nitrate (CAN) and the resulting NH-free β-lactams **229** were reacted with acyl chlorides, chloroformates, and arenesulfonyl chlorides in the presence of an appropriate base to give the corresponding *N*-acyl-, *N*-carbalkoxy, and *N*-arenesulfonyl-β-lactams, **230**, **231**, and **232**, in good yields (Scheme 68) [12]. These β-lactams were converted to the corresponding β-Rf-α-siloxy-β-amino acid methyl esters **233**, **234**, and **235**, respectively, with MeOH, triethylamine, and DMAP at room temperature (Scheme 68) [12]. In a manner similar to the case of 1-Boc-3-hydroxy-β-lactams **219** (Scheme 68), 1-Boc- and 1-Cbz-3-TIPSO-β-lactams **230** reacted with various α- and β-amino acid esters to give the corresponding dipeptides in good to excellent yields [12].

The alkylation of 3-siloxy-β-lactams **216** proceeded with excellent diastereoselectivity in the same manner as that of the Type 1 asymmetric alkylation of 3-imino- and 3-oxazolidinyl-β-lactams (vide supra) to give 3-alkyl-3-siloxy-β-lactams **236** with >99% de, which were converted to the corresponding α-

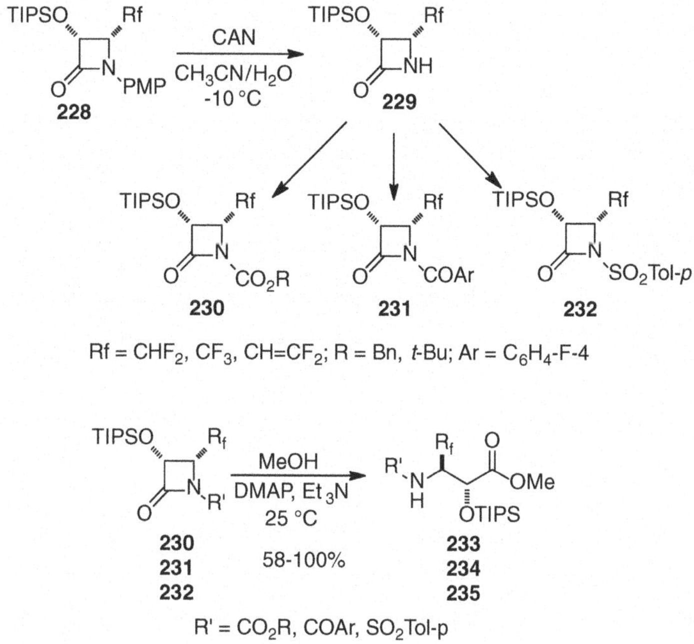

Scheme 68 Preparation of *N*-acyl- and *N*-sulfonyl-β-lactams and their conversion to fluorine-containing α-hydroxy-β-amino acid derivatives

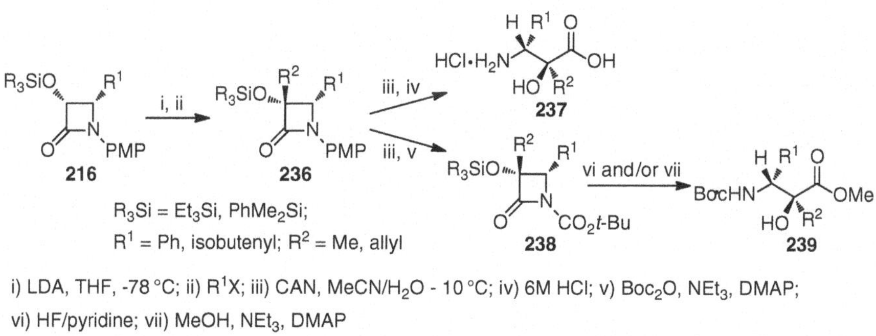

Scheme 69 Synthesis of enaniopure α-alkyl-isoserines and their derivatives via Type 1 alkylation

alkyl-α-hydroxy-β-amino acids **237** via acidic hydrolysis of NH-free β-lactams or *N*-Boc-α-alkyl-α-hydroxy-β-amino acid methyl esters **239** via *N*-Boc-3-alkyl-3-siloxy-β-lactams **238** (Scheme 69) [123]. *N*-Boc-β-lactams **219** and **236a** underwent ring-opening coupling with (*S*)-Leu-OMe and acetate enolate, respectively, to give the corresponding dipeptides **240** and a hydroxy(keto)ethylene dipeptide isostere **241** in high yields (Scheme 70) [123].

Cyclopropanation of 3-hydroxy-4-(2-methylprop-1-enyl)-β-lactam **212** with the Simons–Smith reagent (i.e., Et_2Zn/CH_2I_2) gave 4-[(*S*)-2,2-dimethylcyclopropyl]-

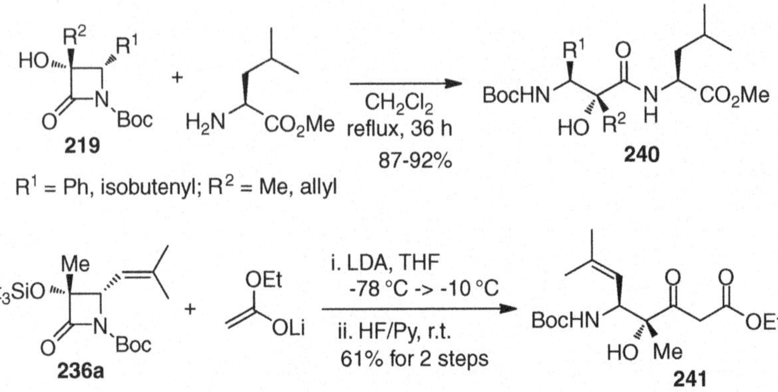

Scheme 70 Synthesis of dipeptides and peptidomimetics via ring-opening coupling of *N*-Boc-β-lactams

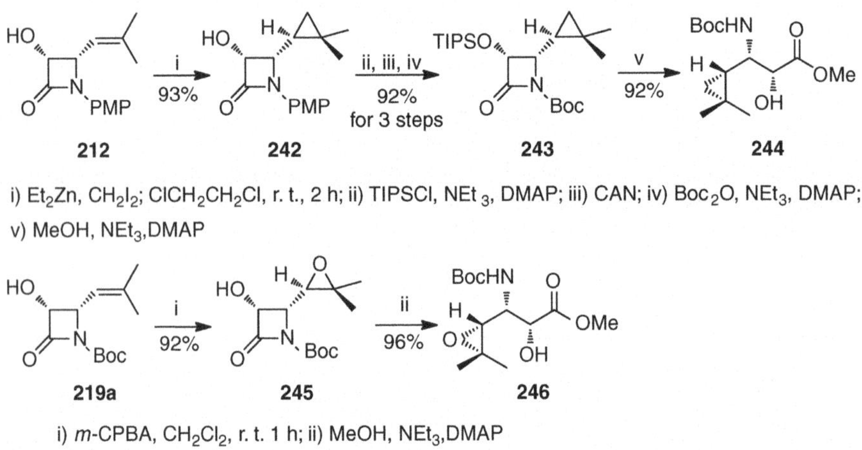

Scheme 71 Synthesis of enantiopure isoserines bearing cyclopropyl and epoxy moieties

β-lactam **242** with 100% de in 93% yield, which was converted to 1-Boc-3-TIPSO-4-(dimethylcyclopropyl)-β-lactam **243** and then to the cyclopropane analog of *N*-Boc-norstatine methyl ester **244** in excellent yield (Scheme 71). Also, the epoxidation of 1-Boc-3-hydroxy-4-(2-methylprop-1-enyl)-β-lactam **219a** with *m*-chloroperbenzoic acid (*m*-CPBA) gave 4-[(*R*)-2-methyl-1,2-epoxypropyl]-β-lactam **245** with 100% de in 92% yield, which was derived to *N*-Boc-epoxynorstatine methyl ester **246** in excellent yield (Scheme 71). It is noteworthy that the chiral 3-hydroxy-β-lactam skeleton served as the excellent stereogenic group for both reactions.

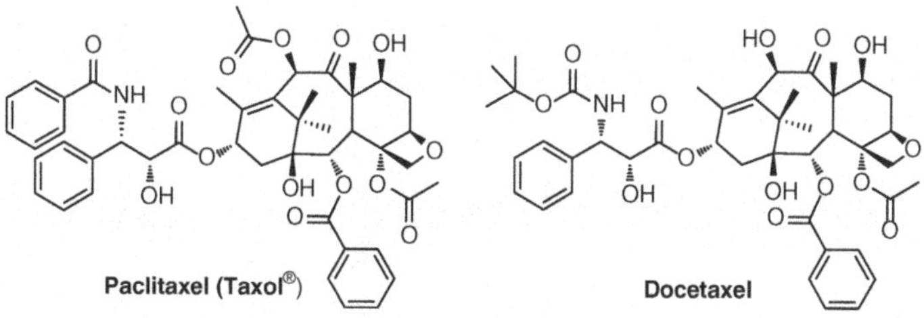

Fig. 10 Chemical structures of paclitaxel and docetaxel

4.3 *Synthesis of Paclitaxel, Docetaxel and New-Generation Taxoids Through Ring-Opening Coupling of N-Acyl-β-Lactams with Baccatin Derivatives*

Paclitaxel (Taxol®), a complex diterpene originally isolated from the bark of *Taxus brevifolia* (Pacific yew), has been approved by the FDA for the treatment of advanced ovarian cancer (1992), breast cancer (1994), AIDS-related Karposi's sarcoma (1997), nonsmall-cell lung cancer (1999), and other cancers [115, 116, 124]. Docetaxel, the first semisynthetic analog ("taxoid") of paclitaxel, was also approved by the FDA for treatment of *cis*-platin-refractory breast cancer in 1996 and has been used extensively for the treatment of various cancers such as breast, ovarian, prostate, and lung cancers [115, 116, 125]. Thus, paclitaxel and docetaxel (Fig. 10) are two of the most important drugs for cancer chemotherapy today. Paclitaxel and docetaxel are potent cytotoxic drugs with a unique mechanism of action, i.e., these drugs act as mitotic spindle poison by accelerating tubulin polymerization to microtubules, but stabilizing the resultant microtubules, which arrests the cell mitosis at the G_2/M phase and causes apoptosis [126–129].

There was a serious supply problem for paclitaxel through isolation from the nonrenewable bark of Pacific yew trees, but this issue was resolved by the discovery of substantial quantity of 10-deacetylbaccatin III (10-DAB) in the leaves of *Taxus baccata* (European yew), which are renewable resources [125, 130]. Since 10-DAB has exactly the same taxane core of paclitaxel, paclitaxel can be constructed in a semisynthetic manner by introducing (2*R*,3*S*)-*N*-benzoylphenylisoserine to the C13 position of properly protected baccatin III. However, the conventional ester formation encountered a serious epimerization problem [130] and thus efficient coupling methods needed to be developed. Then, the β-lactam synthon method provided an exceptionally practical solution, featuring the highly efficient ring-opening coupling of *N*-acyl-β-lactam with a baccatin III derivative [6, 7, 60, 131].

(3*S*,4*R*)-1-Benzoyl-3-EEO-4-phenyl-β-lactam **249a** (EE = 1-ethoxylethyl) was prepared from (3*S*,4*R*)-3-TIPSO-4-phenyl-β-lactam **247** (98% ee) directly obtained through highly efficient chiral ester enolate – *N*-TMS–imine cyclocondensation (Scheme 72) (vide supra) [6, 60].

The reaction of β-lactam **249a** with 7-TES-baccatin III (**250a**) in the presence of a base (DMAP, pyridine at 25°C for 12 h or LiHMDS at −30°C for 30 min) in THF

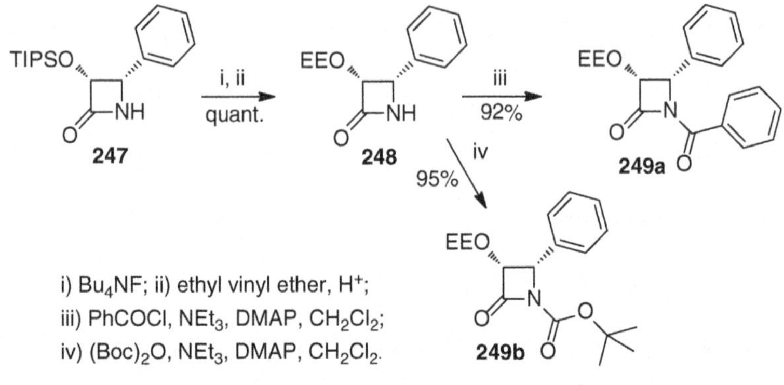

Scheme 72 Preparation of β-lactam synthons for *N*-benzoyl- and *N*-Boc-phenylisoserine moiety for ring-opening coupling with baccatins

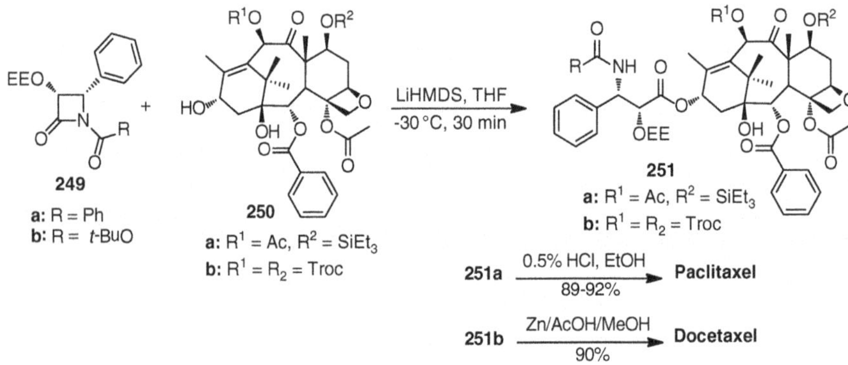

Scheme 73 Highly efficient synthesis of paclitaxel and docetaxel via ring-opening coupling of *N*-acyl-β-lactams with baccatins

gave the ring-opening coupling product **251a**, which was deprotected with 0.5% HCl in ethanol at 0°C to afford enantiopure paclitaxel in 89–92% yield (Scheme 73) [6, 7]. Since there is a kinetic resolution [132] during the ring-opening coupling with the enantiopure baccatin derivative, 1% (3*R*,4*S*)-isomer of the β-lactam **249a** remains unreacted and easily removed during isolation. Thus, **251a** isolated was 100% ee and 100% de. The β-lactam **249** and this ring-opening coupling method were also employed in the total synthesis of paclitaxel [133–136]. In a similar manner, docetaxel was synthesized in 90% yield through the ring-opening coupling of (3*S*,4*R*)-1-Boc-3-TIPSO-4-phenyl-β-lactam **249b** (98% ee) with 7,10-diTroc-10-deacetylbaccatin III (**250b**) (Troc = 2,2,2-trichloroethoxycarbonyl), followed by deprotection of the Troc and EE groups with Zn/AcOH/MeOH [6].

This highly efficient ring-opening coupling method used for the semisynthesis of paclitaxel and docetaxel opened an avenue for the synthesis and biological studies of a variety of paclitaxel congeners, "taxoids" [63, 131, 137–141]. The standard procedure for these couplings has quickly evolved to the use of 1-acyl- or 1-carbalkoxy-3-TIPSO-β-lactams and LiHMDS as the base [7].

Table 4 Selected new generation taxoids

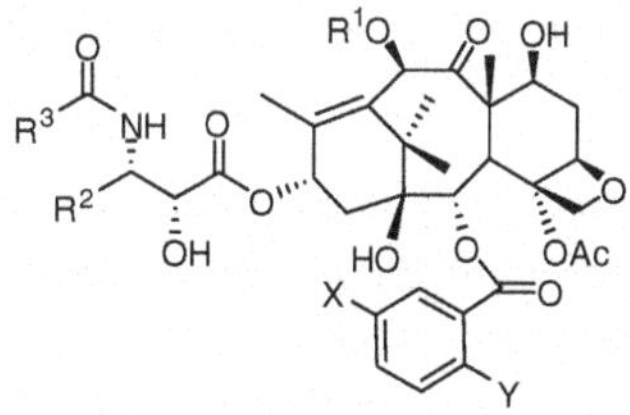

Taxoid	R^1	R^2	R^3	X	Y
Paclitaxel	Ac	Ph	Ph	H	H
Docetaxel	H	Ph	*t*-BuO	H	H
SBT-1213	EtCO	$Me_2C{=}CH$	*t*-BuO	H	H
SBT-1214	*c*-PrCO	$Me_2C{=}CH$	*t*-BuO	H	H
SBT-1216	Me_2NCO	$Me_2C{=}CH$	*t*-BuO	H	H
SBT-11033	EtCO	Me_2CHCH_2	*t*-BuO	MeO	H
SBT-121303	EtCO	$Me_2C{=}CH$	*t*-BuO	MeO	H
SBT-121313	EtCO	$Me_2C{=}CH$	*t*-BuO	MeO	MeO
SBT-121602	Me_2NCO	$Me_2C{=}CH$	*t*-BuO	Me	H
SBT-12854	Me_2NCO	$F_2C{=}CH$	*t*-BuO	H	H
SBT-12823-3	Me_2NCO	CF_3	*t*-BuO	Cl	H
SBT-12855-1	MeOCO	$F_2C{=}CH$	*t*-BuO	MeO	H

Table 5 Cytotoxicity (IC_{50}, nM) of selected new generation taxoids against human cancer cell lines

Taxoid	CFPac-1[a]	DLD-1[b]	HT-29[c]	MCF-7[d]	NCI/ADR[e]	LCC6-MDR[f]
Paclitaxel	68	300	12	1.7	550	346
Docetaxel				1.0	723	120
SBT-1213	4.6	3.9	0.37	0.18	4.0	
SBT-1214	0.38	3.8	0.73	0.2	3.9	
SBT-1216	0.66	5.4	0.052	0.13	7.4	
SBT-11033				0.36	0.61	
SBT-121303	0.89			0.36	0.79	0.90
SBT-121313	0.025	13.2	3.6	0.3		
SBT-121602	0.31	0.46	0.003	0.08		
SBT-12854	0.35	0.25	0.46	0.18		
SBT-12823-3			0.45	0.17	1.87	
SBT-12855-1				0.11	0.92	

[a]Pancreatic cancer cell line
[b]Colon cancer cell line (Pgp+)
[c]Colon cancer cell line (Pgp−)
[d]Breast cancer cell line (Pgp−)
[e]Ovarian cancer cell line (Pgp+)
[f]Breast cancer cell line (Pgp+)

Extensive structure–activity relationship (SAR) studies on taxoids have led to the discovery of highly potent new generation taxoids, which possess 2–3 orders of magnitude higher activity against multidrug resistant (MDR) and paclitaxel-resistant cancer cell lines and exhibit much better efficacy against human tumor xenografts in animal models [63, 139–141]. MDR is caused by the overexpression of P-glycoprotein (Pgp), which acts as an efflux pump for hydrophobic anticancer drugs. Selected new generation taxoids are listed in Table 4 and their potencies in Table 5.

In connection to the elucidation of the bioactive conformation as well as tubulin-bound structure of paclitaxel, conformationally restricted macrocyclic paclitaxel congeners have been designed and synthesized. The synthesis of these macrocyclic taxoids has been accomplished through the combination of the β-lactam synthon method and the Ru-catalyzed ring-closing metathesis (RCM) or intramolecular Heck reaction.

Enantiopure (3*R*,4*S*)-1-(2-alkenylbenzoyl)-3-TESO-4-phenyl-β-lactams **252** were prepared from 3-TIPSO-4-phenyl-β-lactam **247** (vide supra) and coupled to 7-TES-14β-allyloxybaccatin III (**253**) derived from 14β-hydroxybaccatin III [142] to give the corresponding dienyl-taxoids **254** in high yields (Scheme 59). The RCM of *N*-(2-allylbenzoyl)-taxoid **254a** using the first generation Grubbs catalyst at room temperature gave the expected 15-membered macrocyclic taxoid SBT-2053 in high yield after deprotection (Scheme 74) [143a]. However, when the same process was applied to *N*-(2-vinylbenzoyl)-taxoid **254b**, RCM reaction did not occur and an unanticipated novel coupling took place to give 15-membered macrocyclic taxoid

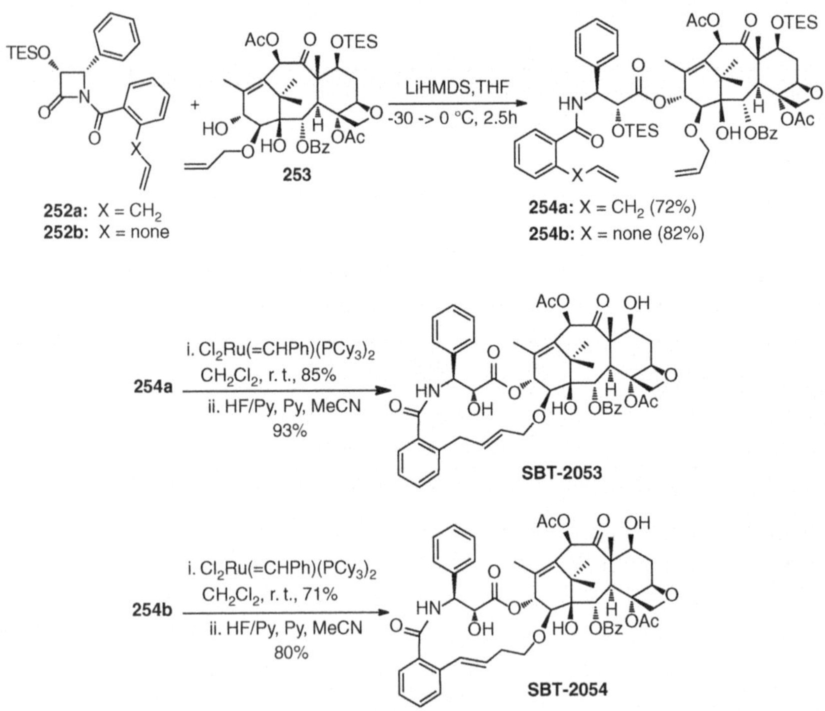

Scheme 74 Synthesis of conformationally restricted macrocyclic paclitaxel congeners via ring-closing metathesis

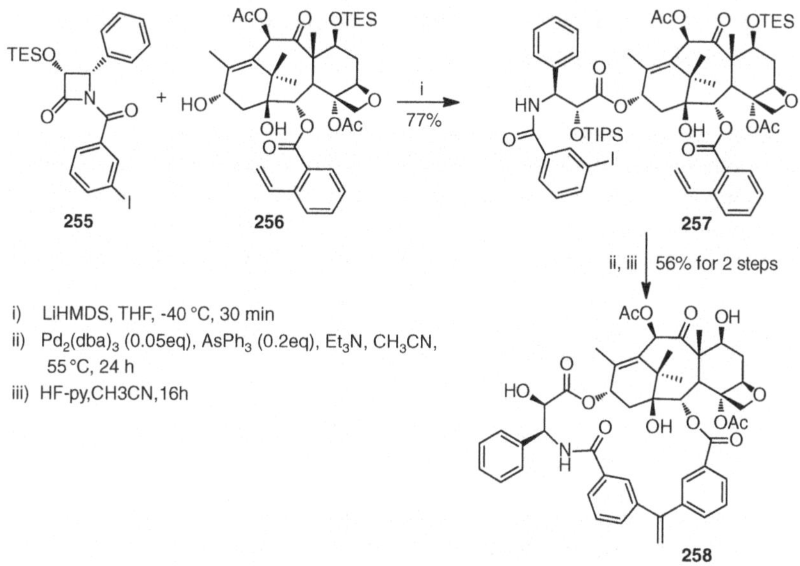

Scheme 75 Synthesis of conformationally restricted paclitaxel congener via Heck reaction

SBT-2054 bearing the double bond conjugated to the benzoyl moiety, in good yield after deprotection (Scheme 74) [143b]. This novel process is mediated by the Ru complex, but not catalytic, and proposed to involve Ru-allyl complexes as key intermediates [143b]. The biological activities of these macrocyclic taxoids, mimicking the tubulin-bound structure of paclitaxel, were examined for their tubulin polymerization and microtubule stabilization potencies as well as cytotoxicities against six human cancer cell lines. Then, both SBT-2053 and SBT-2054 exhibited virtually the same activity as that of paclitaxel in the tubulin polymerization/depolymerization assay and SBT-2054 was found to be as potent as paclitaxel in the cytotoxicity assay, which is the closest conformationally restricted paclitaxel congener to date [143, 144].

(3*R*,4*S*)-1-Boc-3-TESO-4-vinyl-β-lactam and (3*R*,4*S*)-1-Boc-3-TESO-4-allyl-β-lactam derived from (3*R*,4*S*)-1-PMP-3-TIPSO-4-formyl-β-lactam (**222**) were also employed for the preparation of a variety of dienyl-taxoids, through the ring-opening coupling with a number of C2-alkenyl-7-TES-baccatins, which were subjected to the Ru-catalyzed RCM, followed by deprotection to afford the corresponding C2-C3′-linked macrocyclic taxoids in moderate to high overall yields [143c].

3′N–C2 linked macrocyclic taxoid **258** was synthesized through the intramolecular Heck reaction (Scheme 75) [143d]. Enantiopure 1-(3-iodobenzoyl)-3-TESO-4-phenyl-β-lactam **255** was prepared from 3-TIPSO-4-phenyl-β-lactam **247** (vide supra). The ring-opening coupling of **255** with 2-(2-vinylbenzoyl)-7-TES-baccatin III (**256**) gave iodo-vinyl-taxoid **257** in high yield. The Heck reaction of **257** using a Pd(0) complex with $AsPh_3$ as the ligand proceeded at 55°C to form the 19-membered exo-methylene-macrocyclic taxoid **258** in 56% yield after deprotection. Macrocyclic taxoid **258** exhibited the IC_{50} value of 67 nM against LCC6-WT human breast cancer cell line [143d].

4.4 Synthesis of Cryptophycins Through N-Acyl-β-Lactam Macrolactonization

Cryptophycins, macrocyclic cytotoxins produced by cyanobacteria, are tumor-selective tubulin-binding antitumor agents which disrupt cellular mitosis (Fig. 11). These compounds display excellent cytotoxicity against MDR cancer cell lines, and are promising leads for the development of chemotherapeutic agents [145]. A synthetic analog, cryptophycin-52, is in clinical development for treatment of

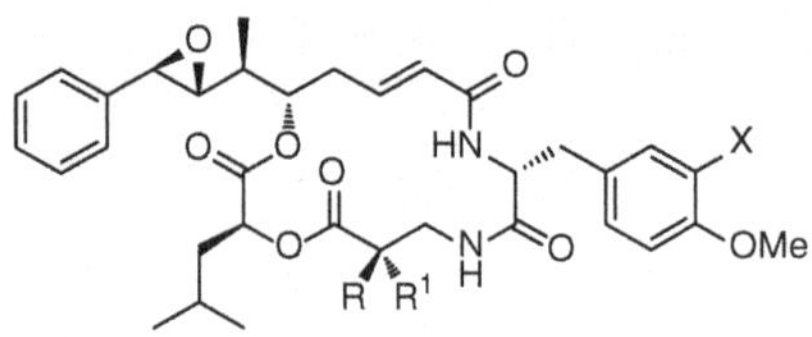

Cryptophycin-1: R = Me, R^1 = H, X = Cl
Cryptophycin-24: R = H, R^1 = H, X = H (**Arenastatin A**)
Cryptophycin-52: R = R^1 = Me, X = Cl
Dechlorocryptophycin-52: R = R^1 = Me, X = H

Fig. 11 Chemical structures of cryptophicins

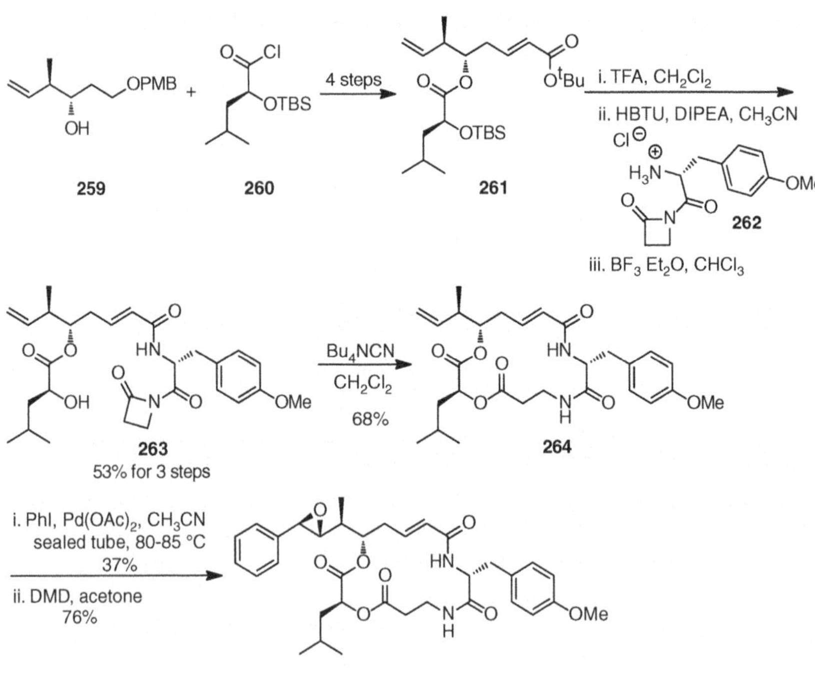

Cryptophycin-24 (Arenastatin A)

Scheme 76 Total synthesis of cryptophycin-24 via β-lactam synthons solid tumors. Cryprophycin-52 possesses extremely potent cytotoxicity(e.g., IC_{50} 0.037 nM for MCF-7) [145].

An efficient and concise route to the macrolide core of cryptophycins has been developed using the *N*-acyl-β-lactam macrolactonization [146, 147]. The synthesis of cryptophycin-24 (Arenastatin A) is illustrated in Scheme 76, as a representative total synthesis of this class of compounds [147]. The β-lactam key intermediate **263** was prepared through the coupling of **261**, obtained from **259** and **260** in four steps, with *N*-(*p*-MeO-Phe)-β-lactam **262**. The macrolactonization of **263** was achieved through cyanide-initiated β-lactam ring opening using Bu_4NCN and subsequent lactonization. The attempted use of NaH and NaHMDS as base for direct ring-opening coupling was unsuccessful, presumably due to the instability of **263** to basic conditions. The C3′-phenyl moiety was introduced using the Heck reaction. Epoxidation of the resulting styryl double bond was carried out with dimethyldioxirane (DMD) to give cryptophycin-24 as a 2:1 (β:α) diastereomeric mixture, which was separated by reverse-phase HPLC. Dechlorocryptophycin-52 was synthesized in the same manner [147].

4.5 Synthesis of (−)-Lankacidin C

Lankacidin C is a 17-membered macrocyclic tetraene isolated from various species of *Streptomyces* [148–150]. Lankacidin C possesses strong antimicrobial activities against a variety of Gram-positive bacteria, including several strains resistant to the conventional macrolide antibiotics. In addition, lankacidin C and its derivatives show considerable in vivo antitumor activity [151].

An intramolecular ring-opening coupling of a β-lactam intermediate **269** has been used as a key step in the total synthesis of (−)-lankacidin C (Scheme 77) [152]. β-Lactam **265**, derived from (*S*)-aspartic acid, served as the C1–C8 synthon in the lankacidin synthesis. The highly stereoselective acylation of the C3 position of **265** through the reaction of the Type 1 lithium enolate (vide supra) with activated ester **266**, which is the C12–C18 synthon, gave C3-acyl-β-lactam **267** in high yield. The reduction of the C18-ketone moiety of **267** with $KHBEt_3$ in ether afforded C18α-alcohol **268** exclusively. Silyl protecting groups of **268** were removed with Bu_4NF to give β-lactam triol **269**, which underwent acid-catalyzed ring-opening coupling with C16-OH, forming δ-lactone, followed by cyclic carbamate formation, linking the NH_2 and C18-OH groups to afford advanced key intermediate **270**. Introduction of the C9–C11 fragment to **270**, followed by macrocyclization completed the total synthesis of (−)-lankacidin C in several steps.

4.6 Synthesis of Combretastatin Mimics

Combretastatins are a group of diarylstilbenes that are isolated from the stem of *Combretum caffrum* (South African Bushwillow tree) [153]. These compounds

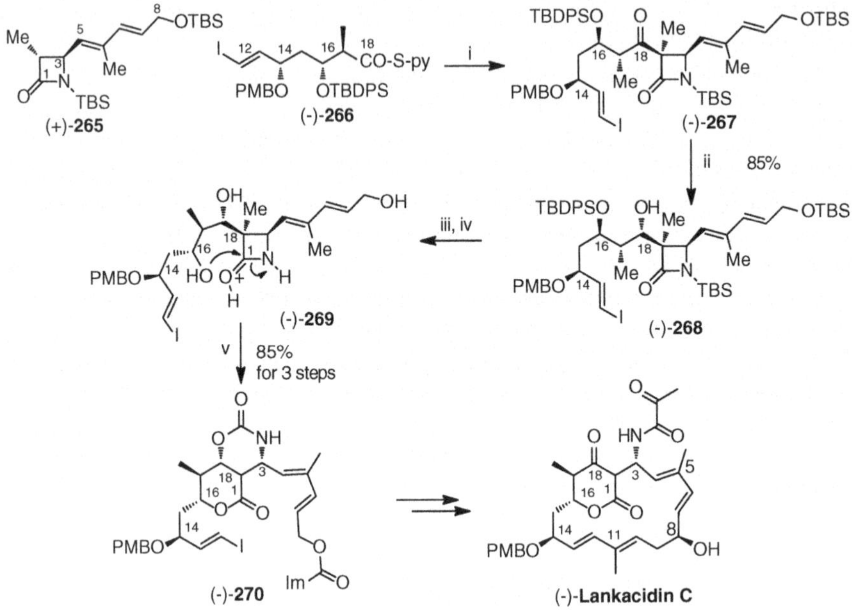

i) LDA, THF, -78 °C, 10 min; ii) KEt_3BH, Et_2O, -78 °C, 10 min; iii) TBAF, THF, r. t., 2 h; iv) MsOH, r. t., 2 h; v) NEt_3, CDI, r. t., 12 h. TBS = *t*-BuMeSi; TBDPS = *t*-$BuPh_2Si$; PMB = *p*-methoxybenzyl

Scheme 77 Total synthesis of (-)-lankacindin C via β-lactam synthons

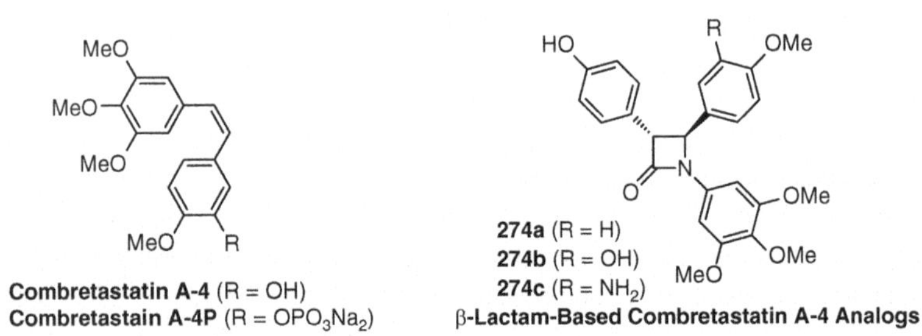

Fig. 12 Combretastatins and β-lactam-based mimics

have potent anticancer activity and share the same tubulin-binding site with colchicine [154]. Among these, combretastatin A-4 is highly potent and its phosphate, combretastatin A-4-P, a water-soluble prodrug, is currently in clinical trials for the treatment of thyroid cancer (Fig. 12) [155, 156]. Since the *cis*-olefin structure of combretastatins is prone to isomerize, which causes the loss of activity, structurally stable combretastatin analogs have obvious advantage. As an approach to develop potent and structurally stable analogs, a series of β-lactam-based combretastatin mimics was designed, synthesized, and examined their activities, and some of these mimics exhibited promising activities [157]. For example, β-lactam mimics **274a**

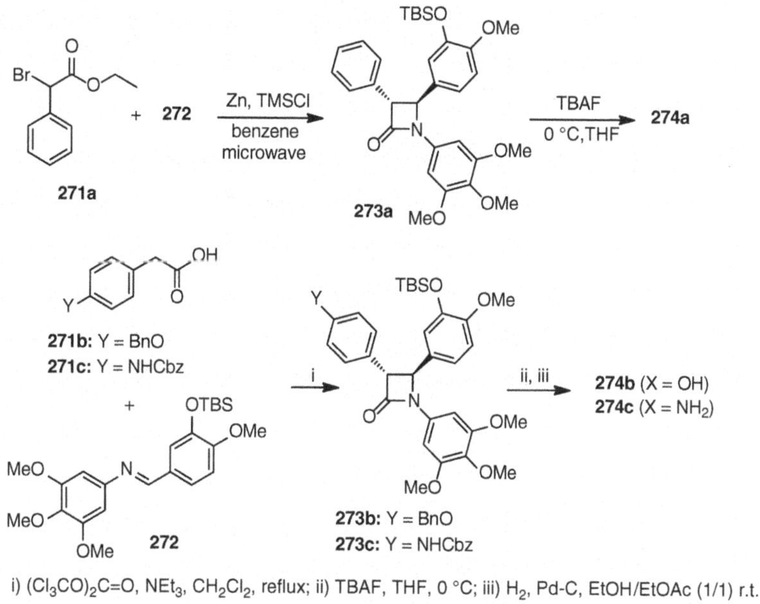

Scheme 78 Synthesis of β-lactam based combretastatin mimics

(IC_{50} 9.6 nM), **274b** (IC_{50} 0.8 nM), and **274c** (IC_{50} 4.5 nM) showed comparable or better potency than combretastatin A-4 (IC_{50} 5.2 nM) against MCF-7 human breast cancer cell line [157]. It is worthy of note that these β-lactam-based analogs did not show significant activity in normal murine breast epithelial cells. In addition, it was shown that compounds **274a** and **274b** inhibited tubulin polymerization with higher efficacy than combretastatin A-4 [157].

The β-lactam analog **274a** was synthesized by using the Reformatsky reaction of phenylbromoacetate **271** with imine **272**, followed by deprotection of the TBS group with Bu_4NF and the benzyl group by hydrogenolysis (Scheme 78) [157]. The analogs **274b** and **274c** were prepared through the Staudinger ketene–imine cycloaddition, followed by deprotection of silyl group and benzyl or carbobenzoxy (Cbz) group (Scheme 78) [157]. Since the chemical yields reported were low, substantial improvement in the syntheses is necessary in the future if these β-lactam analogs move to more advanced phases of drug development.

4.7 *Synthesis of Cholesterol Absorption Inhibitor, Ezetimibe*

Atherosclerotic coronary heart disease (CHD) is the major cause of death in the United States. There is a positive correlation between CHD incidences with elevated serum cholesterol levels. Accordingly, clinical treatment for CHD entails lifestyle changes (diet and exercise) and reduction of serum cholesterol levels.

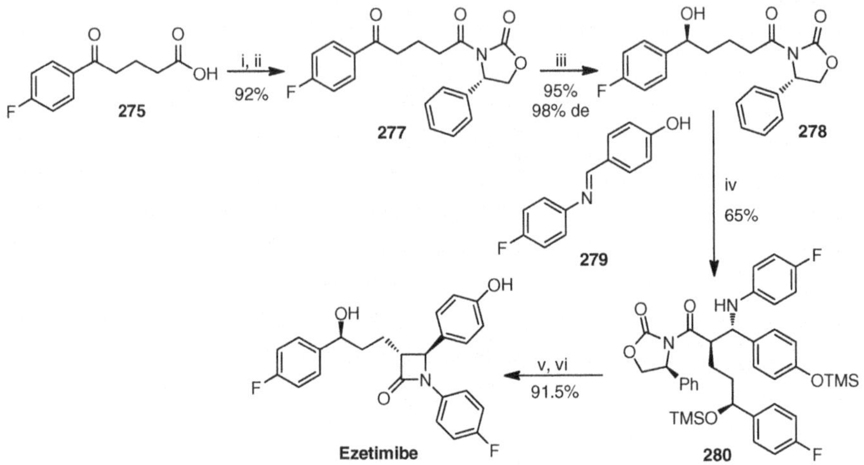

i) pivaloyl chloride, Et_3N; ii) (*S*)-4-phenyloxazolidinone (**276**); MeCBS (3 mol%), iii) BH_3-THF THF; iv) TMSCl, DiPEA, $TiCl_4$, CH_2Cl_2, -25 °C; v) bis-TMS-acetamide, *t*-BuOMe, TBAF.3H_2O; vi) aq. *i*-PrOH, dil. H_2SO_4

Scheme 79 Synthesis of ezetimibe

The major contributor of serum cholesterol is from dietary or intestinal sources. Thus, blocking the intestinal sources of cholesterol should be an effective approach to control the serum cholesterol level. Burnett and others have developed β-lactams which block cholesterol absorption through acyl-CoA:cholesterol *O*-acyltransferase (ACAT) inhibition [158]. Among those β-lactam-based ACAT inhibitors, a *trans*-β-lactam, ezetimibe, was found to be a potent and highly efficacious cholesterol absorption inhibitor and has been approved by the FDA for the treatment of hypercholesterolemia.

A patented asymmetric synthesis of ezetimibe is illustrated in Scheme 79 [158, 159]. In this synthesis, the *S* configuration of the benzylic hydroxyl group in the C3 chain was introduced in the early stage of the synthesis via a Corey–Bakshi–Shibta (CBS) asymmetric reduction of **277**, which was prepared from δ-keto acid **275** and (*S*)-phenyloxazolidinone (**276**). Asymmetric enolate–imine condensation of **278** and **279** gave **280** with high stereoselectivity. Silylation of **280** followed by fluoride-catalyzed cyclization afforded ezetimibe in high yield [159].

An alternative route to ezetimibe was also developed, featuring the one-step highly diastereoselective formation of *trans*-β-lactam **283** through highly diastereo-selective enolate–imine cyclocondensation of (*S*)-hydroxy-γ-lactone (**281**) with imine **282** (see Scheme 65) [160]. Oxidation of diol **283** to the corresponding aldehyde by sodium periodate, followed by Mukaiyama aldol condensation with a silyl enol ester **284** and deprotection gave **285**. Hydrogenation of the double bond and asymmetric CBS reduction of the ketone moiety, followed by the final deprotection of the benzyl group by hydrogenolysis afforded ezetimibe with >99% ee in good overall yield [160] (Scheme 80).

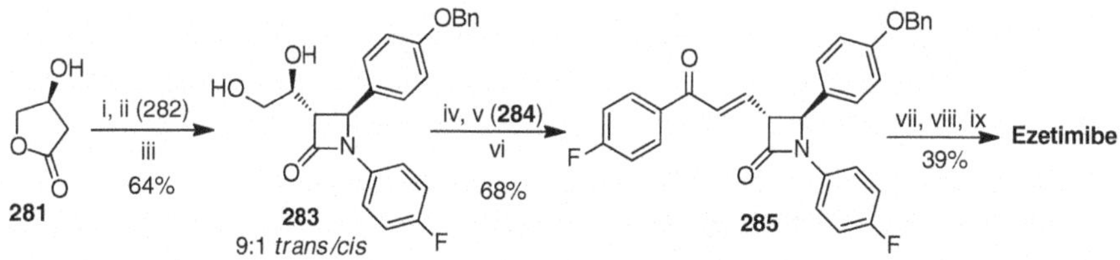

i) LDA; 2) 4-BnO-$C_6H_4CH{=}NC_6H_4$-F-4 (**282**); iii) LiCl; iv) NaIO4; v) (TMSO) (4-F-C_6H_4) $C{=}CH_2$ (**284**); vi) *p*-TsOH; vii) H_2, $(Ph_3P)_3RhCl$; viii) CBS reduction; ix) H_2, Pd-C, EtOH

Scheme 80 Alternative synthetic route to ezetimibe

5 Conclusion

Besides their significance as widely used powerful antibacterial agents, the importance and versatility of β-lactams as key intermediates in organic synthesis have been widely recognized. The β-lactam skeleton provides unique bond angles and configurations of substituents. Selective cleavage of each of the four bonds of the β-lactam skeleton is possible because of its strain energy, which makes this scaffold highly attractive synthetic intermediates or synthons of critical structures. β-Lactams with high enantiopurity can be obtained through asymmetric ester enolate–imine cyclocondensation, diastereoselective ketene–imine cycloaddition reaction coupled with enzymatic resolution, or asymmetric ketene–imine cycloaddition. The chirality of these chiral β-lactams can be transferred directly to an array of useful synthetic intermediates in organic synthesis. The "β-lactam synthon method" has been successfully applied to the synthesis of biologically active oligopeptides, stereoselectively labeled dipeptides, peptidomimetics, antimicrobial, anticancer agents, and complex natural products, including the highly efficient synthesis of paclitaxel and highly potent new generation taxoids. The β-lactam scaffold has also been successfully exploited for the discovery and development of novel anticancer agents and cholesterol-controlling agents. It is highly likely that further development and applications of the β-lactam synthon method will continue to flourish in synthetic organic and medicinal chemistry as well as in chemical biology.

References

1. Georg GI (1992) The organic chemistry of β-lactams. VCH, New York, and references cited therein
2. Ojima I, Inaba S, Yoshida K (1977) New and effective route to β-lactams. The reaction of ketene silyl acetals with Schiff bases promoted by titanium tetrachloride. Tetrahedron Lett 18:3643–3646

3. Hatanaka N, Abe R, Ojima I (1981) β-lactam as synthetic intermediate: synthesis of leucine-enkephalin. Chem Lett 10:1297–1298
4. Ojima I, Shimizu N, Qiu X, Chen H-JC, Nakahashi K (1987) New aspects of β-lactam chemistry: β-lactams as chiral building blocks. Bull Soc Chim France 649–658
5. Ojima I (1992) β-lactam synthon method – enantiomerically pure β-lactams as synthetic intermediates. In: Georg GI (ed) The organic chemistry of β-lactams and β-lactam antibiotics. VCH Publishers, New York, pp 197–255
6. Ojima I, Habus I, Zhao M, Zucco M, Park Y et al (1992) New and efficient approaches to the semisynthesis of taxol and its c-13 side chain analogs by means of β-lactam synthon method. Tetrahedron 48:6985–7012
7. Ojima I (1995) Recent advances in the β-lactam synthon method. Acc Chem Res 28:383–389
8. Ojima I, Delaloge F (1997) Asymmetric synthesis of building-blocks for peptides and peptidomimetics by means of the β-lactam synthon method. Chem Soc Rev 26:377–386
9. Alcaide B, Almendros P (2004) Beta-lactams as versatile synthetic intermediates for the preparation of heterocycles of biological interest. Curr Med Chem 11:1921–1949
10. Deshmukh AR, Bhawal BM, Krishnaswamy D, Govande VV, Shinkre BA et al (2004) Azetidin-2-ones, synthon for biologically important compounds. Curr Med Chem 11:1889–1920
11. Chen J, Kuznetsova L, Ungureanu I, Ojima I (2005) Recent advances in the synthesis of α-hydroxy-β-amino acids and their use in the SAR studies of taxane anticancer agents. In: Juaristi E, Soloshonok V (eds) Enantioselective synthesis of amino acids, 2nd edn. Wiley, New York, pp 447–476
12. Ojima I, Kuznetsova L, Ungureanu IM, Pepe A, Zanardi I et al (2005) Fluoro-β-lactams as useful building blocks for the synthesis of fluorinated amino acids, dipeptides, and taxoids. In: Soloshonok V (ed) ACS Symp Ser 911: Fluorine-containing Synthons. American Chemical Society/Oxford University Press, Washington, DC, pp 544–561
13. Staudinger H (1907) Zur Kenntniss der Ketene. Diphenylketen. Justus Liebigs Ann Chem 356:51–123
14. Cossio FP, Arrieta A, Sierra MA (2008) The mechanism of the ketene-imine (Staudinger) reaction in its centennial: still an unsolved problem? Acc Chem Res 41:925–936
15. Venturini A, Gonzalez J (2006) Mechanistic aspects of the ketene-imine cycloaddition reactions. Mini Rev Med Chem 3:185–194
16. Dumas S, Hegedus LS (1994) Electronic effects on the stereochemical outcome of the photochemical reaction of chromium carbene compleses with imines to form β-lactams. J Org Chem 59:4967–4971
17. Georg GI, Ravikumar VT (1993) Stereocontrolled ketene-imine cycloaddition reactions. In: Georg GI (ed) Organic chemistry β-lactams. VCH Publishing, New York, pp 295–368
18. Hegedus LS, Montgomery J, Narukawa Y, Snustad DC (1991) A contribution to the confusion surrounding the reaction of ketenes with imines to produce β-lactams. A comparison of stereoselectivity dependence on the method of ketene generation: acid chloride/triethylamine s photolysis of chromium carbene complexes. J Am Chem Soc 113:5784–5791
19. Cossio FP, Ugalde JM, Lopez X, Lecea B, Palomo C (1993) A semiempirical theoretical study on hte formation of 5784-5791-lactams from ketenes and imine. J Am Chem Soc 115:995–1004
20. Lopez R, Sordo TL, Sordo JA, Gonzalez J (1993) Torquoelectronic effect in the control of the stereoselectivity of ketene-imine cycloaddition reactions. J Org Chem 58:7036–7037
21. Cossio FP, Arrieta A, Lecea B, Ugalde JM (1994) Chiral control in the staudinger reaction between ketenes and imines. A theoretical SCF-MO study on asymmetric torquoselectivity. J Am Chem Soc 116:2085–2093
22. Wang Y, Liang Y, Jiao L, Du D-M, Xu J (2006) Do reaction conditions affect the stereoselectivity in the Staudinger reaction? J Org Chem 71:6983–6990
23. Aszodi J, Bonnet A, Teusch G (1990) Enantioselective synthesis of a versatile 2-isocepern synthon. Tetrahedron 46:1579–1586

24. Georg G, Wu Z (1994) An investigation of (R)-(+)-1-(1-naphthyl)ethylimines and (R)-(+)-1-(phenyl)ethylimines as chiral templates in the staudinger reaction. Tetrahedron Lett 35:381–384
25. Gunda TE, Sztaricskai F (1997) 2-Amino-1-phenyl-propan-1,3-diol as chiral auxiliary. Application in the synthesis of cis 3-Phthalimido-4-styryl-2-azetidinones. Tetrahedron 53:7985–7998
26. Bose A, Manhas M, van der Veen J, Bari S, Wagle D (1992) Stereoregulated synthesis of β-lactams from Schiff bases derived from threonine esters. Tetrahedron 48:4831–4844
27. Ojima I, Nakahashi K, Brandstadter SM, Hatanaka N (1987) Remarkable effects of lone pair - lone pair interactions on the extremely stereoselective [2+2] cycloaddition of azidoketene to chiral 3-imino-β-lactams. J Am Chem Soc 109:1798–1805
28. Palomo C, Oiarbide M, Esnal A, Landa A, Miranda J et al (1998) practical synthesis of α-amino acid N-carboxy anhydrides of polyhydroxylated α-amino acids from β-lactam frameworks. Model studies toward the synthesis of directly linked peptidyl nucleoside antibiotics. J Org Chem 63:5838–5846
29. Hubschwerlen C, Schmid G (1983) An enantioselective β-lactam synthesis starting from L-(S)-glyceraldehyde acetonide. Helv Chim Acta 66:2206–2209
30. Banik B, Manhas M, Kaluza Z, Barakat J, Bose A (1992) Microwave-induced organic reaction enhancement chemistry.4 convenient synthesis of enantiopure α-hydroxy-β-lactams. Tetrahedron Lett 33:3603–3606
31. Evans D, Williams J (1988) The asymmetric synthesis of β-lactam antibiotics-v. Application of chiral α, β-epoxyimines in ketene-imine cycloaddition reactions leading to homochiral 3-aminoazetidinones. Tetrahedron Lett 29:5065–5068
32. Palomo C, Cossio F, Cuevas C, Leces B, Mielgo A et al (1992) Contribution to the development of new substitution patterns of optically active β-lactams: synthesis of homochiral 4-(1-Aminoalkyl)azetidin-2-ones from N-(tert-Butyloxycarbonyl) α-amino aldehyde-derived imines via asymmetric Staudinger reaction. J Am Chem Soc 114:9360–9369
33. Li L, Thomas SA, Klein LL, Yeung CM, Maring CJ et al (1994) Synthesis and biological evaluation of C-3'-modified analogs of 9(R)-dihydrotaxol. J Med Chem 37:2655–2663
34. Wagle DR, Garai C, Chiang J, Monteleone MG, Kurys BE et al (1988) Studies of lactams. 81. Enantiospecific synthesis and absolute configuration of substituted β-lactams from D-glyceraldehyde acetonide. J Org Chem 53:4227–4236
35. Evans DA, Sjogren EB (1985) The asymmetric synthesis of b-lactam antibiotics-1. Application of chiral oxazolidinones in the Staudinger reaction. Tetrahedron Lett 26:3783–3786
36. Ruhland B, Bhandari A, Gordon E, Gallop M (1996) Solid-supported combinatorial synthesis of structually diverse β-lactams. J Am Chem Soc 118:253–254
37. Delpiccolo C, Mata E (2002) Stereoselective solid-phase synthesis of 3,4-substituted azetidinones as key intermediates for mono- and multicyclic β-lactam antibiotics and enzyme inhibitors. Tetrahedron Asymmetry 13:905–910
38. Delpiccolo CML, Mendez L, Fraga A, Mata EG (2005) Exploring the solid-phase synthesis of 3,4-disubstituted β-lactams: scope and limitations. J Comb Chem 7:331–344
39. Alcaide B, Rodriguez-Vicente A (1999) A convenient trans-stereoselective synthesis of phenanthridine derived 2-azetidinones using the Staudinger ketene-imine cycloaddition. Tetrahedron Lett 40:2005–2006
40. Banik BK, Banik I, Becker FF (2010) Asymmetric synthesis of anticancer β-lactams via Staudinger reaction of chrial ketene from carbohydrate. Euro J Med Chem 45:846–848
41. Guillon CD, Koppel GA, Brownstein MJ, Chaney MO, Ferris CF et al (2007) Azetidinones as vasopressin V1a antagonists. Bioorg Med Chem 15:2054–2080
42. Cremonesi G, Croce P, Fontana F, La Rosa C (2010) Enantiomerically pure polyheterocyclic spiro-β-lactams from trans-4-hydroxy-L-proline. J Org Chem 75:2010–2017
43. Cremonesi G, Croce PD, Fontana F, Fomi A, La Rosa C (2005) Asymmetric synthesis of 1,3-thiazolidine-derived spiro-β-lactams via a Staudinger reaction between chiral ketenes and imines. Tetrahedron Asymmetry 16:3371–3379

44. Chincholkar P, Puranik VG, Deshmukh ARAS (2007) Stereoselective synthesis of spiro-β-lactams using d-(+)-glucose derived chiral pool: remarkable influence of the torquoelectronic effect. Tetrahedron 63:9179–9187
45. Brieva R, Crich JZ, Sih CJ (1993) Chemoenzymic synthesis of the C-13 side chain of taxol: optically active 3-hydroxy-4-phenyl β-lactam derivatives. J Org Chem 58:1068–1075
46. Banik BK, Manhas MS, Bose AK (1994) Stereospecific glycosylation via Ferrier rearrangement for optical resolution. J Org Chem 59:4714–4716
47. Patel RN (1995) Enzymic processes for the resolution of enantiomeric mixtures of compounds useful as intermediates in the preparation of taxanes. Eur Pat Appl 634492 A1: Bristol-Myers Squibb Co., USA
48. Holton RA, Vu P (2001) Enzymatic process for the resolution of enantiomeric mixtures of β-lactams. PCT Int Appl WO 2001/029245: Florida State University, USA
49. Kuznetsova L, Ungureanu I, Pepe A, Zanardi I, Wu X et al (2004) Trifluoromethyl- and difluoromethyl-β-lactams as useful building blocks for the synthesis of fluorinated amino acids, dipeptides, and fluoro-taxoids. J Fluor Chem 125:487–500
50. Kuznetsova LV, Pepe A, Ungureanu IM, Pera P, Bernacki RJ et al (2008) Syntheses and structure-activity relationships of novel 3'-difluoromethyl and 3'-trifluoromethyl-taxoids. J Fluor Chem 129:817–828
51. France S, Weatherwax A, Taggi AE, Lectka T (2004) Advances in the catalytic, asymmetric synthesis of β-lactams. Acc Chem Res 37:592–600
52. Wack H, Drury W, Taggi A, Ferraris D, Lectka T (1999) Nucleophilic metal complexes as acylation catalysts: solvent-dependent "switch" mechanisms leading to the first catalyzed Staudinger reaction. Org Lett 1:1985–1988
53. Taggi A, Hafez A, Wack H, Young B, Drury W et al (2000) Catalytic, asymmetric synthesis of β-lactams. J Am Chem Soc 122:7831–7832
54. Taggi A, Hafez A, Wack H, Young B, Ferraris D et al (2002) The development of the first catalyzed reaction of ketenes and imines: catalytic, asymmetric synthesis of β-lactams. J Am Chem Soc 124:6626–6635
55. France S, Wack H, Hafez A, Taggi A, Witsil D et al (2002) Bifunctional asymmetric catalysis: a tandem nucleophile/lewis acid promoted synthesis of β-lactams. Org Lett 4:1603–1605
56. Zhang Y-R et al (2008) Chiral N-heterocyclic carbene catalyzed staudinger reaction of ketenes with imines: highly enantioselective synthesis of N-Boc-β-lactams. Org Lett 10:277–280
57. Huang X, Chen X, Ye S (2009) Enantioselective synthesis of aza-β-lactams via NHC-catalyzed cycloaddition of ketenes with diazenedicarboxylates. J Org Chem 74:7585–7587
58. Duguet N et al (2008) N-Heterocyclic carbene catalysed β-lactam synthesis. Org Biomol Chem 6:1108–1113
59. Ojima I, Habus I (1990) Asymmetric synthesis of β-lactams by chiral ester enolate – imine condensation. Tetrahedron Lett 31:4289–4292
60. Ojima I, Habus I, Zhao M (1991) Efficient and practical asymmetric synthesis of the taxol C-13 side chain, N-benzoyl-(2R,3S)-3-phenylisoserine, and its analogues via chiral 3-hydroxy-4-aryl-β-lactams through chiral ester Enolate-imine cyclocondensation. J Org Chem 56:1681–1683
61. Ojima I, Park YH, Sun CM, Zhao M, Brigaud T (1992) New and efficient routes to norstatine and its analogs with high enantiomeric purity by β-lactam synthon method. Tetrahedron Lett 33:5739–5742
62. Ojima I (1995) Asymmetric syntheses by means of β-lactam synthon method. In: Hassner A (ed) Advances in asymmetric synthesis. JAI Press, Greenwich, pp 95–146
63. Ojima I, Slater JS, Kuduk SD, Takeuchi CS, Gimi RH et al (1997) Syntheses and structure-activity relationships of taxoids derived from 14β-hydroxy-10-deacetylbaccatin III. J Med Chem 40:267–278

64. Palomo C, Aizpurua JM, Gracenea JJ (1999) Diastereoselective conjugate reduction and enolate trapping with glyoxylate imines. A concise approach to β-lactams that involves a ternary combination of components. J Org Chem 64:1693–1698
65. Alcaide B, Aly M, Rodriguez C, Rodriguez-Vicente A (2000) Base-promoted isomerization of cis-4-formyl-2-azetidinones: chemoselective C4-epimerization vs rearrangement to cyclic enaminones. J Org Chem 65:3453–3459
66. Banfi L, Guanti G, Rasparini M (2003) Intramolecular opening of β-lactams with amines as a strategy toward enzymatically or photochemically triggered activation of lactenediyne prodrugs. Eur J Org Chem 2003:1319–1336
67. Romo D, Rzasa RM, Shea HA, Park K, Langenhan JM et al (1998) Total synthesis and immunosuppressive activity of (−)-pateamine A and related compounds: implementation of a β-lactam-based macrocyclization. J Am Chem Soc 120:12237–12254
68. Del Buttero P, Molteni G, Roncoroni M (2006) Reductive ring opening of 2-azetidinones promoted by sodium borohydride. Tetrahedron Lett 47:2209–2211
69. Ojima I, Wang H, Wang T, Ng EW (1998) New approaches to the asymmetric synthesis of dipeptide isosteres via β-lactam synthon method. Tetrahedron Lett 39:923–926
70. Ojima I, Ng EW, Sun CM (1995) Novel route to hydroxyl(keto)ethylene dipeptide isosteres through the reaction of *N-t*-Boc-β-lactams with enolates. Tetrahedron Lett 36:4547–4550
71. Bose AK, Banik BK, Mathur C, Wagle DR, Manhas MS (2000) Polyhydroxy amino acid derivatives via β-lactams using enantiospecific approaches and microwave techniques. Tetrahedron 56:5603–5619
72. Van Brabandt W, Dejaegher Y, Van Landeghem R, De Kimpe N (2006) Reduction of 4-(haloalkyl)azetidin-2-ones with LiAlH4 as a powerful method for the synthesis of stereodefined aziridines and azetidines. Org Lett 8:1101–1104
73. Vincent G, Williams RM (2007) Asymmetric total synthesis of (−)-cribrostatin 4 (renieramycin H). Angew Chem Int Ed 46:1517–1520
74. Alajarin M, Sanchez-Andrada P (2001) On the mechanism of conversion of N-acyl-4-acyloxy-β-lactams into 2-substituted 1,3-oxazin-6-ones. Can a low-barrier transition state be antiaromatic? J Org Chem 66:8470–8477
75. Ojima I, Zhao M, Yamato T, Nakahashi K, Abe R (1991) Azetidines and bisazetidines. Their synthesis and use as the key intermediates to enantiomerically pure diamines, amino alcohols, and polyamines. J Org Chem 56:5263–5277
76. Ojima I, Shimizu N (1986) Extremely stereoselective and stereospecific reductive cleavage of β-lactams: a highly efficient route to labeled homochiral peptides. J Am Chem Soc 108:3100–3102
77. Ojima I, Chen H-JC (1987) Novel and effective routes to optically pure amino acids, dipeptides, and their derivatives via β-lactams obtained through asymmetric cycloaddition. J Chem Soc Chem Commun 1987:625–626
78. Ojima I, Chen HJ, Qiu X (1988) New approaches to the asymmetric synthesis of non-proteinogenic α-amino acids and dipeptides through chiral β-lactam intermediates. Tetrahedron 44:5307–5318
79. Hatanaka N, Abe R, Ojima I (1982) Synthesis of optically pure enkephalin analog, [D-Ala2, Leu5-ol] enkephalin, using chiral β-lactams as synthetic intermediate. Chem Lett 11:445–448
80. Van Brabandt W, Van Landeghem R, De Kimpe N (2006) Ring transformation of 2-(haloalkyl)azetidines into 3,4-disubstituted pyrrolidines and piperidines. Org Lett 8: 1105–1108
81. Mollet K, Broeckx L, D'hooghe M, De Kimpe N (2011) Synthesis of stereodefined 3,4-disubstituted piperidines through rearrangement of 2-(2-bromo-1,1-dimethylethyl)azetidines. J Org Chem 76:8364–8375
82. Dejaegher Y, Mangelinckx S, De Kimpe N (2002) Rearrangement of 2-aryl-3,3-dichloroazetidines: intermediacy of 2-azetines. J Org Chem 67:2075–2081

83. Alcaide B, Almendros P, Pardo M, Rodriguez-Ranera C, Rodriguez-Vicente A (2003) Lewis acid-promoted intermolecular carbonyl-ene reaction of enantiopure 4-oxoazetidine-2-carbaldehydes. Rapid entry to novel fused polycyclic β-lactams. J Org Chem 68:3106–3111
84. Alcaide B, Almendros P, Cabrero G, Ruiz M (2005) Organocatalytic ring expansion of β-lactams to γ-lactams through a novel N1–C4 bond cleavage. Direct synthesis of enantiopure succinimide derivatives. Org Lett 7:3981–3984
85. Cabell LA, McMurray JS (2002) Two-carbon ring expansion of β-lactams via N(1)—C(4) cleavage reactions. Tetrahedron Lett 43:2491–2493
86. Mandal PK, Cabell LA, McMurray JS (2005) The synthesis of 'tyrosyl' peptidomimetics by acid-catalyzed N(1)–C(4) ring opening of 4-(4′-hydroxyphenyl)-azetidine-2-ones. Tetrahedron Lett 46:3715–3718
87. Palomo C, Aizpurua JM, Ganboa I, Oiardide M (1999) From β-lactams to α- and β-amino acid derived peptides. Amino Acids 16:321–343
88. Paquette LA, Behrens C (1997) A ring-expansion route to analogues of dideoxyhydantocidin. Heterocycles 46:31–35
89. Cossio F, Lopez C, Oiarbide M, Palomo C, Aparicio D et al (1988) Synthetic utility of azetidin-2,3-diones: a new approach to 3-hydroxyethyl-β-lactams and α-amino acid derivatives. Tetrahedron Lett 29:3133–3136
90. Palomo C, Aizpurua JM, Cuevas C, Urchegui R, Linden A (1996) Generation of threonine and azathreonine N-carboxy anhydrides from α-hydroxy β-lactams promoted by 2,2,6,6-tetramethylpiperidinyl-l-oxyl (TEMPO) in combination with sodium hypochlorite. J Org Chem 61:4400–4404
91. Durham T, Miller M (2003) Enantioselective synthesis of α-amino acids from n-tosyloxy β-lactams derived from β-keto esters. J Org Chem 68:27–34
92. Alcaide B, Almendros P, Aragoncillo C (2002) A novel one-step approach for the preparation of α-amino acids, α-amino amides, and dipeptides from azetidine-2,3-diones. Chem Eur J 8:3646–3652
93. Alcaide B, Martin-Cantalejo Y, Rodriguez-Lopez J, Sierra M (1993) New reactivity patterns of the l-lactam ring: tandem C3-C4 bond breakage-rearrangement of 4-acyl- or 4-imino-3,3-dimethoxy-2-azetidinones promoted by stannous chloride ($SnCl_2$-$2H_2O$). J Org Chem 58:4767–4770
94. Alcaide B, Almendros P, Cabrero G, Callejo R, Ruiz M et al (2010) Ring expansion versus cyclization in 4-oxoazetidine-2-carbaldehydes catalyzed by molecular iodine: experimental and theoretical study in concert. Adv Synth Catal 352:1688–1700
95. Saito T, Kobayashi S, Ohgaki M, Wada M, Nagahiro C (2002) Diene-transmissive hetero Diels–Alder reaction of cross-conjugated azatrienes with ketenes: a novel and efficient, stereo-controlled synthetic method for hexahydroquinolinones. Tetrahedron Lett 43: 2627–2631
96. Almendros P, Aragoncillo C, Cabrero G, Callejo R, Carrascosa R et al (2010) Thermal Cope reaction of 2-azetidinone-tethered 1,5-dienes: synthesis of tetrahydroazocinones. Arkivoc 2010:74–92
97. Alcaide B, Rodriguez-Campos IM, Rodriguez-Lopez J, Rodriguez-Vicente A (1999) Stereoselective synthesis of fused bicyclic β-lactams through radical cyclization of enyne-2-azetidinones. J Org Chem 64:5377–5387
98. Souers AJ, Ellman JA (2001) β-Turn mimetic library synthesis: scaffolds and applications. Tetrahedron 57:7431–7448
99. Palomo C, Aizpurua JM, Benito A, Miranda JI, Fratila RM et al (2003) Development of a new family of conformationally restricted peptides as potent nucleators of β-turns. design, synthesis, structure, and biological evaluation of a β-lactam peptide analogue of melanostatin. J Am Chem Soc 125:16243–16260
100. Palomo C, Aizpurua JM, Ganboa I, Benito A, Cuerdo L et al (2004) Synthesis of type II β-turn surrogate dipeptides based on syn-α-amino-α, β-dialkyl-β-lactams. Org Lett 6:4443–4446

101. Palomo C, Aizpurua JM, Balentova E, Jimenez A, Oyarbide J et al (2007) Synthesis of β-lactam scaffolds for ditopic peptidomimetics. Org Lett 9:101–104
102. Jasko'lski M, Tomasselli AG, Sawyer TK, Staples DG, Heinrikson RL et al (1991) Synthesis and biological evaluation of a series of HIV-1 protease inhibitors. Biochemistry 30:1600–1609
103. Burgess K (2001) Solid-phase syntheses of β-turn analogues to mimic or disrupt protein-protein interactions. Acc Chem Res 34:826–835
104. Arnold U, Hinderaker MP, Nilsson BL, Huck BR, Gellman SH et al (2002) A reverse turn structure induced by a D, L-alpha-aminoxy acid dimer. J Am Chem Soc 124:8522–8523
105. Aizpurua JM, Ganboa JI, Palomo C, Loinaz I, Oyarbide J et al (2011) Cyclic RGD b-lactam peptidomimetics induce differential gene expression in human endothelial cells. Chembiochem 12:401–405
106. Freidinger RM, Veber DF, Perlow DS, Brooks JR, Saperstein R (1980) Bioactive conformation of luteinizing hormone-releasing hormone: evidence from a conformationally constrained analog. Science 210:656–658
107. Freidinger RM, Perlow DS, Veber DF (1982) Protected lactam-bridged dipeptides for use as conformational constraints in peptides. J Org Chem 47:104–109
108. Freidinger RM (1985) Synthesis of γ-lactam-constrained tryptophyllysine derivatives. J Org Chem 50:3631–3633
109. Jung M (1985) In: Barrett G (ed) Chemistry and biochemistry of amino acids. Chapman and Hall, New York
110. Ojima I, Chen HJ, Nakahashi K (1988) Asymmetric synthesis with chiral β-lactams. α-substituted aromatic α-amino acids and their derivatives through highly stereoselective alkylations. J Am Chem Soc 110:278–281
111. Ojima I, Komata T, Qiu X (1990) Asymmetric alkylations of a phenylalanylglycinate equivalent. Novel route to dipeptides bearing α-alkyl-α-amino acid residues. J Am Chem Soc 112:110–114
112. Thaisrivongs S, Pals DT, Kroll LT, Turner SR, Han FS (1987) Renin inhibitors. Design of angiotensinogen transition-state analogues containing novel. (2*R*,3*R*,4*R*,5*S*)-5-amino-3,4-dihydroxy-2-isopropyl-7-methyloctanoic acid. J Med Chem 30:976–982
113. Huff JR (1991) HIV protease: a novel chemotherapeutic target for AIDS. J Med Chem 34:2305–2314
114. Okino TM, Matsuda H, Murakami M, Yamaguchi K (1993) Microginin, an angiotensin-converting enzyme inhibitor from the blue-green alga *Microcystis aeruginosa*. Tetrahedron Lett 34:501–504
115. Rowinsky EK (1997) The development and clinical utility of the taxane class of antimicrotubule chemotherapy agents. Ann Rev Med 48:353–374
116. Kingston DGI (2009) Tubulin-interactive natural products as anticancer agents. J Nat Prod 72:507–515
117. Ojima I, Sun CM, Park YH (1994) New and efficient coupling method for the synthesis of peptides bearing the norstatine residue and their analogs. J Org Chem 59:1249–1250
118. Ojima I, Wang T, Ng EW (1998) New approaches to the asymmetric synthesis of dipeptide isosteres via β-lactam synthon method. Tetrahedron Lett 39:923–926
119. Palomo C, Aizpurua JM, Cuevas C (1994) A route to dipeptides containing β-amino-α-hydroxy acid fragments by coupling of N-boc-β-lactams with α-amino esters. Application to the synthesis of (−)-bestatin. J Chem Soc Chem Commun 1994:1957–1958
120. Kukhar VP, Soloshonok VA (1994) Fluorine containing amino acids: synthesis and properties. Wiley, Chichester
121. Ojima I (1996) Biomedical frontiers of fluorine chemistry. In: McCarthy J, Welch J (eds) ACS Symposium Series 639, Washington DC
122. Ojima I, Kuznetsova LV, Sun L (2007) Organofluorine chemistry at the biomedical interface – a case study on fluoro-taxoid anticancer agents. In: Soloshonok V, Mikami K, Yamazaki T, Welch JT, Honek J (eds) ACS Sym Ser 949: Current fluoroorganic chemistry new synthetic

directions, technologies, materials and biological applications. American Chemical Society/ Oxford University Press, Washington, DC, pp 288–304
123. Ojima I, Wang T, Delaloge F (1998) Extremely stereoselective alkylation of 3-siloxy-β-lactams and its applications to the asymmetric syntheses of novel 2-alkylisoserines, their dipeptides, and taxoids. Tetrahedron Lett 39:3663–3666
124. Suffness M (1995) Taxol: science and applications. CRC, New York
125. Guénard D, Guéritte-Vogelein F, Potier P (1993) Taxol and taxotere: discovery, chemistry, and structure-activity relationships. Acc Chem Res 26:160–167
126. Schiff PB, Fant J, Horwitz SB (1979) Promotion of microtubule assembly *in vitro* by taxol. Nature 277:665–667
127. Schiff PB, Horwitz SB (1980) Taxol stabilizes microtubules in mouse fibroblast cells. Proc Natl Acad Sci U S A 77:1561–1565
128. Jordan MA (2002) Mechanism of action of antitumor drugs that interact with microtubules and tubulin. Curr Med Chem: Anti-Cancer Agents 2:1–17
129. Jordan M, Wilson L (2004) Microtubules as a target for anticancer drugs. Nat Rev Cancer 4:253–265
130. Denis J-N, Greene AE, Guénard D, Guéritte-Voegelein F, Mangatal L et al (1988) Highly efficient, practical approach to natural taxol. J Am Chem Soc 110:5917–5919
131. Holton RA, Biediger RJ, Boatman PD (1995) Semisynthesis of taxol and taxotere. In: Suffness M (ed) Taxol®: science and applications. CRC, New York, pp 97–121
132. Lin S, Geng X, Qu C, Tynebor R, Gallagher D et al (2000) Synthesis of highly potent second-generation taxoids through effective kinetic resolution coupling of racemic β-lactams with baccatins. Chirality 12:431–441
133. Holton RA, Kim H-B, Somoza C, Liang F, Biediger RJ et al (1994) First total synthesis of taxol. 2. completion of the C and D rings. J Am Chem Soc 116:1599–1600
134. Nicolaou KC, Ueno H, Liu J-J, Nantermet PG, Yang Z et al (1995) Total synthesis of taxol. 4. The final stages and completion of the synthesis. J Am Chem Soc 117:653–659
135. Danishefsky S, Masters J, Young W, Link J, Snyder L et al (1996) Total synthesis of baccatin III and taxol. J Am Chem Soc 118:2843–2859
136. Wender PA, Badham NF, Conway SP, Floreancig PE, Glass TE et al (1997) The pinene path to taxanes. 6. A concise stereocontrolled synthesis of taxol. J Am Chem Soc 119:2757–2758
137. Ojima I, Sun CM, Zucco M, Park YH, Duclos O et al (1993) A highly efficient route to taxotère by the β-lactam synthon method. Tetrahedron Lett 34:4149–4152
138. Georg GI, Boge TC, Cheruvallath ZS, Clowers JS, Harriman GCB et al (1995) The medicinal chemistry of taxol. In: Suffness M (ed) Taxol®: science and applications. CRC, New York, pp 317–375
139. Ojima I, Slater JC, Michaud E, Kuduk SD, Bounaud P-Y et al (1996) Syntheses and structure-activity relationships of the second generation antitumor taxoids. Exceptional activity against drug-resistant cancer cells. J Med Chem 39:3889–3896
140. Ojima I, Chen J, Sun L, Borella CP, Wang T et al (2008) Design, synthesis and biological evaluation of new generation taxoids. J Med Chem 51:3203–3221
141. Ojima I, Das M (2009) Recent advances in the chemistry and biology of new generation taxoids. J Nat Prod 72:554–565
142. Appendino G, Gariboldi P, Gabetta B, Pace R, Bombardelli E et al (1992) 14β-Hydroxy-10-deacetylbaccatin III, a New Taxane from Himalayan Yew (Taxus Wallichiana Zucc.). J Chem Soc, Perkins Trans 1 2925–2929
143. (a) Geney, R., Sun, L., Pera, P., Bernacki, R.J., Xia, S., Horwitz, S.B., Simmerling, C.L., Ojima, I. (2005). Use of the tubulin-bound paclitaxel conformation for structure-based rational drug design. Chem & Biol 12, 339-348 (b) Sun L, Geng X, Geney R, Li Y, Li Z et al (2008) Design, synthesis and biological evaluation of novel C14-C3'BzN–linked macrocyclic taxoids. J Org Chem 73:9584–9593 (c) Ojima, I., Lin, S., Inoue, T., Miller, M.L., Borella, C.P., Geng, X., Walsh, J.J. (2000) Macrocycle Formation by Ring-Closing Metathesis. Application to the Syntheses of Novel Macrocyclic Taxoids. J Am Chem Soc

122: 5343–5353 (d) Geng, X., Miller, M., Lin, S., Ojima, I. (2003) Synthesis of Novel C2-C3′ N-Linked Macrocyclic Taxoids by Means of Highly Regioselective Heck Macrocyclization. Org Lett 5: 3733–3736

144. Sun L, Simmerling C, Ojima I (2009) Recent advances in the study of the bioactive conformation of taxol. ChemMedChem 4:719–731
145. Eggen MJ, Georg GI (2002) The cryptophycins: their synthesis and anticancer activity. Med Res Rev 22:85–101
146. Eggen M, Nair S, Georg G (2001) Rapid entry into the cryptophycin core via an acyl-β-lactam macrolactonization: total synthesis of cryptophycin-24. Org Lett 3:1813–1815
147. Vidya R, Eggen MJ, Nair SK, Georg GI, Himes RH (2003) Synthesis of cryptophycins via an N-acyl-β-lactam macrolactonization. J Org Chem 68:9687–9693
148. Keller-Schierlein K, Roncari G (1964) Stoffwechselprodukte von Mikroorganismen 46. Mitteilung Die Konstitution des Lankamycins. Helv Chim Acta 47:78–103
149. Sakamoto JMJ, Kondo S, Yumoto H, Arishima M (1962) Bundlins A and B, two antibiotics produced by *Stereptomyces grieseofuscus nov. sp.* J Antibiot Ser A 15:98–102
150. Harada S, Higashide E, Fugono T, Kishi T (1969) Isolation and structures of T-2636 antibiotics. Tetrahedron Lett 27:2239–2244
151. Ootsu K, Matsumoto T, Harada S, Kishi T (1975) Antitumor and immunosuppressive activities of lankacidin-group antibiotics: structure-activity relationships. Cancer Chemother Rep, Part 1 59:919–928
152. Kende AS, Liu K, Kaldor I, Dorey G, Koch K (1995) Total synthesis of the macrolide antitumor antibiotic lankacidin C. J Am Chem Soc 117:8258–8270
153. Watt JM, Gerdina M (1962) The medicinal and poisonous plants of southern and eastern Africa. E. & S. Livingstone Ltd., Edinburgh
154. Cragg G, Kingston D, Newman D (2005) Anticancer agents from natural products. CRC, Boca Raton
155. Cooney M, Ortiz J, Bukowski R, Remick S (2005) Novel vascular targeting/disrupting agents:combretastatin A4 phosphate and related compounds. Curr Oncol Rep 7:90–95
156. Young SL, Chaplin DJ (2004) Combretastatin A4 phosphate: background and current clinical status. Expert Opin Invest Drugs 13:1171–1182
157. O'Boyle N, Carr M, Greene L, Bergin O, Nathwani S et al (2010) Synthesis and evaluation of azetidinone analogues of combretastatin A-4 as tubulin targeting agents. J Med Chem 53:8569–8584
158. Burnett D (2004) β-Lactam cholesterol absorption inhibitors. Curr Med Chem 11:1873–1887
159. Tiruvettipuram KT, Fu X, Tann CH, McAllister TL, Chiu JS et al (2001) Process for the synthesis of azetidinones (Schering Corporation)
160. Wu G, Wong YS, Chen X, Ding Z (1999) A novel one-step diastereo- and enantioselective formation of trans-azetidinones and its application to the total synthesis of cholesterol absorption inhibitors. J Org Chem 64:3714–3718

Top Heterocycl Chem (2013) 30: 65–84
DOI: 10.1007/7081_2012_87

Published online: 7 August 2012

β-Lactams from Fischer Carbene Complexes: Scope, Limitations, and Reaction Mechanism

Israel Fernández and Miguel A. Sierra

Abstract The irradiation (visible light) of Fischer carbene complexes promotes the reversible insertion of a *cis*-carbonyl ligand into the metal–carbene carbon bond in a process known as photocarbonylation. This reaction gives rise to the formation of ketene-like complexes which are able to react with nucleophiles to yield a great variety of reaction products. When the nucleophile is an imine, β-lactams are formed in good to excellent yields and with high diastereoselectivities. The scope and limitations of this synthetically useful transformation as well as its reaction mechanism are considered herein.

Keywords Fischer Carbene complexes · β-lactams · Photochemistry · Reaction mechanisms · Metallocenes · DFT calculations

Contents

I. Fernández and M.A. Sierra (✉)
Departamento de Química Orgánica I, Facultad de Ciencias Químicas,
Universidad Complutense de Madrid, 28040 Madrid, Spain
e-mail: sierraor@quim.ucm.es

1 Introduction

The chemistry of Fischer carbene complexes has been largely developed since the first synthesis of the pentacarbonyl[methoxy(methyl)carbene] tungsten(0) complex by Fischer and Maasböl in 1964 [1]. Due to their special electronic features and multifunctional structure, most of the investigations performed so far have mainly focused on their applications in organic synthesis (selected reviews on the chemistry of Fischer carbenes [2–10]). Nowadays, Fischer carbene complexes have shown themselves to be very efficient and versatile starting materials to carry out a wide variety of organic transformations under mild reaction conditions.

Representative examples of the rich and versatile chemistry of this family of organometallic complexes are the useful Dötz-benzannulation reaction [11], which produces substituted phenols by reaction with alkynes and carbon monoxide, cycloaddition reactions such as cyclopropanation [12–15], Diels–Alder reaction ([16–19]; for a recent computational study, see [20]) or dipolar cycloadditions ([21–24]; for a recent computational study, see [25]), and the catalytic transmetallation to late transition metals [26, 27], which enhances the reactivity of the carbene complexes leading to a great number of different reaction products.

The discovery by Hegedus and McGuire at the beginning of the 1980s that alkoxychromium(0) carbene complexes can be transformed into β-lactams when irradiated (visible light) in the presence of imines marked the beginning of a key reaction in organometallic chemistry [28]. This reaction, which is known as photocarbonylation, gives rise to ketene-like intermediates which show all the advantages of ketenes without their shortcomings, i.e., dimerization, formation of undesired adducts, and so forth. Depending on the nucleophile added to the reaction medium, this process allows the easy and efficient access to a wide variety of compounds such as β-lactams, cyclobutanones, amino acids and peptides, poly-nuclear hydrocarbons, or β-lactones [29] (Scheme 1). This synthetically powerful photoreactivity has no parallel in the photochemistry of any other class of organometallic compounds and competes in efficiency and exceeds in versatility with many well-established and synthetically useful photochemical organic reactions [30–32].

Herein, the scope, limitations, and mechanism of the photocarbonylation reaction of group 6 Fischer carbene complexes in the presence of imines to produce β-lactams are summarized [33].

2 Synthesis of β-Lactams from Fischer Carbene Complexes

2.1 Scope and Limitations

The irradiation of group 6 Fischer metal–carbene complexes with visible light in the presence of a wide range of imines (including heterocyclic imines) produces β-lactams in good to excellent yields [28, 34–36]. In contrast to the analogous

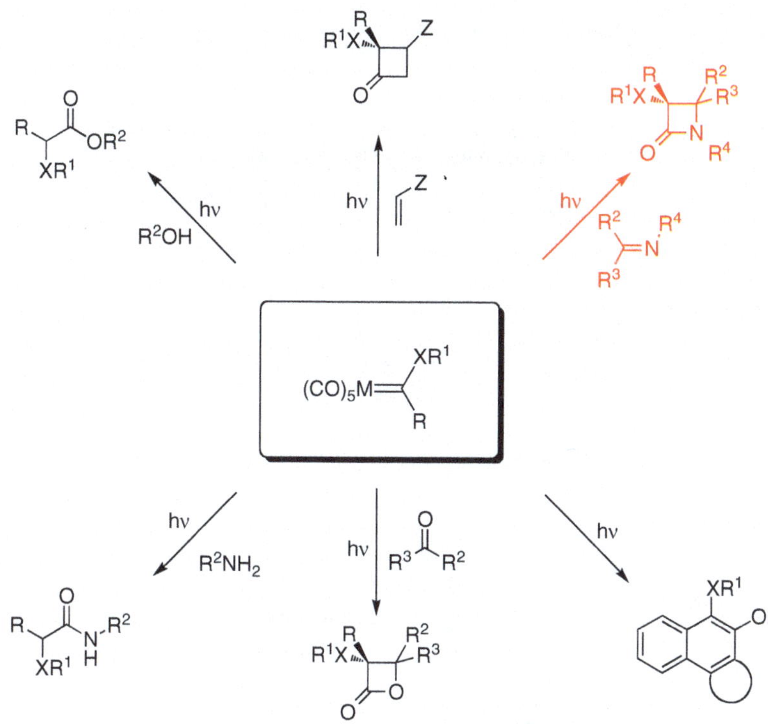

Scheme 1 The synthetic usefulness of group 6 (Fischer) carbene complexes

Staudinger process, the photoreaction of these organometallic compounds is free of the typical by-products formed when ketenes derived from acyl chlorides are involved. Moreover, the transformation is highly diastereoselective, usually giving rise to a single diastereoisomer (Scheme 2).

The usefulness of this process toward the synthesis of new β-lactams is clearly demonstrated in the reactions depicted in Schemes 3 and 4. Thus, the family of β-lactams *azapenams* [37–39] and *bis-azapenams* [40] are easily produced by the reaction of mono- or bis-alkoxycarbene complexes with *N*-protected imidizolines followed by deprotection of the Cbz group.

Nevertheless, most biologically active β-lactams lack an alkyl and an alkoxy group α to the carbonyl group of the 2-azetidinone ring, but rather they possess an amino group and a hydrogen atom. Therefore, aminocarbene complexes must be used to produce such β-lactams. Usually, complexes where R^1 = H, which are readily prepared by the reaction of $M_2Cr(CO)_5$ with iminium chlorides [41] or with amides with the addition of trimethylsilyl chloride [42, 43] can undergo an efficient photoreaction with a wide range of imines to produce α-amino β-lactams in good to excellent yields and with high diastereoselectivity [44] (Scheme 5). However, aminocarbene complexes with R^1 = alkyl or aryl formed β-lactams in only low yields.

Attempts to induce asymmetry into the reaction of chromium alkoxycarbene complexes with imines met with mixed results. In most cases chiral auxiliaries attached

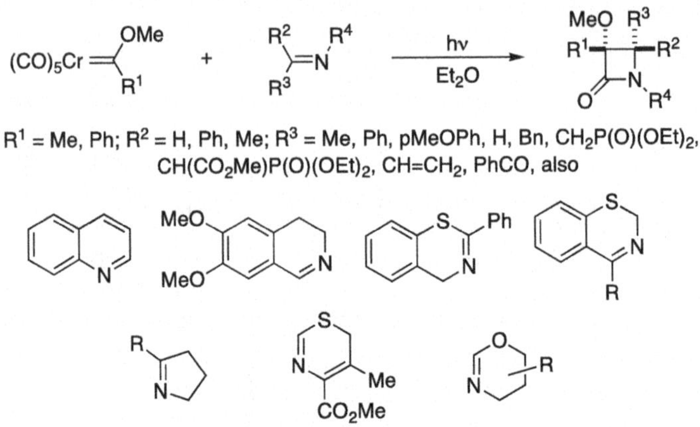

Scheme 2 The scope of the reaction between chromium(0) carbene complexes and imines to yield β-lactams

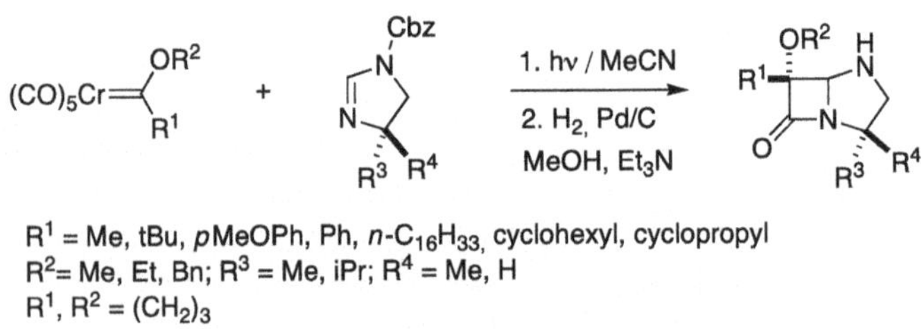

Scheme 3 The synthesis of azapenams from imidizolines

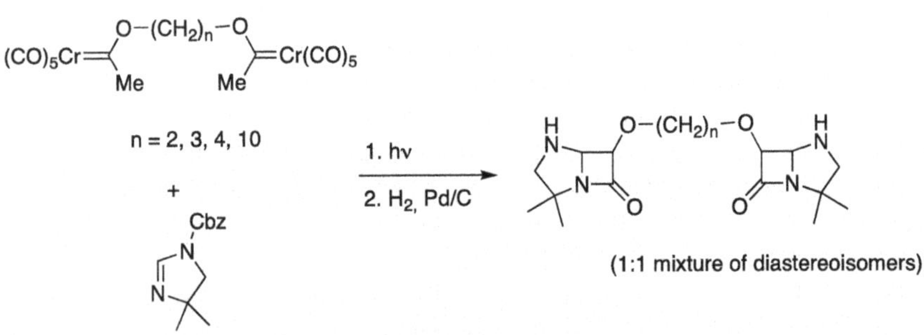

Scheme 4 The synthesis of bis-β-lactams from bis-carbene complexes

to the nitrogen of the imine were ineffective, resulting in low to moderate diastereoselectivity [35]. However, when rigid, cyclic imines such as thiazolines (Scheme 6) [35] and imidazolines [37] were used, very high (essentially 100%) *d.e.* were observed. Although the formed β-lactams were not of biological interest, they provided chiral templates for the synthesis of highly functionalized quaternary systems (Scheme 7) [45].

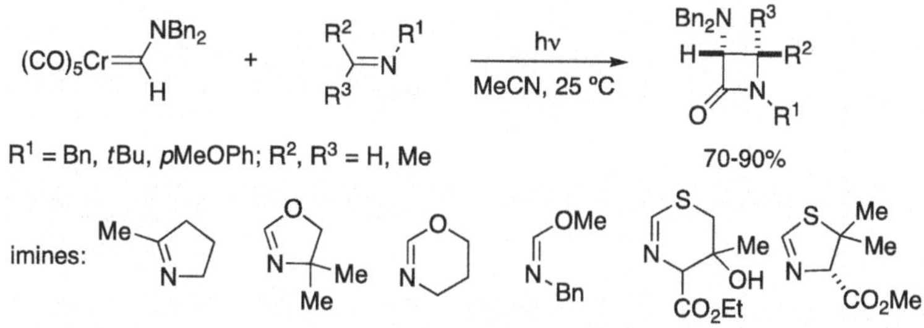

Scheme 5 The synthesis of 3-amino-β-lactams from chromium(0) aminocarbene complexes

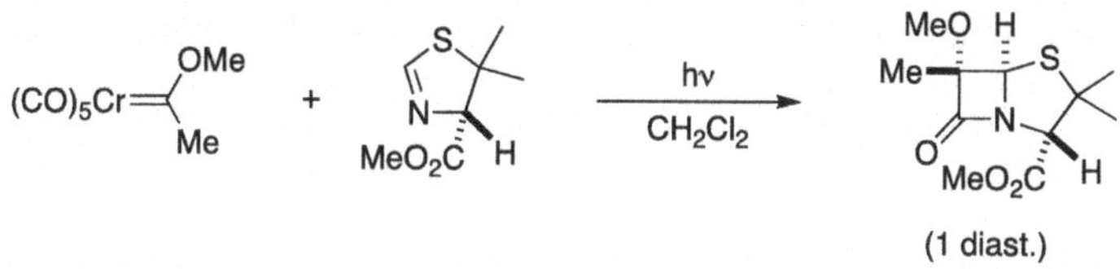

Scheme 6 The synthesis of chiral bicyclic-β-lactams from cyclic chiral thiazolines carboxylates

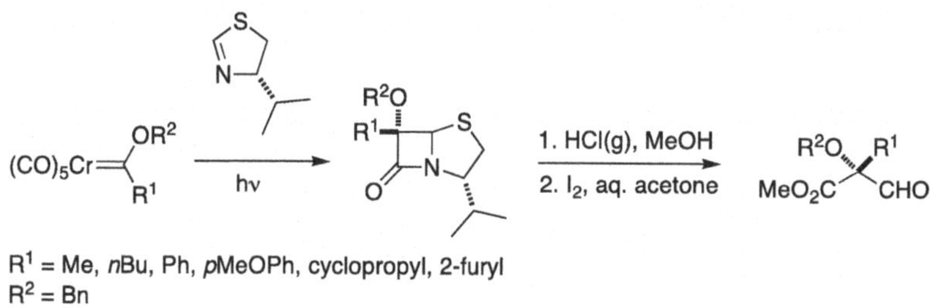

Scheme 7 The synthesis of chiral bicyclic-β-lactams from cyclic chiral thiazolines

Chiral auxiliary oxazolidine was also directly attached to the carbene ligand [46]. The photochemically produced oxazolidine metal–ketene complex gave excellent chemical yields of single diastereoisomers of β-lactams with imidates, oxazines, thiazines, and cyclic and acyclic aliphatic imines, but only modest yields of mixtures of *cis* and *trans* diastereoisomers of β-lactams with imines derived from benzaldehyde and cinnamaldehyde (Scheme 8). In sharp contrast, the related nonorganometallic oxazolidinone ketene gave excellent chemical yields of single *cis* diastereoisomers of β-lactams with imines of benzaldehyde and cinnamaldehyde, but very low yields of β-lactams with other imines. This different behavior clearly exemplifies the dramatically different albeit complementary reactivity of these closely related intermediates ([47]; for the related oxazolidinone ketene see [48]).

Finally, there is a single example of the intramolecular cyclization of chromium (0) carbene complex bearing imino tethers [49]. The preparation of these complexes is achieved in two-steps from simple chromium(0) complexes and their cyclization

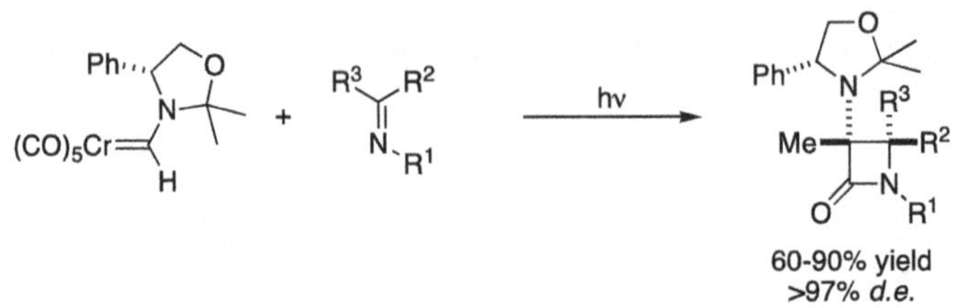

Scheme 8 The synthesis of chiral β-lactams from chiral aminocarbene complexes

yields bicyclic anti-Bredt γ-lactams instead of the expected β-lactams. This is probably due to the opening of these unstable intermediates or, alternatively, by a competitive evolution of the intermediate zwitterion. This zwitterion is derived from the intramolecular attack of the imine nitrogen on the photogenerated ketene (Scheme 9).

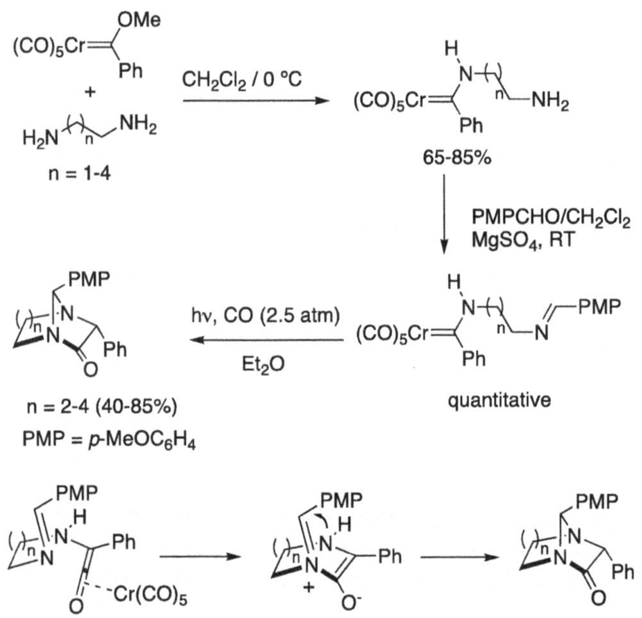

Scheme 9 The intramolecular cyclization of chromium(0) carbene complexes having imino tetheres

2.2 *Synthesis of β-Lactams from Chromium(0) Carbene Complexes and Subsequent Manipulation of the Four-Membered Ring*

The use of appropriate imine substrates in the chromium(0) carbene-mediated photocycloaddition allows for ulterior manipulation of the four-membered ring.

Thus, malonoimides have been prepared by reaction of iminodithiocarbonates and chromium(0) carbene complexes followed by oxidation with *N*-bromosuccinimide (NBS) [50] (Scheme 10).

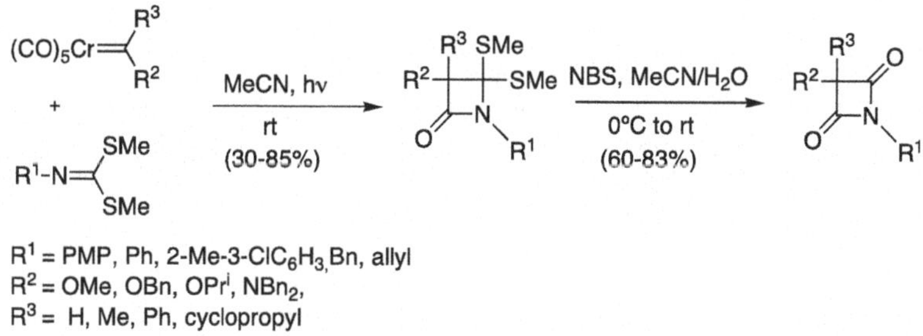

Scheme 10 The two-step synthesis of malonoimides from iminodithiocarbonates

These iminodithiocarbonates also behave as formaldehyde imine equivalents using the sequence chromium(0) carbene photocycloaddition-nickel boride desulfuration [51]. This sequence allows the preparation of either chiral or racemic 4-unsubstituted-2-azetidinones in two steps and highly overall yields (Scheme 11).

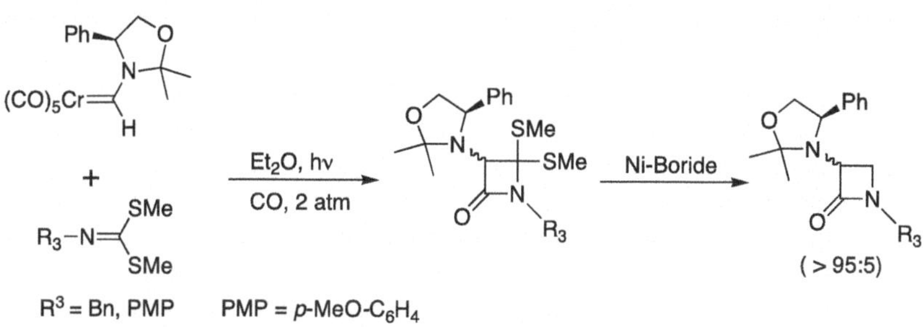

Scheme 11 The synthesis of chiral 4-unsubstituted-β-lactams from chiral aminocarbene chromium(0) complexes and iminodithiocarbonates

Alternatively, the chromium(0) carbene moiety can be incorporated in a preformed 2-azetidinone ring. This methodology based mainly on solvolysis reactions of chromium(0) alkoxycarbene complexes have been used in the preparation of peptide containing 2-azetidinones, including penicillin and cephalosporin derivatives [52] (Scheme 12).

2.3 Synthesis of Metallocenyl-β-Lactams from Bimetallic Carbenes

The synthesis of novel β-lactams which incorporate organometallic moieties in their structures has attracted much attention in the recent years due to their novel and

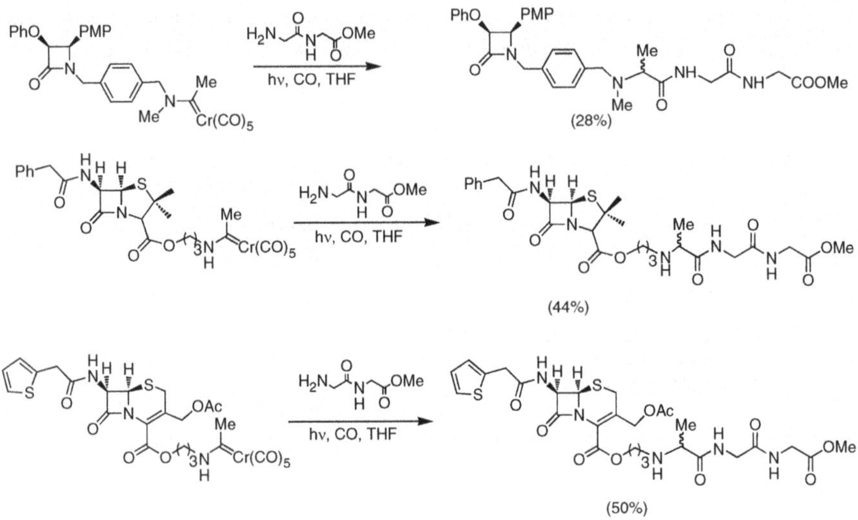

Scheme 12 The reactivity of 2-azetidinones having pendant chromium(0) carbenes

interesting biological properties ([53–62]). As instance, almost all of the ferrocenyl-penicillins and cephalosporins exhibited antibiotic activity, some being highly active, while others proved to be potent β-lactamase inhibitors. Despite that, 2-azetidinones having organometallic fragments attached to the four-membered ring are scarce. The use of Fischer carbene complexes is a useful alternative toward the synthesis of new organometallic β-lactams.

We pioneered the use of the photocarbonylation reaction to access novel ferrocenyl-β-lactams [63]. Thus, alkoxychromium(0) carbene complexes react smoothly with ferrocene imines to place ferrocene substituents at the N1, C4, or simultaneously at N1 and C4 positions of the β-lactam ring (Scheme 13). Interestingly, when the ferrocene unit was introduced in the carbene ligand, the reaction yield dramatically decreases to 23% for aminoferrocenyl-chromium(0) complex or no reaction is observed when the ferrocenyl substituent is directly attached to the carbene carbon atom (Scheme 14).

In contrast, the analogues mono- or bis-ruthenocyl-carbene complexes are photoreactive and readily produce the expected β-lactams [64] (Scheme 15). To our knowledge, these are the unique examples of a 2-azetidinone bearing a metallocene directly attached to the C3 of the four-membered ring. Reasons for the differential behavior of ruthenocenyl- vs. ferrocenyl-substituted carbene complexes are found in the structure of the corresponding excited states formed upon irradiation (see below). Using this useful methodology, the preparation of the first 6-ruthenocenyl-substituted penicillin by reaction of ruthenocenyl-substituted carbene complex and thiazoline was achieved [64] (Scheme 15). This metallapenicillin might exhibit potential biological activity.

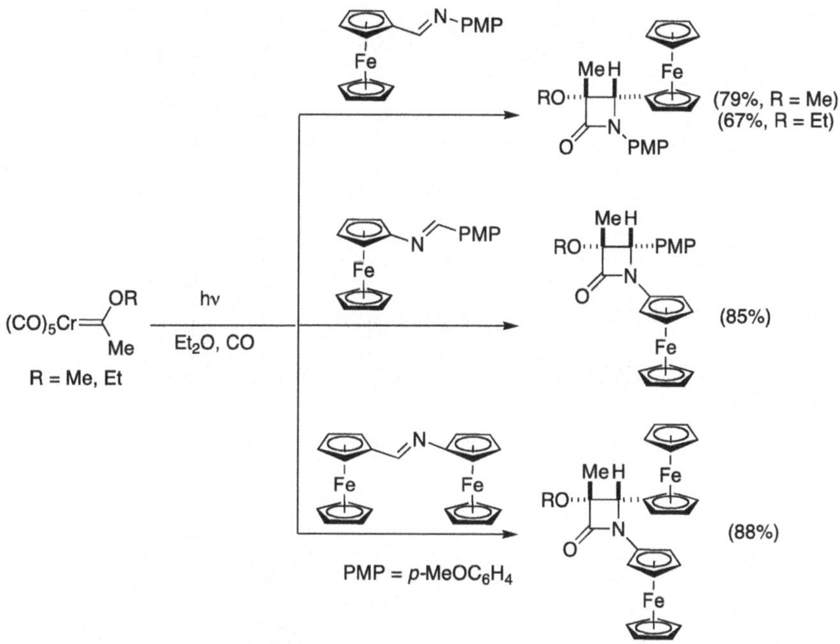

Scheme 13 Ferrocenyl-substituted β-lactams

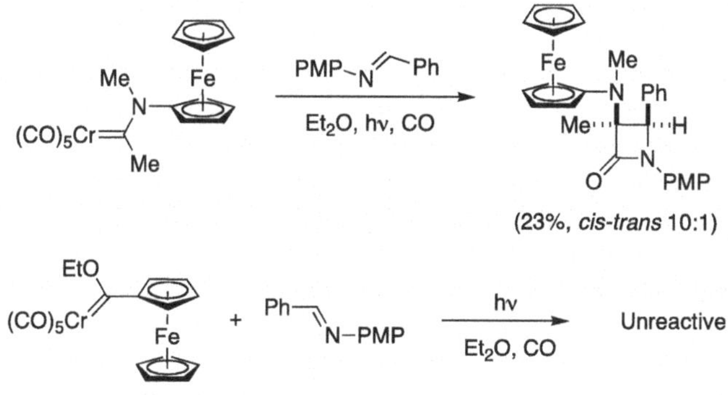

Scheme 14 The anomalous behavior of ferrocene-substituted chromium(0) carbene complexes

3 Reaction Mechanism

It is well established that not all carbene complexes can undergo the photocarbonylation reaction and produce metallaketenes able to react with imines to yield β-lactams. In fact, those group 6 Fischer carbenes complexes based on tungsten(0) instead of chromium(0) or molybdenum(0) do not photocarbonylate. This is mainly due to the higher back-donation of tungsten(0) compared to its group 6 congeners

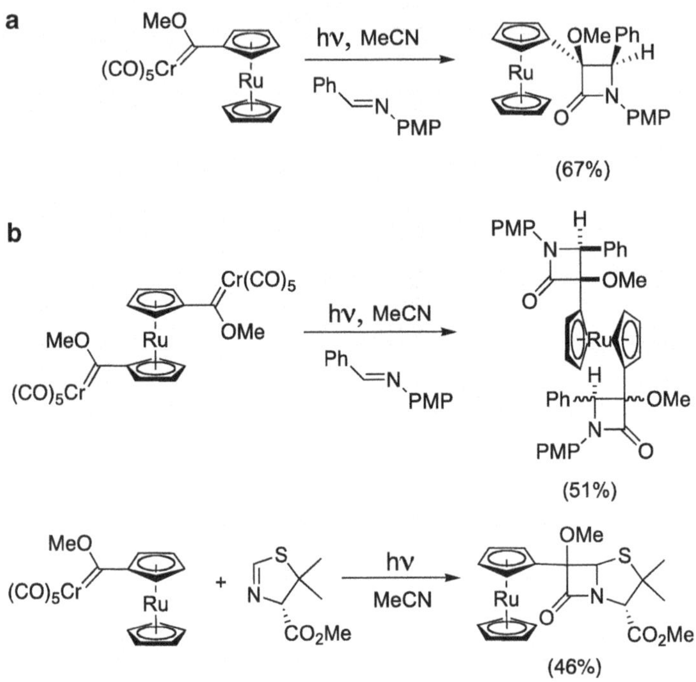

Scheme 15 The synthesis of 3-ruthenocenyl β-lactams

which makes the metal–carbene and metal–carbonyl bonds stronger inhibiting the photoinsertion of the *cis*-CO into the metal–carbene bond (see below). Similarly, chromium(0) complexes bearing σ-donating ligands (i.e., PBu_3 or diphos) are not photoreactive either [65]. In contrast, soft σ-donor ligands (like PPh_3 or phosphites) can form β-lactams in moderate to good yields [65, 66].

As stated above, the nature of the substituents attached to carbene carbon atom has a strong influence on the photoreactivity of Fischer carbenes. Thus, alkoxycarbenes with alkenyl or alkynyl groups attached to the carbene or aminocarbenes, where the R^1 substituent is different from hydrogen (see above), are not photoactive at all or lead to very low conversions [28]. Therefore, it seems that the electronic properties of the carbene ligand (mainly, the occupation of the "empty" p_z-atomic orbital of the carbene carbon atom) is the main factor controlling the photoreactivity of Fischer carbene complexes. By the proper selection of the ligands surrounding the metal center and the substituents at the carbene ligand, the photoreactivity of the complexes can be indeed modulated and processes different to the photocarbonylation reaction such us stepwise 1,2-dyotropic rearrangements, ([67, 68]; for a recent review on dyotropic rearrangements, see [69]), α-fragmentations [70] or photoslippage [71] can occur [33].

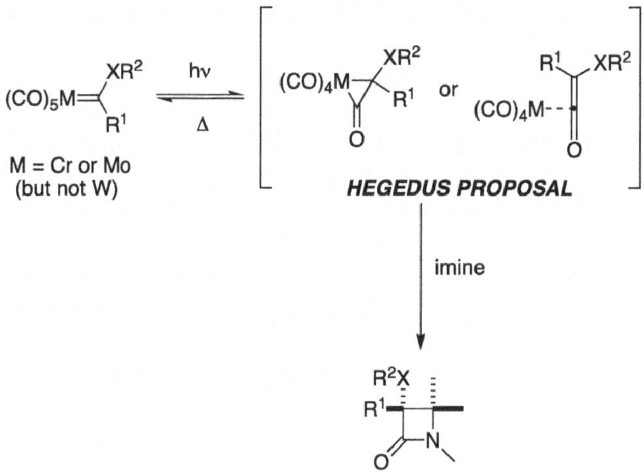

Scheme 16 The postulated pathway from chromium(0) carbene complexes to β-lactams

3.1 The Photocarbonylation Process

The photochemistry of chromium(0) and molybdenum(0) carbene complexes (but not tungsten(0) carbene complexes) relies on the reversible insertion of a *cis*-CO ligand into the M=C bond to yield metal-coordinated ketenes or metallacyclopropanone species. This mechanism was proposed by Hegedus in 1988 [72] (Scheme 16). Despite the close parallelism of the ketenes derived from the irradiation of group 6 metal–carbene complexes and free ketenes (for a recent review in the state of the art of the research on the mechanism of the Staudinger reaction, see [73]), all the efforts directed toward the isolation or detection of these elusive intermediates have been fruitless so far.

By means of a combination of experimental and computational tools [65, 74], it was found that the irradiation of alkoxychromium(0) carbene complexes, either in the LF band followed by relaxation to the MLCT band, or directly in the MLCT band, results in the excitation of these complexes to the S_1 excited state, which readily decays to the triplet T_1 state by intersystem crossing (ISC) due to spin–orbit coupling (this is a general phenomenon in group 6 metal carbonyl complexes; see [32] and [75].) These triplet species (**1(T₁)**) have a chromacyclopropanone structure whose unpaired electrons are mainly localized in the metal fragment and in the former carbene carbon atom and therefore *corresponds with the chromacyclopropanone proposed by Hegedus and co-workers* to explain the reaction products obtained in the photochemical reaction of chromium(0) carbene complexes [28, 72]. These coordinatively unsaturated complexes change their multiplicity prior to evolving the ketene-derived products. This is accomplished by filling the corresponding free coordination site with a molecule of a coordinating solvent in the apical position to form **2(S_0)** complex (Fig. 1). To experimentally confirm this theoretical prediction, the photolysis of pentacarbonyl[ethoxymethyl] carbene-chromium(0) complex with the imine PhCH=N(p-OMeC_6H_4) was carried out in solvents with different coordinative ability. It was found that the higher the donor number of the solvent (i.e., tetrahydrofuran or acetonitrile) the higher were the conversions of complex **1** into the corresponding β-lactam [74].

The geometrical features and the Natural Bond Orbital (NBO) analyses carried out on these newly formed species (**2(S$_0$)**) show that they possess a structure which corresponds to a ketene species coordinated to chromium with a highly polarized Cr–C (former carbene carbon atom) bond. Therefore, complexes **2(S$_0$)** can be viewed as acylchromate complexes from which ketene-derived products can now be formed in the presence of nucleophiles on the S_0 hypersurface, while in their absence, these species revert to the starting carbene complex in a highly exothermic process (Fig. 1).

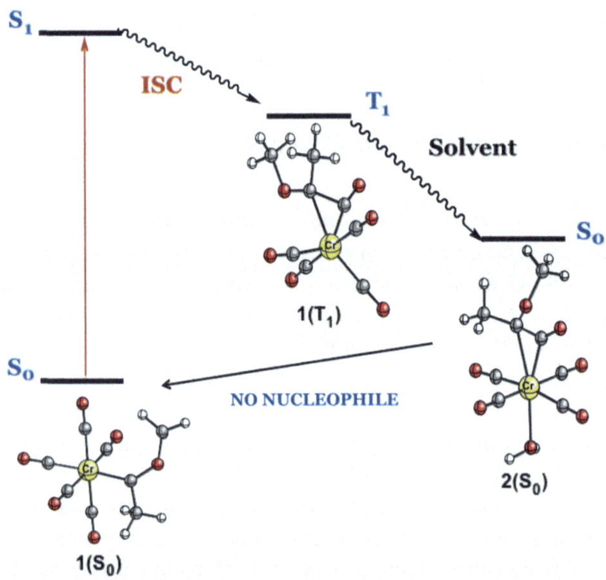

Fig. 1 The computed mechanism for the photocarbonylation of chromium(0) carbene complexes

The solvent-induced T_1–S_0 crossing of chromoketene **1(T$_1$)** to the acylchromate **2(S$_0$)** was further analyzed in detail (for a recent example on Lewis-base-induced triplet-singlet crossing in organometallic compounds, see [76]; for a review on this problem, see [77]). Relaxed scans of this complex at different Cr–OH_2 distances, where the OH_2 ligand models an ethereal solvent (Fig. 2), showed that the T_1 state exhibits a very shallow Morse-like curve, whose minimum is located at $r = 3.746$ Å, r being the Cr–OH_2 distance (Fig. 2). The harmonic analysis of this structure shows two low-frequency vibrations associated with stretching of the Cr–O interaction and symmetrical bending of two carbonyls to achieve the octahedral coordination. This result, together with the negligible energy difference between both

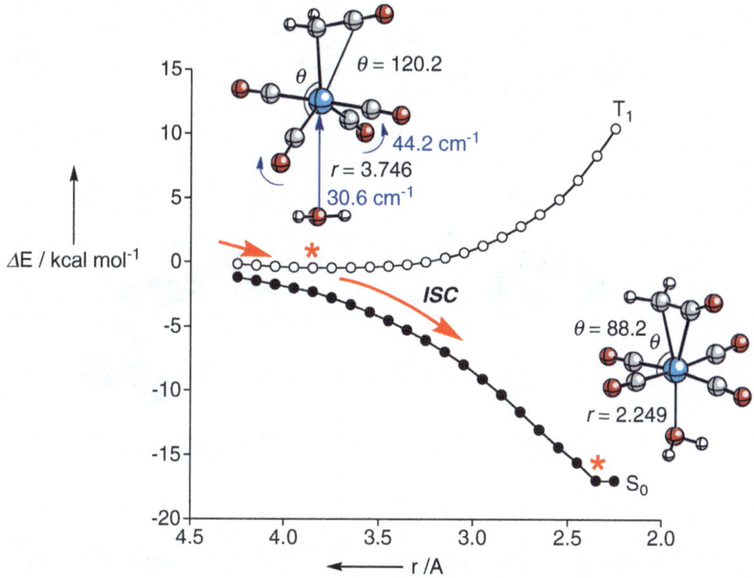

Fig. 2 The Loose-Bolt mechanism for the ISC in chromium(0) carbene photochemistry

states at long Cr–OH_2 distances, indicates that the available thermal energy is enough for the molecule to go from the T_1 to the S_0 potential energy surface. This radiationless ISC does not take place through a narrow conical intersection (a photochemical funnel) but occurs at *r*-values larger than 4.0 Å along an energy plateau shared by both spin states. This process, which is called the Loose-Bolt effect in organic chemistry [32], has no known precedents in the photochemistry of metal complexes.

The formation of the metallacyclopropanone complex in the triplet state is decisive for the production of β-lactams in the presence of imines. DFT calculations clearly show that the most stable triplet species of the analogous tungsten(0) alkoxycarbene complex does not have the required metallacyclopropanone structure. Thus, the computed bond lengths between the carbene carbon atom and the nearest C(=O) groups are 2.716 Å and 2.567 Å, respectively, without any measurable bond order between both atom pairs [74]. Therefore, no ketene-like species are formed by excitation of an alkoxy-pentacarbonyltungsten(0) carbene complex which is in good accord with the experimentally observed lack of photochemical reactivity of tungsten(0) carbene complexes.

Similarly, the computational data show the coexistence of a very low-energy noncarbonylated triplet for ferrocenyl-substituted carbene complex (**T1**) and a high-energy metallacyclopropanone triplet (**T2**, Fig. 3a). This implies that irradiation of this complex leads to the formation of a triplet species, which does not possess the structure required to react with nucleophiles [64]. In sharp contrast, two nearly degenerated triplet states coexist in its ruthenocenyl counterpart (Fig. 3b).

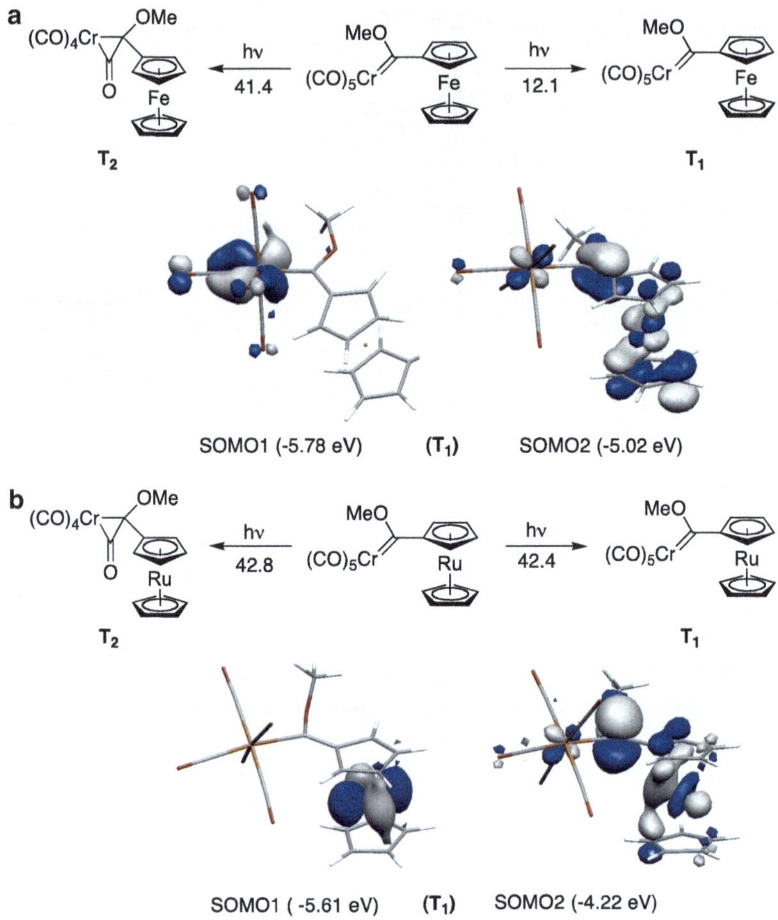

Fig. 3 Computed most stable excited states of ferrocenyl- and ruthenocenyl-substituted Fischer carbene complexes

Simple visual inspection of the corresponding SOMOs of **T1** species (Fig. 3) indicates clear differences in the electronic structures of ferrocenyl- and ruthenocenyl-substituted carbenes, which are translated into a differential photoreactivity (see above). One of these two triplets is a carbonylated metallacyclopropanone species (similar to **1(T1)** in Fig. 1), which can react with imines to form the observed β-lactams [64] (Schemes 14 and 15).

3.2 *The Reaction with Imines to Yield β-Lactams*

Having established a reasonable model for the photocarbonylation step of chromium(0) carbene complexes, the addition of an imine to the coordinated ketene was

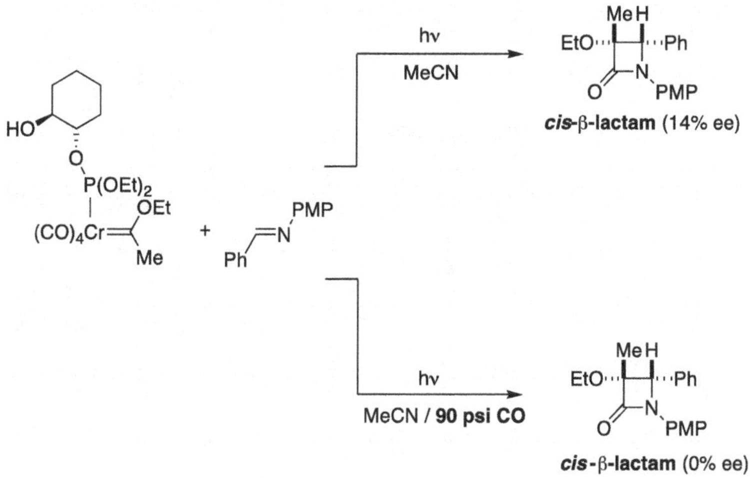

Scheme 17 The enantioselection in the reaction of chromium(0) carbene complexes having chiral ligands as a probe for the β-lactam formation mechanism

studied next. As above-mentioned, this reaction forms 2-azetidinones in good yields and it is usually free of by-products [28]. The reaction is highly diastereoselective since Fischer carbene complexes form β-lactams where the larger substituent of the ketene is placed *cis* to the *anti*-substituent of the imine. This contrasteric bias has been claimed to be due to the presence of the metal moiety during the cycloaddition [28].

The participation of the metal in the transformation of the chromium complex into the final products and not only in the first carbonylation step was experimentally confirmed by irradiating the chiral complex depicted in Scheme 17 in the presence of an imine. This reaction yielded a mixture of the corresponding *cis*/*trans*-β-lactams in a 4.0:1 ratio with a 66% of conversion [66]. After separation of both diastereoisomers, a 14% *e.e.* for *cis*-diastereomer was measured by ^{1}H NMR in the presence of Eu(hfc)$_3$. Even with the observed low *e.e.*, which can be very likely ascribed to the fact that the inducing center is located far away from the emerging chiral centers, the formation of chiral products is compatible with a mechanism in which the metal is present in the enantio-discriminating step, namely the conrotatory ring closure of the zwitterion ([73]). Nevertheless, there is a possibility of organocatalysis by the free chiral phosphine which would be responsible for the observed *e.e.* (related examples [78–80]). To safely disregard this process, the removal of the metal-moiety from the reactive system was ensured by working under moderate CO-pressures (90 psi). Thus, the irradiation of the same chiral complex and imine in MeCN under 90 psi of CO gave the mixture of the corresponding *cis*/*trans*-β-lactams in a 3.8:1 ratio and in 70% isolated yield (Scheme 17). Similarly, the isomers were separated and the *cis*-isomer was analyzed again by ^{1}H NMR in the presence of Eu(hfc)$_3$. The *e.e.* was 0% within

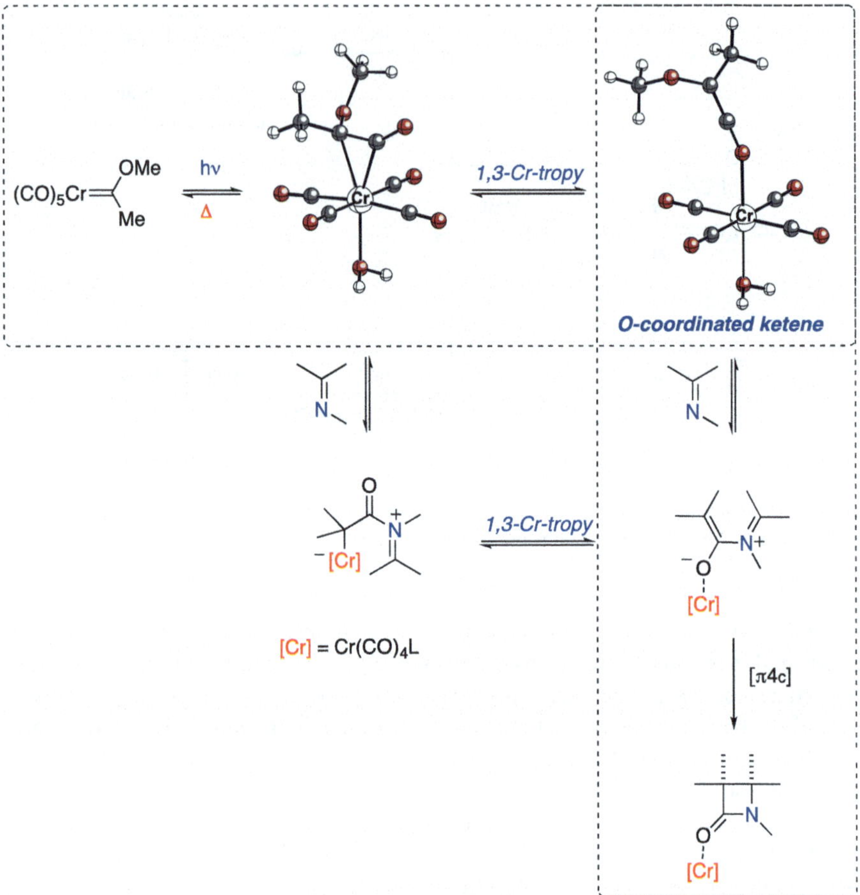

Scheme 18 Pausible reaction pathways in the reaction between chromium(0) Fischer carbene complexes and imines

the experimental error, thus confirming that no asymmetric induction was obtained if the metal moiety is removed from the reactive system.

Two related reaction mechanisms can be envisaged for this transformation: (1) the nucleophilic attack of the lone-pair of the nitrogen atom of the imine to the carbonyl group of the acyl-chromate **2(S$_0$)** followed by a 1,3-metallatropic process forming a chromium enolate–iminium zwitterionic complex which produces the final cycloadduct through a conrotatory ring closure, or alternatively, (2) the 1,3-metallatropic process can occur first from the acyl-chromate to produce the corresponding O-coordinated ketene, which produces the final cycloadduct after nucleophilic attack of the imine and subsequent four electron conrotatory electrocyclation, following the standard mechanism for the Staudinger reaction (see [73]) (Scheme 18).

It has been suggested that the 1,3-metallatropy must occur before the nucleophilic attack of the imine [66]. This rearrangement, albeit slightly endergonic (see Fig. 4), takes place due to a more favorable HOMO(imine)–LUMO+1(oxygen-coordinated ketene) interaction which constitutes the driving force of the transformation. Moreover, the NBO-charge analyses carried out on both key intermediates show that the carbon atom (former carbonyl ligand), which suffers the nucleophilic attack by the nitrogen atom of the imine, bears a more positive charge in the oxygen coordinated ketene ($\Delta q = +\,0.11$ au) thus indicating a higher electrophilic character for this species (Fig. 4).

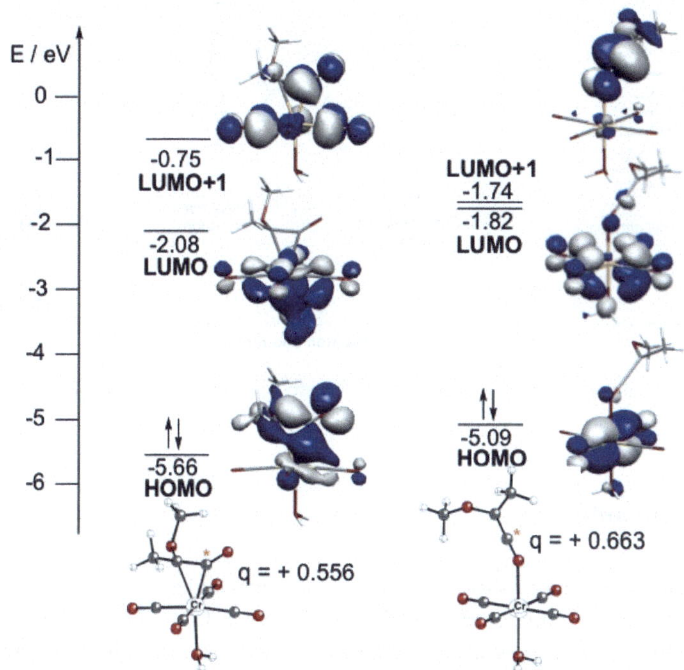

Fig. 4 Molecular orbitals of the metallacyclopropanone (*left*) and oxygen-coordinated ketene (*right*) complexes

Finally, it was also found that the *cis–trans* stereoselection is produced during the nucleophilic addition of the nitrogen atom of the imine to the oxygen-coordinated ketene, since there are no significant differences between the classical and the metallated processes in the four-electron conrotatory electrocyclization step leading to the final 2-azetidinones [66]. The complete reaction profile for the reaction of complex **1** and MeHC=NMe to form the final β-lactam is shown in Fig. 5.

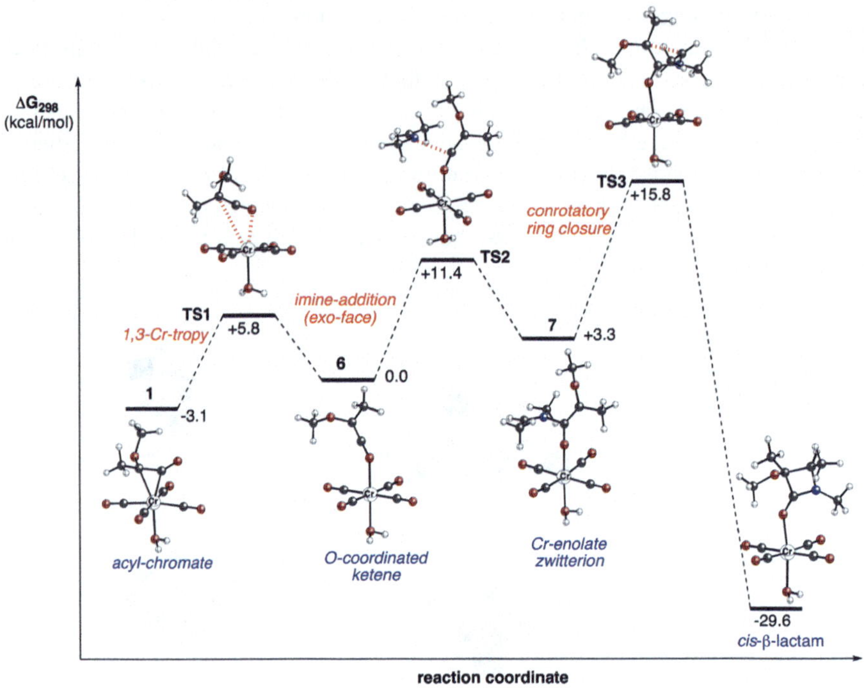

Fig. 5 Computed reaction profile of the process involving complex **1** and MeHC=NMe imine

4 Conclusions

In this chapter, we have shown that the photocarbonylation reaction of Fischer carbene complexes in the presence of imines is a powerful tool toward the synthesis of β-lactams. This process allows the easy access to 2-azetidinones, which are difficult to obtain using standard methodologies with very good reaction yields and excellent diastereoselectivities. From a reaction mechanism point of view, the irradiation of Fischer carbene complexes promotes the insertion of a *cis*-carbonyl ligand into the metal–carbene bond to form a metallacyclopropanone intermediate in the triplet excited state which produces ketenes coordinated to metal. From the latter species and in the singlet hypersurface, the reaction with imines occurs. Different to the closely related Staudinger reaction between ketenes and imines, the *cis–trans* stereoselection is produced during the nucleophilic addition of the nitrogen atom of the imine to the oxygen-coordinated ketene, since there are no significant differences between the classical and the metallated processes in the final four-electron conrotatory electrocyclization step leading to the final β-lactams.

References

1. Fischer EO, Maasböl A (1964) Angew Chem Int Ed 3:580
2. Wulff WD (1995) In: Abel EW, Stone FGA, Wilkinson G (eds) Comprehensive organometallic chemistry II. Pergamon, New York
3. Herndon JW (2000) Tetrahedron 56:1257
4. Barluenga J, Fañanás FJ (2000) Tetrahedron 56:4597
5. Sierra MA (2000) Chem Rev 100:3591
6. de Meijere A, Schirmer H, Duestch M (2000) Angew Chem Int Ed 39:3964
7. Barluenga J, Santamaría J, Tomás M (2004) Chem Rev 104:2259
8. Sierra MA, Gómez-Gallego M, Martínez-Álvarez R (2007) Chem Eur J 13:736
9. Sierra MA, Fernández I, Cossío FP (2008) Chem Commun 4671
10. Dötz KH, Stendel J Jr (2009) Chem Rev 109:3227
11. Dötz KH (1975) Angew Chem Int Ed 14:644
12. Casey CP, Tuinstra HE, Saeman MC (1976) J Am Chem Soc 98:608
13. Casey CP, Albin LD, Burkhardt TJ (1977) J Am Chem Soc 99:2533
14. Harvey DF, Brown MF (1990) J Am Chem Soc 112:7806
15. Harvey DF, Lund KP (1991) J Am Chem Soc 113:8916
16. Dötz KH, Kuhn W, Müller G, Huber B, Alt HG (1986) Angew Chem Int Ed 25:812
17. Wulff WD, Yang DC, Murray CK (1988) J Am Chem Soc 110:2653
18. Wulff WD, Yang DC, Murray CK (1988) Pure Appl Chem 60:137
19. Wulff WD, Bauta WE, Kaesler RW, Lankford PJ, Miller RA, Murray CK, Yang DC (1990) J Am Chem Soc 112:3642
20. Fernández I, Sierra MA, Cossío FP (2008) J Org Chem 73:2083
21. Chan KS, Wulff WD (1986) J Am Chem Soc 108:5229
22. Chan KS, Yeung ML, Chan W, Wang R, Mak TCW (1995) J Org Chem 60:1741
23. Yeung ML, Li WK, Liu HJ, Wang Y, Chan KS (1998) J Org Chem 63:7670
24. Barluenga J, Aznar F, Palomero MA (2001) Chem Eur J 7:5318
25. Fernández I, Sierra MA, Cossío FP (2006) J Org Chem 71:6178
26. Gómez-Gallego M, Mancheño MJ, Sierra MA (2005) Acc Chem Res 38:44
27. Fernández I, Mancheño MJ, Vicente R, López LA, Sierra MA (2008) Chem Eur J 14:11222
28. McGuire MA, Hegedus LS (1982) J Am Chem Soc 104:5538
29. Hegedus LS (1997) Tetrahedron 53:4105
30. Hoffmann N (2008) Chem Rev 108:1052
31. Gilbert A, Baggott J (1991) Essentials of molecular photochemistry. Blackwell Science, Oxford
32. Turro NJ (1991) Modern molecular photochemistry. University Science Books, Sausalito
33. Fernández I, Cossío FP, Sierra MA (2011) Acc Chem Res. 44:479
34. Hafner A, Hegedus LS, de Weck G, Hawkins B, Dötz KH (1988) J Am Chem Soc 110:8413
35. Hegedus LS, McGuire MA, Schultze LM, Yijun C, Anderson OP (1984) J Am Chem Soc 106:2680
36. Hegedus LS, Schultze LM, Toro J, Yijun C (1985) Tetrahedron 41:5833
37. Betschart C, Hegedus LS (1992) J Am Chem Soc 114:5010
38. Hegedus LS, Moser WH (1994) J Org Chem 59:7779
39. Es-Sayed M, Heiner T, de Meijere A (1993) Synlett 57
40. Dumas S, Lastra E, Hegedus LS (1995) J Am Chem Soc 117:3368
41. Borel C, Hegedus LS, Krebs J, Satoh Y (1987) J Am Chem Soc 109:1101
42. Imwinkelried R, Hegedus LS (1988) Organometallics 7:702
43. Schwindt MA, Lejon T, Hegedus LS (1990) Organometallics 9:2814
44. Hegedus LS, D'Andrea S (1998) J Org Chem 53:3113
45. Thompson DK, Suzuki N, Hegedus LS, Satoh Y (1992) J Org Chem 57:1461
46. Hegedus LS, Imwinkelried R, Alarid-Sargent M, Dvorak D, Satoh Y (1990) J Am Chem Soc 112:1109

47. Hegedus LS, Montgomery J, Narukawa Y, Snustad DC (1991) J Am Chem Soc 113:5784
48. Evans DA, Sjogren EB (1985) Tetrahedron Lett 26:3783
49. Alcaide B, Casarrubios L, Domínguez G, Sierra MA (1995) J Am Chem Soc 117:5604
50. Alcaide B, Dominguez G, Plumet J, Sierra MA (1992) J Org Chem 57:447
51. Alcaide B, Casarrubios L, Domínguez G, Sierra MA (1994) J Org Chem 59:7934
52. Sierra MA, Rodríguez-Fernández M, Casarrubios L, Gómez-Gallego M, Mancheño MJ (2009) Eur J Org Chem 2998
53. Edwards EI, Epton R, Marr G (1975) J Organomet Chem 85:C23
54. Edwards EI, Epton R, Marr G (1976) J Organomet Chem 107:351
55. Scutaru D, Mazilu I, Vata M, Tataru L, Vlase A (1991) J Organomet Chem 401:87
56. Simionescu C, Lixandru T, Tataru L, Mazilu I, Vata M, Luca S (1983) J Organomet Chem 252: C43
57. Alcaide B, Pérez-Castells J, Sánchez-Vigo B, Sierra MA (1994) Chem Commun 587
58. Baldoli C, Del Buttero P, Licandro E, Maiorana S, Papagni A (1994) Synlett 183
59. Baldoli C, Del Buttero P, Licandro E, Maiorana S, Papagni A (1994) Tetrahedron Asymmetry 5:809
60. Baldoli C, Del Buttero P, Licandro E, Papagni A (1996) Tetrahedron 52:4849
61. Tong Y-Y, Frausto da Silva JJR, Pombeiro AJL, Wagner G, Herrmann R (1998) J Organomet Chem 552:17
62. Ghatak A, Becker FF, Banik BK (2000) Heterocycles 53:2769
63. Sierra MA, Mancheño MJ, Vicente R, Gómez-Gallego M (2001) J Org Chem 66:8920
64. Lage ML, Fernández I, Mancheño MJ, Gómez-Gallego M, Sierra MA (2009) Chem Eur J 15:593
65. Arrieta A, Cossío FP, Fernández I, Gómez-Gallego M, Lecea B, Mancheño MJ, Sierra MA (2000) J Am Chem Soc 122:11509
66. Fernández I, Sierra MA, Mancheño MJ, Gómez-Gallego M, Cossío FP (2008) J Am Chem Soc 130:13892
67. Sierra MA, Fernández I, Mancheño MJ, Gómez-Gallego M, Torres MR, Cossío FP, Arrieta A, Lecea B, Poveda A, Jiménez-Barbero J (2003) J Am Chem Soc 125:9572
68. Fernández I, Sierra MA, Gómez-Gallego M, Macheño MJ, Cossío FP (2006) Angew Chem Int Ed 45:125
69. Fernández I, Cossío FP, Sierra MA (2009) Chem Rev 109:6687
70. Fernández I, Sierra MA, Mancheño MJ, Gómez-Gallego M, Cossío FP (2008) Eur J Inorg Chem 2454
71. Lage ML, Fernández I, Mancheño MJ, Gómez-Gallego M, Sierra MA (2010) Chem Eur J 16:6616
72. Hegedus LS, de Weck G, D'Andrea S (1988) J Am Chem Soc 110:2122
73. Cossío FP, Arrieta A, Sierra MA (2008) Acc Chem Res 41:925
74. Fernández I, Sierra MA, Gómez-Gallego M, Mancheño MJ, Cossío FP (2005) Chem Eur J 11:5988
75. Wrighton M (1974) Chem Rev 74:401
76. Iwakura I, Ikeno T, Yamada T (2005) Angew Chem Int Ed 44:2524
77. Poli R, Harvey JN (2003) Chem Soc Rev 32:1
78. France S, Shah MH, Weatherwax A, Wack H, Roth JP, Lectka T (2005) J Am Chem Soc 127:1206
79. Lee EC, Hodous BL, Bergin E, Shih C, Fu GC (2005) J Am Chem Soc 127:11586
80. France S, Weatherwax A, Taggi AE, Lectka T (2004) Acc Chem Res 37:592

Top Heterocycl Chem (2013) 30: 85–110
DOI: 10.1007/7081_2012_85

Published online: 7 August 2012

Synthesis of β-Lactams Through Alkyne–Nitrone Cycloadditions

Bablee Mandal and Basudeb Basu

Abstract The [3+2] cycloaddition reaction between an alkyne and nitrone in the presence of copper catalyst leading to the formation of β-lactam is known as the Kinugasa reaction. Over the years, the Kinugasa reaction has been established as a robust and versatile route for the synthesis of β-lactam derivatives in both *cis* and *trans* configuration. This review has focused on various approaches highlighting recent stereoselective syntheses along with plausible mechanisms and other salient features.

Keywords 1,3-Dipolar cycloadditions · Alkynes · Kinugasa reaction · Nitrones · β-Lactams

Nitrogen-containing heterocycles are ubiquitous in nature and found in almost every sphere of life. They are present in vitamins, alkaloids, DNA/RNA bases, antibiotics, etc. For the synthesis of these kinds of molecules, β-lactams are considered ideal building blocks. Though synthesized by Staudinger [1] in 1907, investigations on β-lactam got a high impetus only after the phenomenal discovery of penicillin by Fleming in 1929. Currently thousands of chiral compounds containing β-lactam moiety are known which have high efficacy and safe toxicological profile. Till date, penicillin and its derivatives are still the most commonly used antibiotics. The various families of β-lactam antibiotics differ in their spectrum of antibacterial activity and in their susceptibility to β-lactamase enzymes, which constitute the most common form of resistance to β-lactam.

B. Mandal
Department of Chemistry, Surya Sen College, Siliguri 734004, India

B. Basu (✉)
Department of Chemistry, North Bengal University, Darjeeling 734013, India
e-mail: basu_nbu@hotmail.com

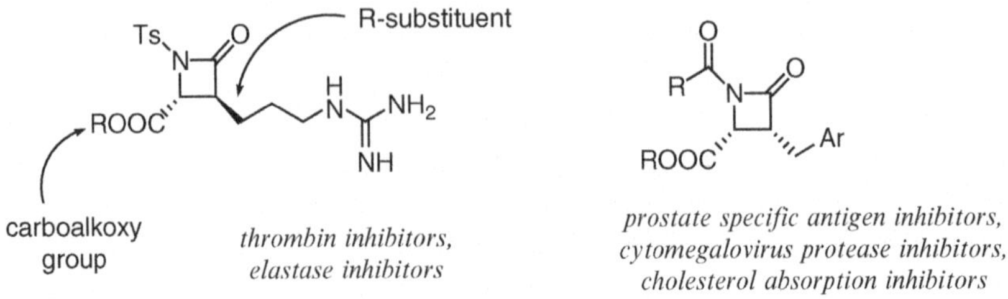

Fig. 1 General structures for serine-protease β-lactam inhibitors

The 2-azetidinone nucleus has been recognized as the central motif of the β-lactam antibiotics. Apart from the immense use of the β-lactam moiety as bioactive agents, many important nonantibiotic uses have been developed in recent years. Some of the most notable recent discoveries concern applications such as serine protease inhibitors (Fig. 1) [2] of elastase [3], cytomegalovirus protease [4], thrombin [5], prostate-specific antigen [6], β-lactamase [7], and cell metastasis [8] and as inhibitors of acyl-CoA cholesterol acyl transferase [9].

Besides their biological relevance, the importance of β-lactams as potential synthetic intermediates has also been recognized in organic synthesis [10, 11]. They are used in the stereo-controlled synthesis of complex organic compounds [10, 12, 13]. They are used as potential synthons for bioactive natural products [14], e.g., the side chain of taxol [15, 16] (the anticancer drug) and bestatin [17] (peptide enzyme inhibitor).

The importance of β-lactams as synthetic intermediates has been widely recognized in organic synthesis because ring cleavage of any of the four single bonds of the β-lactam system is enhanced by ring strain [18]. Selective bond cleavage of the strained β-lactam ring coupled with further interesting synthetic transformations renders these fascinating molecules as powerful synthetic building blocks [13]. 2-Azetidinones have been used as precursors for the preparation of α- and β-amino acids, alkaloids, different-sized heterocycles, taxoids, and other types of compounds of biological and medicinal interest.

Consequently, β-lactam synthesis has been the subject of intensive research all over the world and numerous methods have been reported till date [19] (Fig. 2). Among them the most exploited methods are the ketene–imine cycloaddition (Staudinger reaction; Eqn. 1) [20–23], the copper-catalyzed [3+2] cycloaddition of an alkyne with a nitrone (Kinugasa reaction; Eqn. 2) [24–29], and the condensation of ester enolates with imines (Gilman–Speeter reaction; Eqn. 3) [30–35]. In this review, we aim to consider the various exploitations that utilize the Kinugasa reaction for β-lactam synthesis.

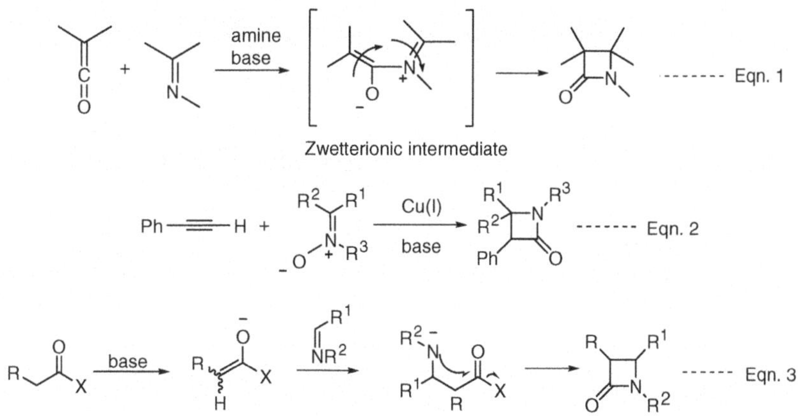

Schematic representation of the most important methods of β-lactam synthesis

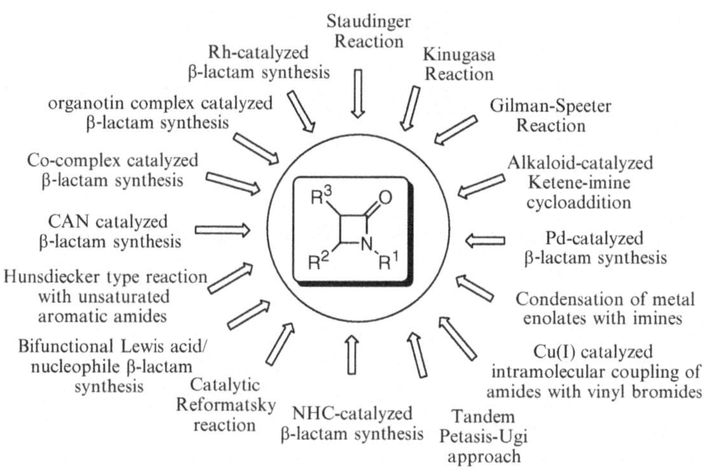

Fig. 2 Various approaches for β-lactam synthesis

The Kinugasa reaction is a simple [3+2] cycloaddition between a nitrone and in situ generated Cu(I)–acetylide species producing both *cis* and *trans* isomers (Scheme 1) [24]. The reaction offers several advantages, which include mild reaction conditions and the availability of a large repertoire of alkynes and nitrones. Crafting of these functionalities on two arms of the same molecule to facilitate an intramolecular reaction is comparatively easier than the widely used Staudinger reaction, which requires the use of an acyl halide, a more reactive functionality.

In general, the reaction proceeds in the presence of an organic base (Scheme 2) [36]. According to the mechanism proposed by Ding and Irwin [25], the initially formed copper–alkyne π-complex undergoes deprotonation. Such activated triple bond is subjected to the 1,3-dipolar cycloaddition with a nitrone to provide a

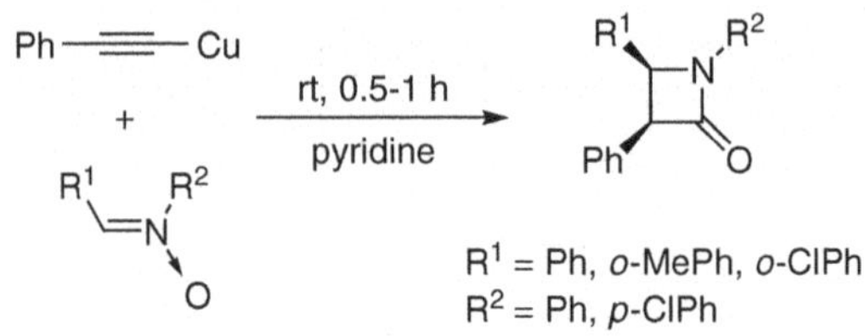

Scheme 1 β-Lactam synthesis by Kinugasa reaction

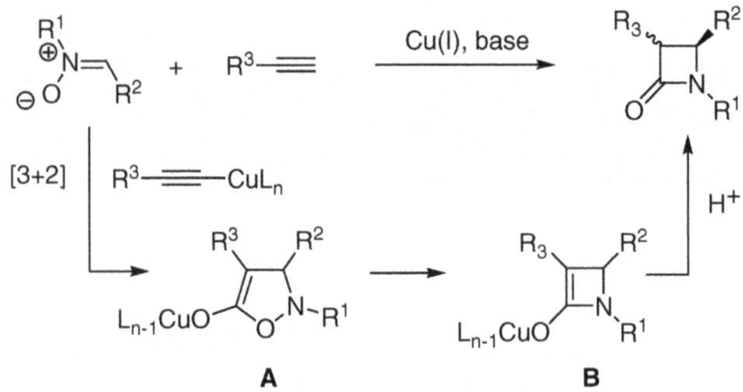

Scheme 2 Plausible mechanism of Kinugasa reaction

five-membered isoxazoline **A**. Rearrangement of the isoxazoline copper complex **A** to the copper enolate **B** and its subsequent protonation leads to the formation of the β-lactam ring.

The evidence in favor of the mechanism was revealed in a work by Miura et al. [26] who in their course of study, found that the absence of CuI does not yield any azetidin-2-one. This goes to suggest that the copper(I) acetylide coordinated by the ligand is involved as the key intermediate.

In a further study, the same group [27] reported that in a reaction between phenyl acetylene and α,*N*-diphenylnitrones, a mixture of four compounds are obtained (Scheme 3) and the yield of the desired product can be increased just by altering the ligand.

The use of dppe ligand afforded the azenyne **3** selectively, whereas using pyridine at 0°C gave the azetidin-2-one **4** as the major product. Again if either of the phenyl rings of the nitrone has a substituent at 4-position, then the reaction takes place at room temperature giving a mixture of **4** and **5**. The ratio of **4** to **5** increases as the electron-withdrawing nature of the substituent increases. The azetidin-2-one is always obtained as a mixture of *cis* and *trans* isomer, the ratios ranging from 1:1 to 2:3.

With the aim to improve the selectivity, Fu et al. [37] used a new C_2-symmetric planar-chiral bis(azaferrocene) ligand **7** to generate β-lactams with good enantiomeric excess and *cis* diastereoselection (Scheme 4).

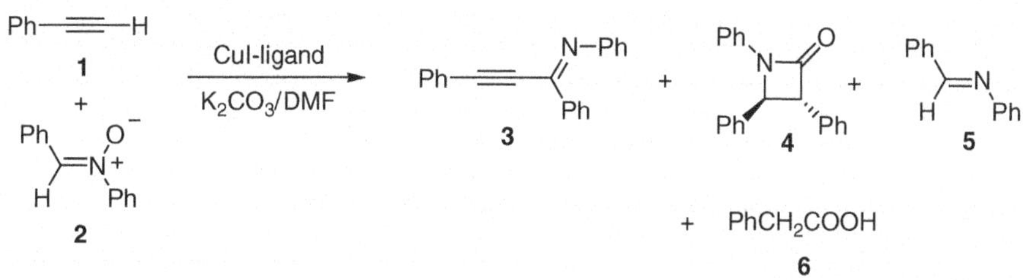

Scheme 3 Reaction of phenylacetylene **1** with α,*N*-diphenylnitrone **2**

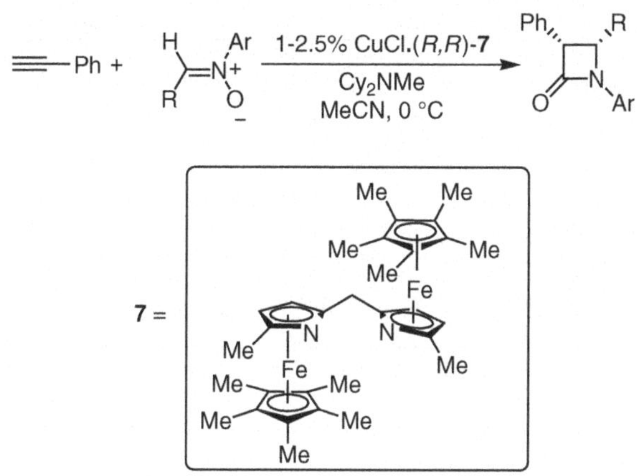

Scheme 4 Kinugasa reaction using bis(azaferrocene) ligand **7**

When the R is Ph, generation of the β-lactam proceeded with excellent *cis* diastereoselectivity irrespective of the nature of the aromatic ring (95:5). With regard to enantioselectivity, the more electron-rich the aromatic group, the higher was the ee (Table 1). Again lowering the reaction temperature improved the yields as well as the stereoselectivity (Table 1, entry 5). Whereas keeping the Ar group fixed as *para*-methoxyphenyl and bringing in the variations in the R-groups, higher ee was observed for electron-deficient groups (Table 2).

Again when phenyl acetylene was replaced by alkyl acetylene, the stereoselection was lowered. But alkenyl substituted alkynes furnish good de and ee (Table 3). In these cases, alkyl **8** and acyl **9** substituted nitrones were taken.

Efforts were now directed to achieve better success in the asymmetric version of the Kinugasa reaction. Basak et al. [38] reported a chiral auxiliary based approach, which proceeds with high enantiospecificity. They crafted the ligand on one of the components of the Kinugasa reaction, namely, the acetylene, so that the reaction proceeds through an intramolecularly chelated copper complex. A homochiral *N*-propargyl oxazolidinone was used for the purpose and it was hoped that the side chain would control the approach of the nitrone. The proposed mechanism is shown in Fig. 3. The two substituents (benzyl and R) being on opposite faces in intermediate **D** will lower its energy, than **E**. Thus, there will be a preference for the

Table 1 Catalytic asymmetric synthesis of β-lactams: variation of the *N*-substituent

Entry	Ar	*cis*:*trans*	% ee, *cis*	Isolated yield *cis* isomer (%)
1	Ph	95:5	77	69
2	4-(OMe)C_6H_4	95:5	85	53
3	4-BrC_6H_4	94:6	72	74
4	4-(EtO_2C)C_6H_4	94:6	67	79
5[a]	4-(EtO_2C)C_6H_4	95:5	71	91

[a]Run at −20°C

Table 2 Catalytic asymmetric synthesis of β-lactams: variation of the R-group

Entry	R	*cis*:*trans*	% ee, *cis*	Isolated yield *cis* isomer (%)
1	Ph	95:5	85	53
2	4-(F_3C)C_6H_4	93:7	90	50
3[a]	4-(OMe)C_6H_4	93:7	83	46
4	Cy	93:7	89	57
5	PhCO	91:9	72	42

[a]Run at room temperature

Table 3 Catalytic asymmetric synthesis of β-lactams: scope with respect to the alkyne component

Entry	Nitrone	R	*cis*:*trans*	% ee, *cis*	Isolated yield *cis* isomer (%)
1[a]	**8**	Ph	>95:5	92	65
2[a]	**8**	4-(F_3C)C_6H_4	>95:5	93	57
3[a]	**8**	4-(OMe)C_6H_4	92:8	91	60
4[a]	**8**	$PhCH_2$	71:29	73	43
5[b]	**9**	Ph	90:10	90	56
6[b]	**9**	1-Cyclohexenyl	90:10	91	45

[a]Run at −20°C
[b]Run at −40°C

chirality at C-4 of the resulting isooxazoline. This, in turn, will fix the chirality at C-3 of the *cis* and the *trans* azetidinone.

Thus the alkyne, (4*S*)-benzyl-3-propargyl oxazolidone, prepared from (*S*)-phenylalanine underwent reaction with various nitrones (Scheme 5), giving a mixture of one *cis* **12** and one *trans* **13** isomer, which could easily be separated by column chromatography. The reactions proceeded with good yields and excellent stereochemical control.

Intramolecular Kinugasa reactions are in principle ideal for the generation of monocyclic and polycyclic β-lactams. Fu et al. [39] made an endeavor to study

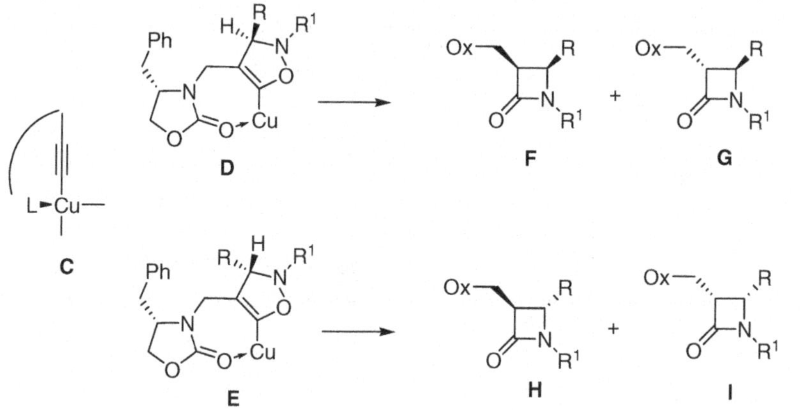

Fig. 3 Control over the approach of nitrone by a homochiral *N*-propargyl oxazolidinone

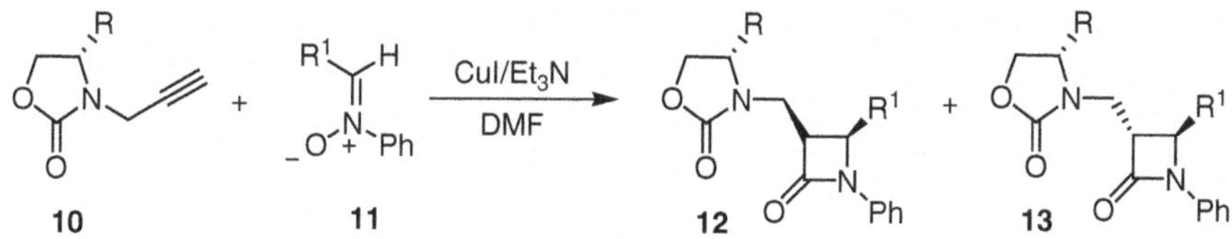

R = Ph, CH_2Ph
R^1 = Ph, thiophenyl, furyl, *p*-anisyl

Scheme 5 Use of chiral oxazolidones in Kinugasa reaction

these reactions and successfully came up with a protocol for preparing enantiomerically enriched bi- and polycyclic β-lactams. The group demonstrated that a copper/phosphaferrocene–oxazoline catalyst mediates asymmetric intramolecular Kinugasa reactions to produce two new rings with very good stereoselectivity (Scheme 6).

Thus, they explored the application of their technique to the synthesis of a range of tricyclic compounds containing a 6,4 or a 7,4 ring system (Table 4).

Thus, the intramolecular Kinugasa reaction methodology was further extended by the development of conditions that allow α-allylated β-lactams to be prepared with the same catalyst. The intermediate **B** (Scheme 2) could be intercepted by an electrophile thereby generating a quaternary stereocenter [40, 41]. This group successfully exploited this scope thereby enhancing the utility of Kinugasa reaction. In the presence of a mixture of a silyl enol ether and KOAc as the base [rather than $(C_6H_{11})_2NMe$], alkyne–nitrone **14** underwent cyclization followed by α-alkylation with good stereoselectivity and in good yield (85% ee and 76% yield) (Scheme 7). The protocol when applied to heterocyclic substrates also yielded

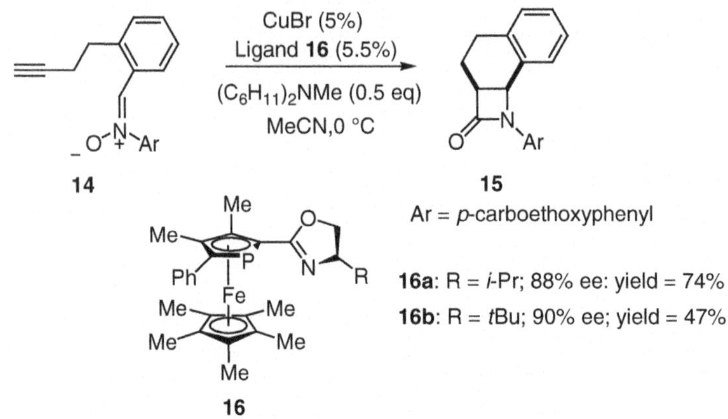

Scheme 6 Intramolecular Kinugasa reaction generating tricyclic β-lactam framework

fruitful results (Scheme 8). The substrate **20** was converted into the desired enantioenriched β-lactam in 90% ee and 70% chemical yield.

The above reaction sequence is quite impressive because of the formation of two carbon–carbon bonds, a carbon–nitrogen bond, a carbocyclic ring, a β-lactam, a carbonyl group, a tertiary stereocenter, and an all-carbon quaternary stereocenter.

The group [42] also showed the use of nitrones in the synthesis of potential antihypercholesterolemic and antibacterial mono and tricyclic β-lactams. They elaborated the hydroxyethyl group at C-3 of monocyclic β-lactams by a series of reactions to the appropriate side chain meant for acting as cholesterol absorption inhibitor without perturbing the sensitive β-lactam moiety. In addition, novel tricyclic β-lactam was also synthesized using the nitrone cycloaddition approach.

Tang et al. [43] designed a pseudo C_3-symmetric tris(oxazoline) ligand **22** and applied it successfully for asymmetric Kinugasa reaction (Scheme 9).

A study of various organobases revealed that secondary amines afforded the desired products with better diastereoselectivity and enantioselectivity. Obviously, bulkier amines gave better diastereoselection. Screening experiments showed dicyclohexyl amine to be the best choice. Under the conditions, the electronic character of the α-aryl group on nitrones had almost no effect on the enantioselection. But the substituents on the N-atom showed profound influence on both the yield and the selection. Electron-rich ones increased enantioselectivity but decreased the reactivity, whereas electron-deficient ones though increased yield reduced the enantioselectivity. Thus the method provided a facile access to *cis* β-lactams with good enantioselectivity. A repertoire of copper-bis(oxazoline)-catalyzed synthesis of β-lactams is already encompassed in a review by Evans et al. [44]. The topic includes enantioselective reaction of alkynes with nitrones.

A variety of synthetic applications of *N*-Boc allylic amines were reported by Whisler and Beak [45]. One of the applications involve a synthetic sequence that involves a stereo-controlled intramolecular nitrone–olefin dipolar cycloaddition

Table 4 Copper-catalyzed intramolecular Kinugasa reactions in the presence of planar-chiral phosphaferrocene–oxazoline ligands: enantioselective synthesis of two new rings

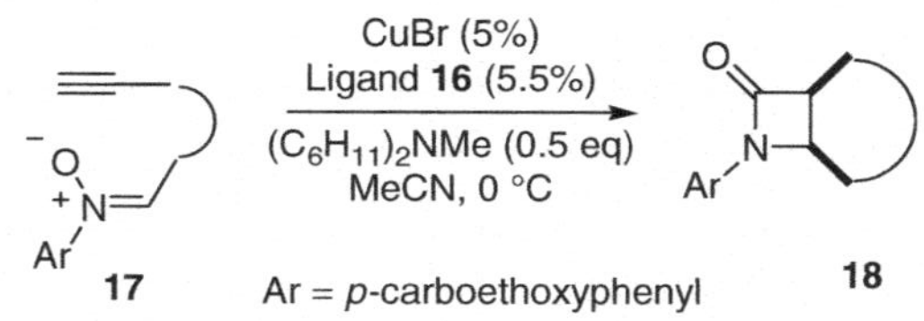

Entry	Product **18**	Ligand	ee (%)	Yield (%)
1	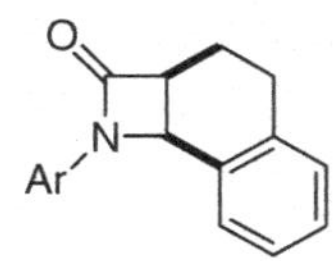	16a	88	74
2	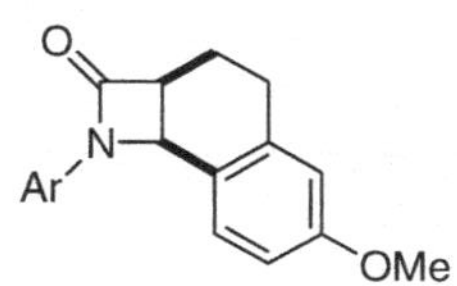	16a	86	60
3[a]	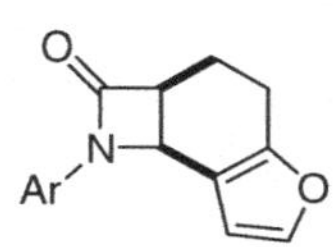	16a	90	46

(continued)

Table 4 (continued)

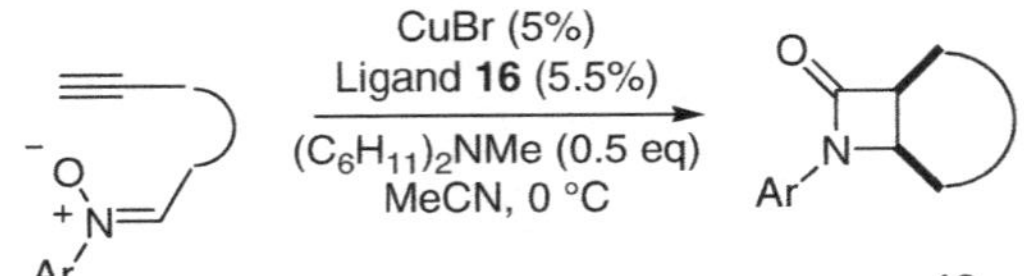

17 Ar = *p*-carboethoxyphenyl **18**

Entry	Product **18**	Ligand	ee (%)	Yield (%)
4[a]	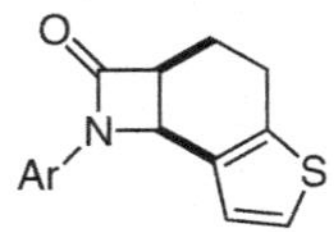	16b	90	64
5[a]	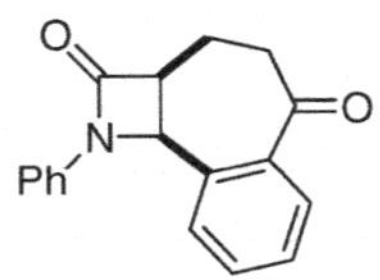	16b	85	53
6	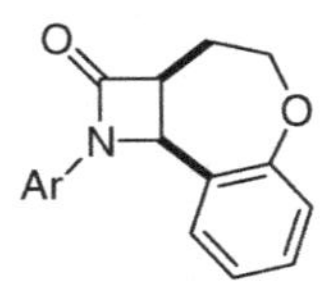	16b	91	68

[a]Run at room temperature

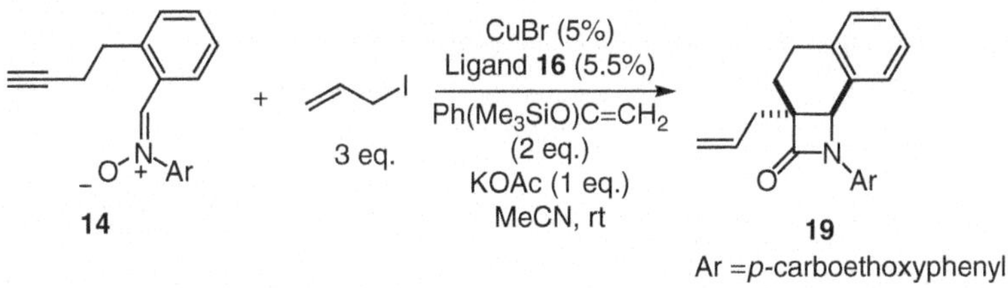

Scheme 7 Intramolecular Kinugasa reaction using silyl enol ether and KOAc as the base

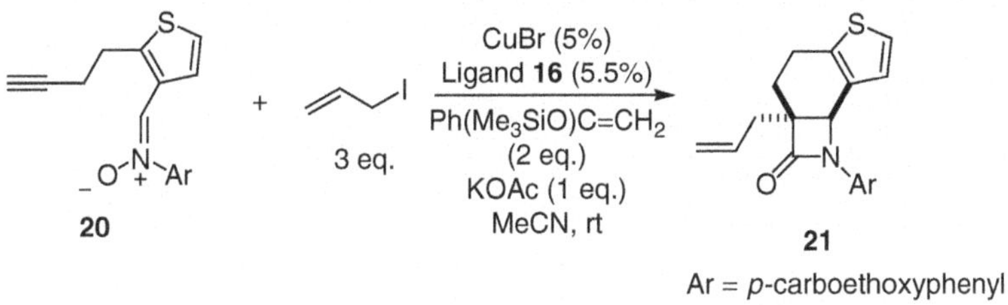

Scheme 8 Intramolecular Kinugasa reaction using heterocyclic substrate

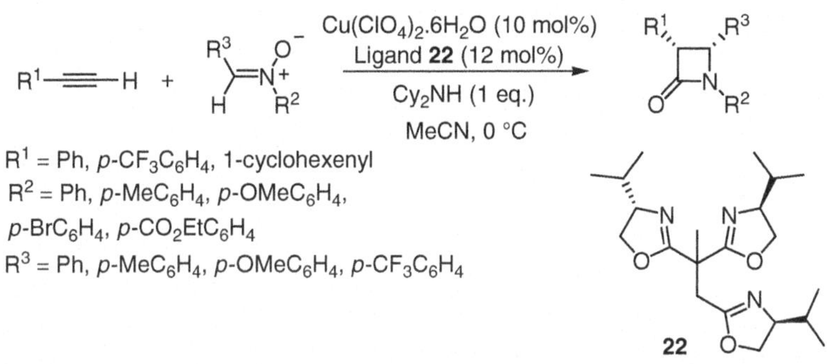

Scheme 9 Kinugasa reaction using a pseudo C_3-symmetric tris(oxazoline) ligand

(compound **25**), which has been developed for the preparation of enantioenriched 2-formyl-4-phenyl-1-aminocyclopentanes from one β-allyl-substituted aldehyde. Further manipulations allow access to an enantiomerically enriched β-lactam **32** (Scheme 10).

The 5-spirocyclopropane isoxazolidines **36** can be easily synthesized by 1,3-dipolar cycloaddition of nitrones and methylenecyclopropane derivatives [46, 47]. Cycloadducts **36** can be selectively converted into β-lactam derivatives through ring contraction and concomitant extrusion of ethylene by heating at 70–110°C in the presence of a protic acid [48, 49]. Cordero et al. [50] synthesized a new series of tri- and tetracyclic spirocyclopropane isoxazolidines **36** starting from the 1-vinylcyclopropyl tosylate **33,** which underwent Pd(0)-catalyzed nucleophilic substitution of α-amino and α-hydroxy acid derivatives to afford exclusively

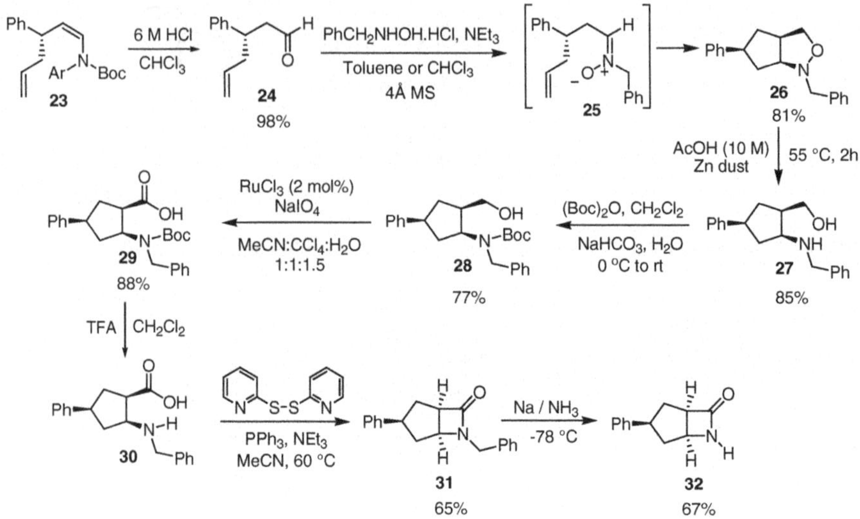

Scheme 10 Kinugasa reaction starting from *N*-Boc allylic amines

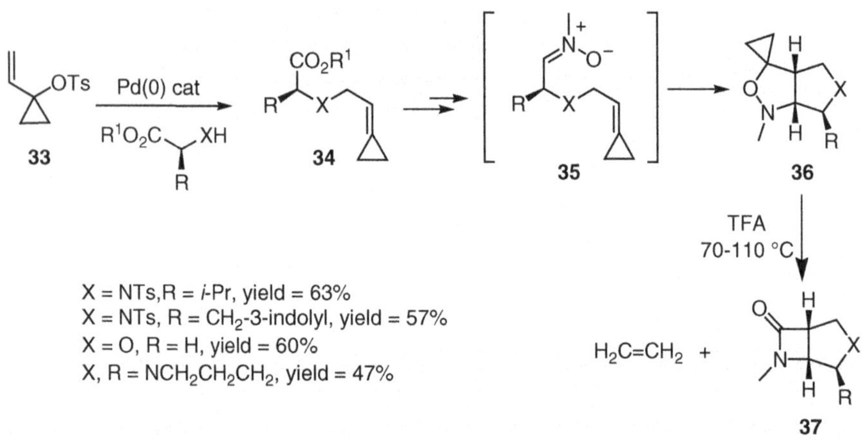

Scheme 11 Synthesis of β-lactams starting from 1-vinylcyclopropyl tosylate

alkylidenecyclopropanes **34**. The esters **34** were easily converted into the corresponding nitrones **35** which spontaneously evolved to **36**. In the presence of trifluoroacetic acid (TFA), compound **36** undergoes a clean reaction to afford 6-azabicyclo[3.2.0]-heptan-7-ones, **37** with conservation of the relative and absolute configuration (Scheme 11). The yields presented in Scheme 11 refer to those of the β-lactams **37**.

In a preliminary attempt to rationalize the mechanism of β-lactam formation from 5-spirocyclopropane isoxazolidines **36**, the authors discarded the possible

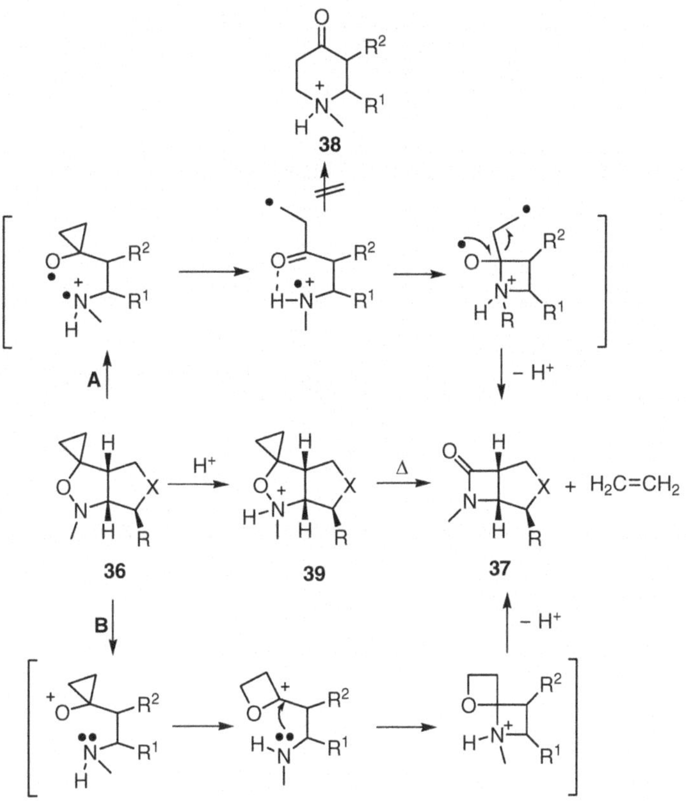

Scheme 12 Possible mechanistic pathway of formation of β-lactam from isoxazolidine

involvement of piperidin-4-ones **38**, as reaction intermediates, when it was observed that they are stable under the acidic rearrangement conditions. Further, the protonation of the isoxazolidine nitrogen atom was unequivocally demonstrated by the significant downfield shift of signals in the ^{1}H NMR spectra of adducts **36** in the presence of protic acids, proving it to be the initial step of the formation of β-lactams **37**. The protonated isoxazolidine **39**, then, undergoes a thermally induced cleavage of the weak protonated N–O bond that might occur either in a homo- (Scheme 12, path A) or in a heterolytic (Scheme 12, path B) mode.

Basak and Ghosh have developed a variant of the Kinugasa reaction that allows for straightforward synthesis of 3-exomethylene β-lactams [51]. The reaction of propargylic alcohol with nitrones in the presence of CuI/L-proline resulted in good yields of the desired 3-exomethylene β-lactams **40** (Scheme 13). The authors reported that the use of DMSO as solvent is critical to the formation of the compound **40** favorably over the corresponding *cis* β-lactam **41**. Furthermore, the authors explain that the use of common amines as bases in this reaction lead only to

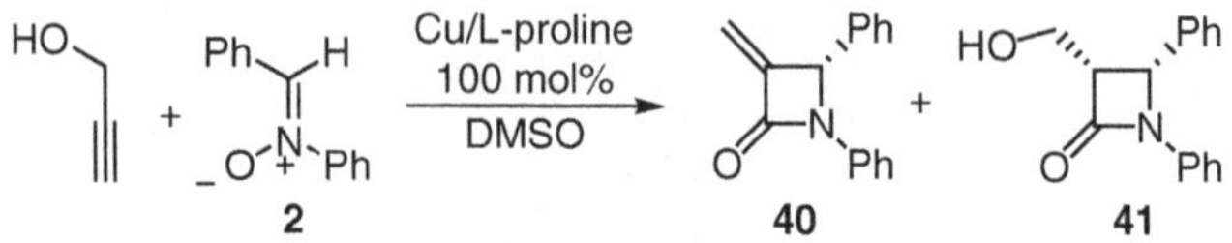

Scheme 13 β-Lactam synthesis by the reaction of propargyl alcohol with nitrones

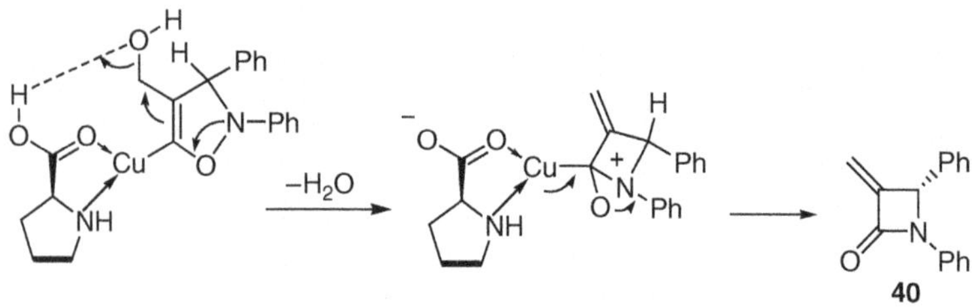

Scheme 14 Proposed mechanism for the formation of 3-exomethylene β-lactams

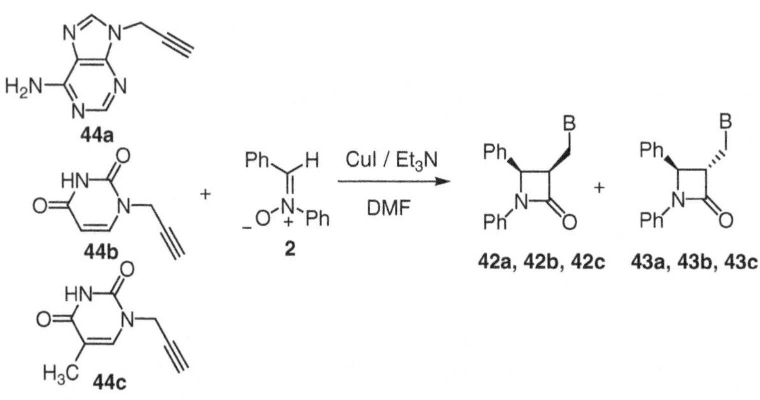

Scheme 15 Synthesis of several β-lactam nucleoside chimeras

formation of the *cis*-β-lactam, and that the amphoteric nature of the amino acid is necessary to promote elimination and formation of the 3-exomethylene β-lactam. A reasonable explanation as suggested by the authors for their experimental observation is depicted in Scheme 14.

Several β-lactam nucleoside chimeras **42** (*cis*) and **43** (*trans*) were synthesized by Basak et al. [52] from the corresponding *N*-propargyl nucleobases **44** via Kinugasa reaction in moderate yields. The reaction was carried out using propargyl nucleobases and diphenyl nitrone in the presence of CuI and Et_3N in DMF (Scheme 15). The reaction time in all the three cases was 36 h and the combined

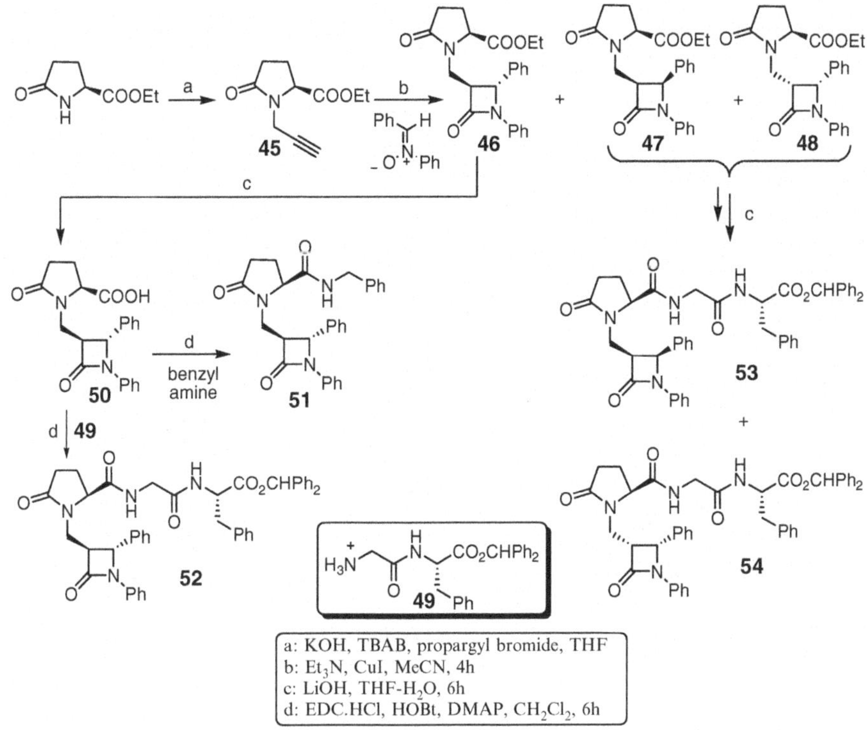

Scheme 16 Synthesis of tripeptide using Kinugasa reaction

yield from the adenine propargyl base **44a** is 64% and from uracil **44b** and thymine **44c** is 60%.

With the aim to synthesize tripeptides **52–54**, Basak et al. [53] applied the Kinugasa reaction as their method of choice. Their synthetic strategy is outlined in Scheme 16. The propargyl ethyl pyroglutamate **45** was prepared and the Kinugasa reaction between **45** and the diphenyl nitrone was carried out in CH_3CN solution in the presence of organic base triethylamine and cuprous iodide. Interestingly, the reaction produced three diastereomers: one *trans* isomer **46** and a pair of *cis* isomers **47** and **48**. It appears that only one of the *cis* isomers, **48**, has epimerized to the *trans* compound, the other *cis* isomer **47** being configurationally more stable is resistant to epimerization. Since the two *cis* isomers have similar polarity on silica gel, the peptide synthesis was carried out using the mixture.

Intramolecular Kinugasa reaction was also explored by Basak et al. [54] as it has the possibility of synthesizing β-lactam fused enediynes. This class of molecule has gained importance because of the ability of the β-lactam ring to act as a molecular lock [55–57] in stabilizing the otherwise unstable enediyne moiety. The group first attempted the synthesis of the enediynes **59** and **60** which involved the following steps: (1) the construction of the acyclic enediyne framework by Sonogashira coupling, (2) O-propargylation, (3) functional group modification to generate the

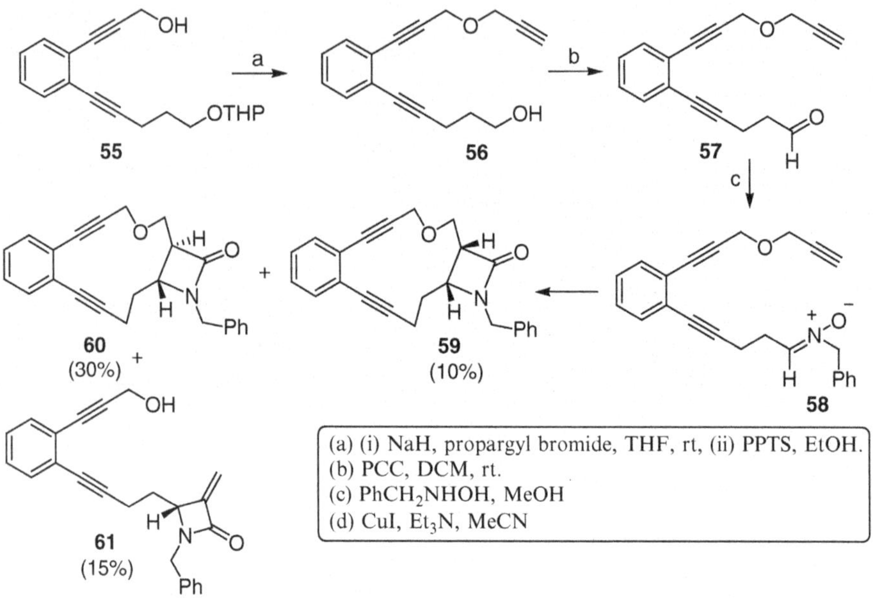

Scheme 17 Intramolecular Kinugasa reaction for synthesizing β-lactam fused enediynes

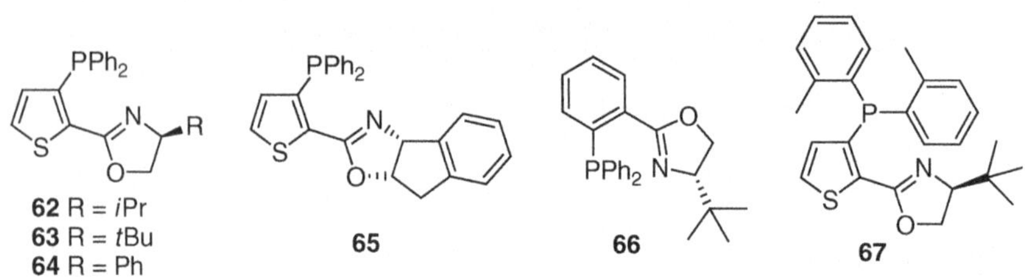

Fig. 4 HETPHOX ligands

nitrone, and (4) an intramolecular Kinugasa reaction. The nitrone **56** essentially existing in the *Z*-form was subjected to Kinugasa reaction conditions with some modification (Scheme 17). The desired β-lactams **59** and **60** were isolated by careful chromatography over silica gel using hexane–ethyl acetate of increasing polarity as eluent. The lactam **61** is produced during the collapse of the isooxazoline intermediate formed by initial cycloaddition [51].

β-Lactam synthesis mimicking the click chemistry conditions was also reported by Basak et al. [58]. Various monocyclic β-lactams, both *cis* and *trans*, were successfully prepared via Kinugasa reaction, involving a Cu(I)-catalyzed cycloaddition between a nitrone and a terminal acetylene.

The HETPHOX ligands (Fig. 4) that were developed independently by Tietze and Cozzi compare favorably with other P,N ligands in a wide range of asymmetric

Table 5 HETPHOX/Cu(I)-mediated synthesis of β-lactam via the Kinugasa reaction

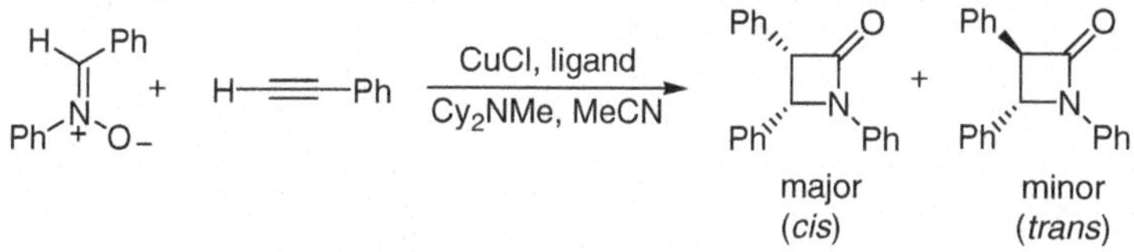

Entry	Ligand	Time (h)	Conversion	*cis:trans*	ee
1	64	72	40	89:11	12
2	64	120	58	91:9	12
3	62	120	43	93:7	16
4	63	120	73	91:9	37
5	65	120	72	93:7	16
6	66	120	85	93:7	22
7	67	120	44	90:10	55

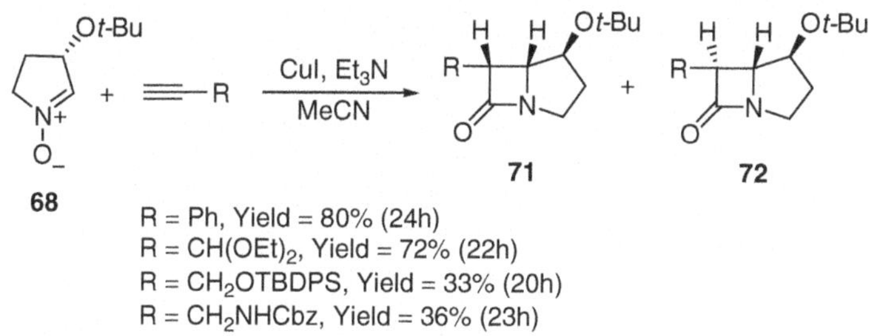

Scheme 18 Diastereoselective synthesis of carbapenams via Kinugasa reaction of nitrone **68**

transformations. Guiry et al. [59] are the first to investigate these HETPHOX ligands in asymmetric intermolecular Kinugasa reaction (Table 5). The *cis:trans* ratios were determined by ^{1}H NMR and the enantiomeric excess by chiral-phase HPLC analyses.

Chmielewski et al. [60] carried out the Kinugasa reaction with cyclic nitrones thereby establishing a facile approach to carbapenams. They have performed their studies using non-racemic cyclic nitrones derived from *S*-malic (**68** and **69**) and *L*-tartaric acid (**70**).

Studies carried out with the nitrone **68** and terminal alkynes afforded the 5,6-*cis*-penams (**71**) as the major product preferentially over the other diastereomer **72**. A variety of terminal acetylenes were used (Scheme 18) and the reaction time ranged from 20 to 24 h.

The stereochemical outcome of the Kinugasa reaction is possibly controlled by the initial cycloaddition step that leads to the formation of the isoxazoline intermediate. The cycloaddition step determines the configuration at the bridgehead carbon atom. The two possible approaches are depicted in Fig. 5. The approach of the acetylide to the *si* side of the nitrone (*syn* to *t*-BuO) is disfavored due to the steric

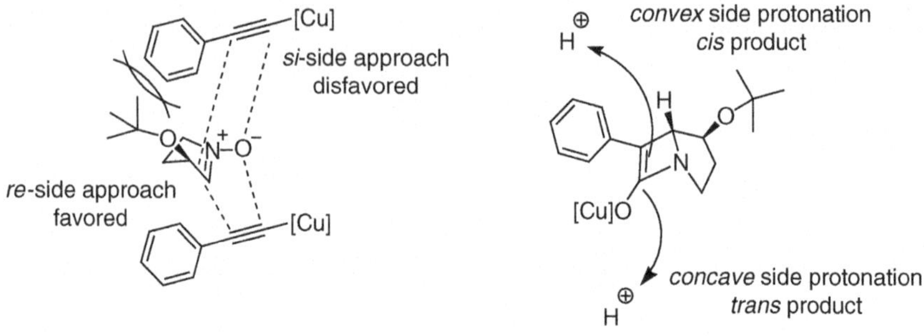

Fig. 5 Mechanism defining the diastereoselection

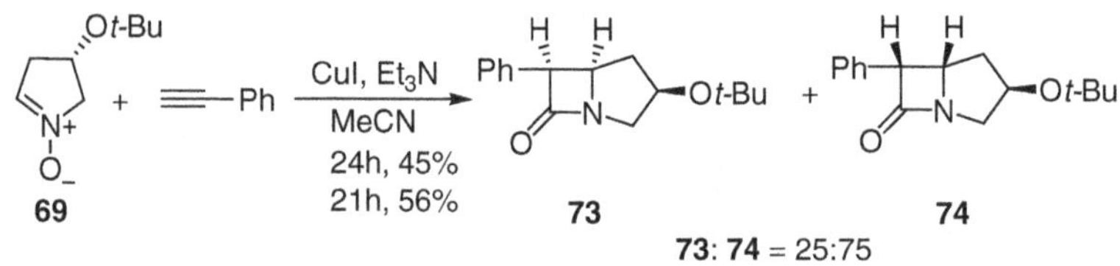

Scheme 19 Diastereoselective synthesis of carbapenams via Kinugasa reaction of nitrone **69**

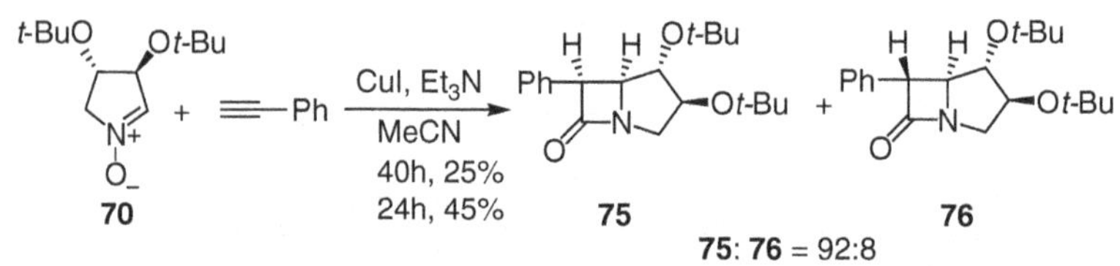

Scheme 20 Diastereoselective synthesis of carbapenams via Kinugasa reaction of nitrone **70**

interactions. On the contrary, the *re* side of the nitrone lacks the steric hindrance, thus favoring the attack of the acetylide. As a result, the cycloaddition step proceeds with a high degree of diastereoselectivity, giving products with a *cis* arrangement of the *t*-BuO group and the bridgehead proton.

The shift of the *t*-BuO group from C-3 to C-4 of the nitrone moiety **69**, however, did change the observed diastereoselectivity significantly (Scheme 19). However, the introduction of a second *t*-BuO group to the nitrone **70** did not affect the stereochemical outcome of the cycloaddition step, and the 5,6-*cis* penam **75** was formed as the major product (Scheme 20).

In a further communication, the same group [61] reported the Kinugasa reaction involving chiral, optically pure acetylenes and five-membered ring nitrones, both nonchiral or with a stereogenic center.

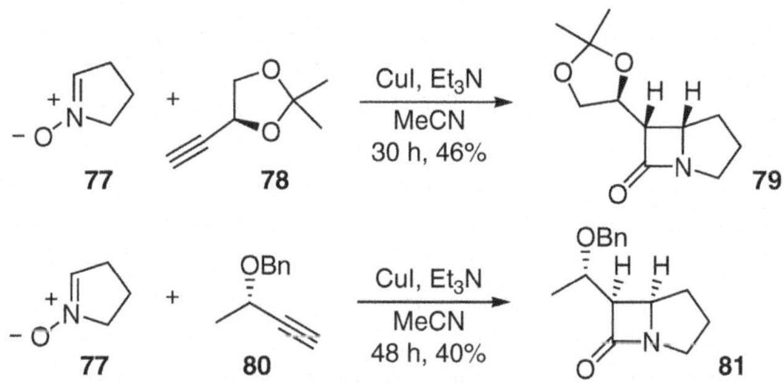

Scheme 21 Kinugasa reaction between chiral acetylene and nonchiral nitrone

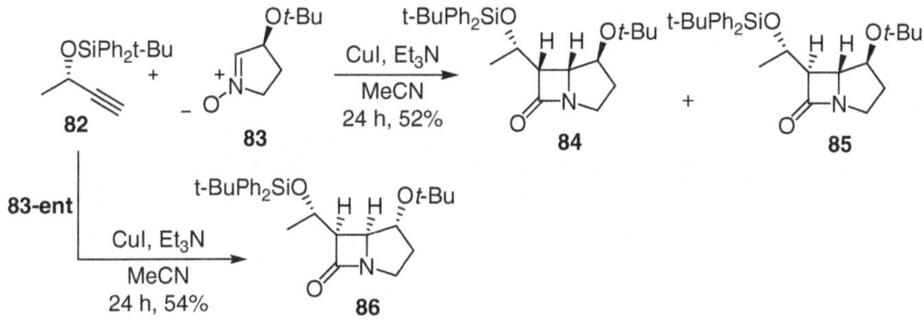

Scheme 22 Kinugasa reaction between chiral nitrone and chiral acetylene

As shown in Scheme 21, reactions between nitrone **77** and chiral acetylenes **78** and **80** afforded corresponding carbapenams **79** and **81**, respectively, with excellent diastereoselectivity but with moderate yield (~50%). The relative configuration of H-5 and H-6 protons in both products was proven using $^3J_{H5-H6}$ coupling constants and was found to be *cis* for both of them.

In case where both reactants (acetylene and nitrone) are chiral (Scheme 22), the first 1,3-dipolar cycloaddition step involving nitrone **83/83-*ent*** proceeds exclusively *anti* to the *tert*-butoxy group. The reaction of **82** with the nitrone **83** led to two carbapenams **84** and **85** in a 9:1 ratio, respectively, whereas nitrones' enantiomeric form (**83-*ent***) afforded solely the *cis* isomer **86**.

In another report, Chmielewski et al. [36] demonstrated that reactions of acetylenes derived from glyceraldehyde and propargyl aldehyde (Scheme 23) show remarkable reactivity in Kinugasa cycloaddition/rearrangement cascade process catalyzed by Cu(I) ion. Reactions proceed by formation of a rigid dinuclear copper(I) complex in which each copper ion is coordinated to one or both oxygen atoms in the acetylene molecule and to the triple bonds. Their studies showed that the effectiveness of acetylene components in the Kinugasa reaction can be improved by the addition of 1,10-phenatroline.

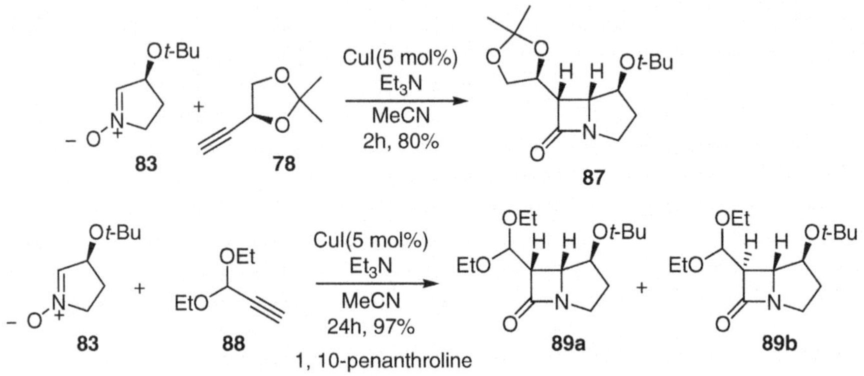

Scheme 23 Direct catalytic synthesis of carbapenams via Kinugasa reaction

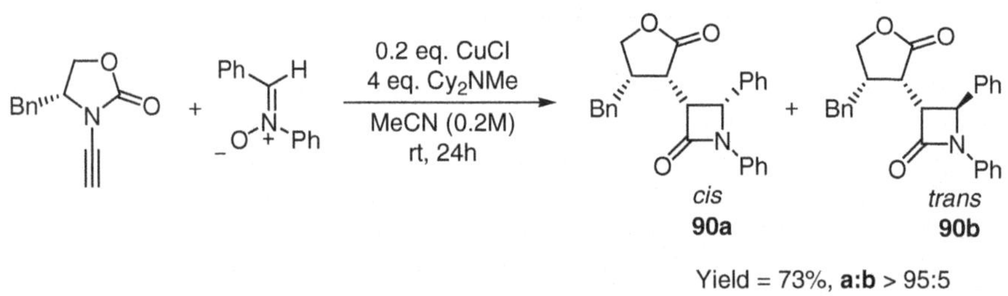

Scheme 24 Stereoselective synthesis of chiral α-amino-β-lactams via the Kinugasa reaction employing ynamides

Another highly stereoselective synthesis of chiral α-amino-β-lactam through an ynamide-Kinugasa reaction (Scheme 24) has been described by Hsung et al. [62].

The major product is *cis* (**90a**), which is unambiguously determined by X-ray structural analysis. The minor isomer **90b** was assigned as *trans* initially based on proton coupling constants and was later confirmed by NOE experiments.

The authors demonstrated an immediate application of this reaction in the preparation of chiral α-amino-β-lactams **93** (Scheme 25).

Saito et al. [63] have reported an enantioselective Kinugasa reaction of nitrones with terminal alkynes in the presence of 20 mol% of IndaBox–$Cu(OTf)_2$ and di-sec-butylamine (1.5 eq.) (Scheme 26) which they claim produced β-lactams with the highest level of enantiomeric excesses among the catalytic enantioselective Kinugasa reactions reported so far. The authors report an enantioselectivity of 85%, the *cis*:*trans* ratio being 84:16.

They have demonstrated a wide variety of reactions using both aliphatic and aromatic terminal alkynes. Varying the substituents in the nitrone also successfully resulted in products with high enantioselectivity.

Further "on water" application of Kinugasa reaction was reported by Pezacki et al. [64] demonstrating a micelle-promoted, copper-catalyzed multicomponent

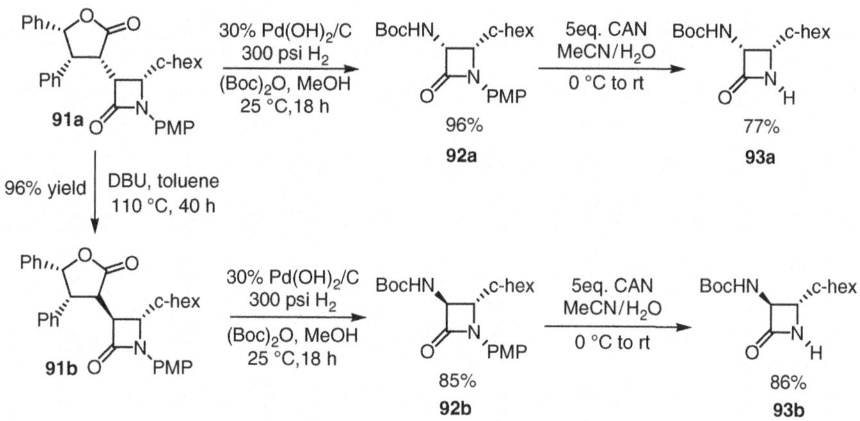

Scheme 25 Synthesis of chiral α-amino-β-lactams

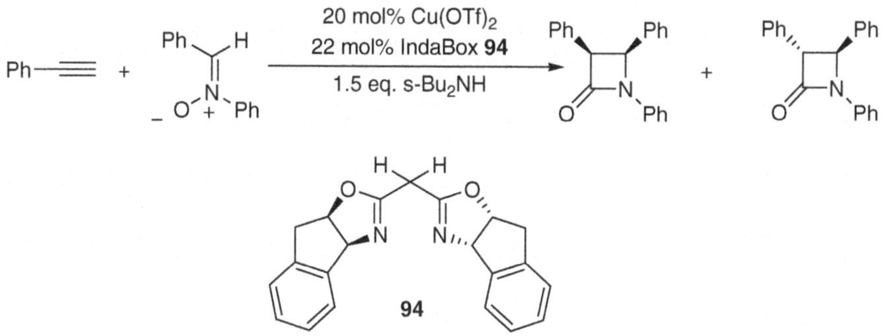

Scheme 26 Enantioselective Kinugasa reaction of nitrones with terminal alkynes using catalytic IndaBox–Cu$(OTf)_2$

Kinugasa reaction. Reactions were performed in a "single pot" for a series of in situ generated C,N-diphenylnitrones with Cu(I) phenylacetylide providing β-lactams in the yield range of 45–85%. The multicomponent process proceeds by a two-step reaction sequence involving the micelle-promoted nitrone formation from substituted benzaldehydes and *N*-phenylhydroxyl amine followed by the in situ 1,3-dipolar cycloaddition and rearrangement reaction with Cu(I) phenylacetylide.

The authors found that cooling the in situ-generated C,N-diphenylnitrone at 0°C led to better yields of β-lactam. The cooled solution was then reacted with phenylacetylene (0.2 mmol), (+)-sodium-L-ascorbate (0.08 mmol), and Cu$(SO)_4$ (0.04 mmol) in the dark. The latter reaction also contained an excess of pyridine (1.6 mmol) and 0.1 mmol of ethanolamine (ETA), buffered at pH 10. This mixture was allowed to react for 30 min at 0°C and then stirred at room temperature for 10 h. The reaction mixture consisted of essentially three products (Scheme 27), the

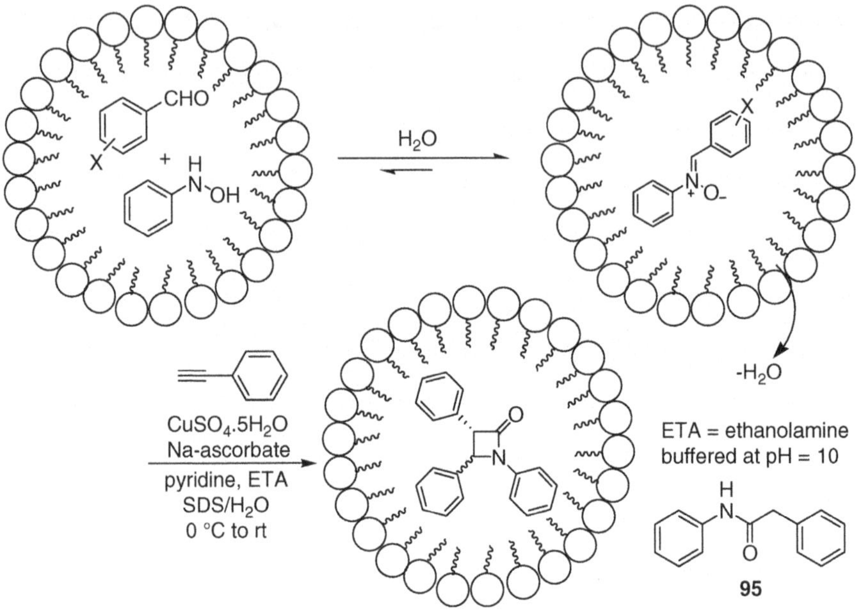

Scheme 27 Micelle-promoted multicomponent Kinugasa reactions in aqueous media

cis and *trans*-β-lactams in 46% overall yield and an amide **95** in 36% yield, the structure of which was confirmed by X-ray crystallography.

In order to account for the formation of the amide **95**, the authors performed a series of reactions and proposed that a ketene is formed, which gives **95** in water according to the pathway outlined in Scheme 28.

The reaction was proved to be tolerant to substituents at the α-aryl position of the nitrone, and the highest yields of β-lactams were obtained when electron-withdrawing substituents were employed. Thus, the reaction provides a convenient method for the construction of the β-lactam ring in aqueous media.

Khangarot and Kaliappan [65] disclosed an application Kinugasa reaction from sugar-derived cyclic nitrones and alkynes for the synthesis of chiral β-lactams (Scheme 29). The sugar-derived β-lactams were prepared by the reaction between cyclic nitrones, derived from different sugars, tartrate ester, and alkynes obtained from different sugars. The authors proposed that the addition of a sugar unit to both the templates may have a significant effect in improving the bioavailability of these chiral β-lactams. Their investigation showed the Glaser coupling product as a minor side product in some cases.

The stereochemistry of the β-lactams was tentatively assigned on the basis of previous reports [36, 60, 61]. The acetylene–copper complex approaches the nitrone *anti* to the substituent close to the 1,3-dipole, and the protonation of the copper enolate occurs from the less hindered convex side of the β-lactam skeleton, which leads exclusively to *cis*-β-lactams (Fig. 6).

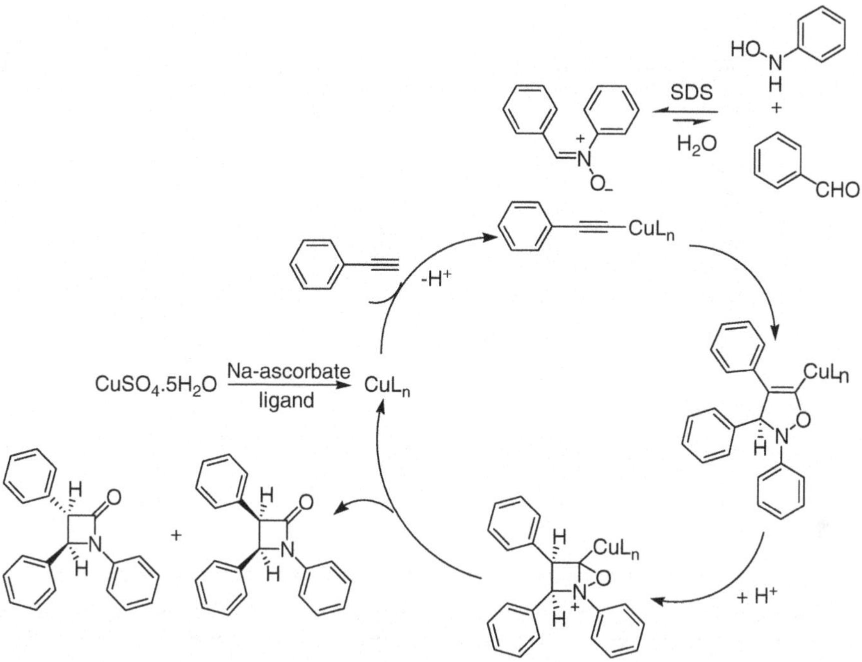

Scheme 28 Mechanism of micelle-promoted Kinugasa reaction in aqueous media

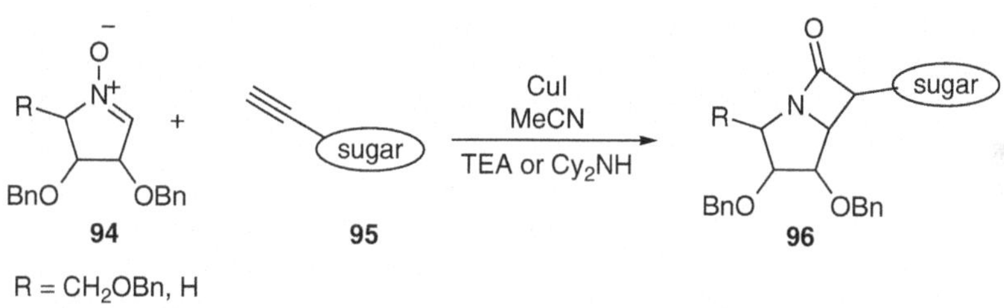

Scheme 29 Kinugasa reaction using sugar-derived cyclic nitrones and alkynes

The stereogenic centers in the alkynes do not influence the stereochemical outcome of the reaction unless the nitrones are achiral or the chiral center in the nitrone is not close to the 1,3-dipole [66]. Thus, the reaction methodology offers a means to synthesize a range of chiral sugar-derived β-lactams that may act as precursors to several potential water-soluble β-lactams.

4-Oxoazetidine-2-carbaldehydes or 4-formyl-β-lactams can be considered both as protected α-amino aldehydes and masked β-amino acids. These substrates are very useful for the preparation of substances of biological interest, including α-amino acids, β-amino acids, amino sugars, polycyclic-β-lactams, alkaloids, and complex natural products. Alcaide and Almendros [67] have composed a review in

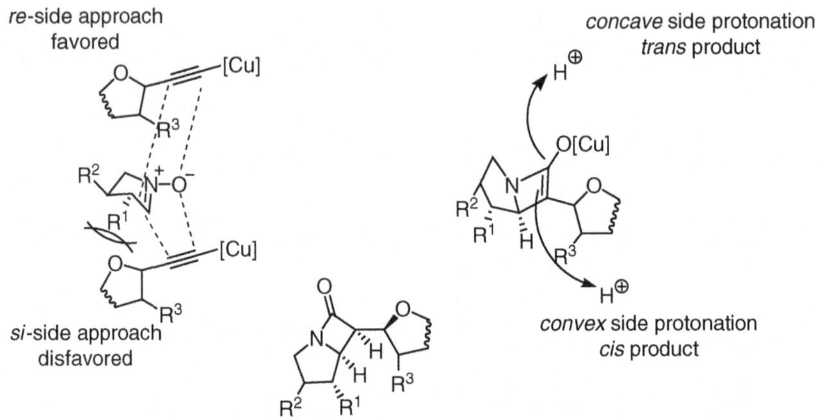

Fig. 6 Stereochemical outcome of the Kinugasa reaction

this regard which has highlighted the usefulness of 4-oxoazetidine-2-carbaldehydes as building blocks in various stereo-controlled syntheses with particular emphasis on diastereoselective processes. Since then there has been numerous works in this area involving enantioselective reaction of alkynes with nitrones [68–71]. But we have kept this particular area outside the scope of our review.

In this review, we have tried to draw attention to the various techniques developed for the synthesis of β-lactams using [3+2] cycloaddition reaction between an alkyne and nitrone, known as the Kinugasa reaction. This mild approach to the generation of β-lactam derivatives is attractive owing to its convergence, its high functional group tolerance, and the ready availability and stability of alkynes and nitrones. Though the reaction gained popularity a long time after its inception in 1972, there have been volumes of reports in the latest years showing variations in the reaction conditions, substrates, and ligands. Hence, we believe that this area has a great potential and further research is likely to bring forth many interesting undiscovered avenues of the Kinugasa reaction.

References

1. Staudinger H (1908) Liebigs Ann Chem 356:51
2. Wilmouth RC, Kassamally S, Westwood NJ, Sheppard RJ, Claridge TD, Alpin RT, Wright PA, Pritchard GJ, Schofield CJ (1999) Biochemistry 38:7989
3. Skiles JW, Sorcek R, Jacober S, Miao C, Mui PW, McNeil D, Rosenthal AS (1993) Bioorg Med Chem Lett 3:773
4. Ogilvie WW, Yoakim C, Hache FDB, Lagace L, Naud J, O'Meara JA, Deziel R (1999) Bioorg Med Chem 7:1521

5. Han WT, Trehan AK, Wright JJK, Federici ME, Seiler SM, Meanwell NA (1995) Bioorg Med Chem 3:1123
6. Addington RM, Baldwin JE, Chen B, Cooper SL, McCoull W, Pritchard GJ, Howe TJ (1997) Bioorg Med Chem Lett 7:1689
7. Danelon GO, Mata EG, Mascaretti OA, Girardini J, Marro M, Roveri OA (1995) Bioorg Med Chem Lett 5:2037
8. Tsuruoka T, Nakabayahi S, Fukuyasu H, Ishii Y, Tsuruoka T, Yamamoto H, Inouye S, Kondo S (1989) Eur Pat Appl 31pp
9. Dugar S, Yumibe N, Clader JW, Vizziano M, Huie K, van Heek M, Compton DS, Davis HR Jr (1996) Bioorg Med Chem Lett 6:1271
10. Ojima I, Delaloge F (1997) Chem Soc Rev 26:377
11. Manhas MS, Wagle DR, Chiang J, Bose AK (1988) Heterocycles 27:1755
12. Juaristi E (ed) (1997) Enantioselective synthesis of β-amino acids. Wiley-VCH, New York
13. Alcaide B, Almendros P, Aragoncillo C (2007) Chem Rev 107:4437
14. Ojima I (1995) Acc Chem Res 28:383
15. Brieva R, Crich JZ, Sih CJ (1993) J Org Chem 58:1068
16. Ojima I, Habus I, Zhao M, Zucco M, Park YH, Sun CM, Brigaud T (1992) Tetrahedron 48:6985
17. Palemo C, Arrieta A, Cossio FP, Aizpurua JM, Mielgo A, Aurrekotxea N (1990) Tetrahedron Lett 31:6429
18. Mandal B, Ghosh P, Basu B (2010) Top Heterocycl Chem 22:261
19. France S, Weatherwax A, Taggi AE, Lectka T (2004) Acc Chem Res 37:592
20. Taggi AE, Hafez AM, Wack H, Young B, Drury WJ III, Lectka T (2000) J Am Chem Soc 122:7831
21. Taggi AE, Hafez AM, Wack H, Young B, Ferraris D, Lectka T (2002) J Am Chem Soc 124:6626
22. Testero SA, Mata EG (2006) Org Lett 8:4783
23. Méndez L, Testero SA, Mata EG (2007) J Comb Chem 9:189
24. Kinugasa M, Hashimoto S (1972) J Chem Soc Chem Commun 466
25. Ding LK, Irwin WJ (1976) J Chem Soc Perkin Trans 1:2382
26. Okuro K, Enna M, Miura M, Nomura M (1993) J Chem Soc Chem Commun 1107
27. Miura M, Enna M, Okuro K, Nomura M (1995) J Org Chem 60:4999
28. Pal R, Ghosh SC, Chandra K, Basak A (2007) Synlett 15:2321
29. Marco-Contelles J (2004) Angew Chem Int Ed 43:2198
30. Gilman H, Speeter M (1943) J Am Chem Soc 65:2255
31. Brown MJ (1989) Heterocycles 29:2225
32. Hart DJ, Ha DC (1989) Chem Rev 89:1447
33. Fujieda H, Kanai M, Kambara T, Iida A, Tomioka K (1997) J Am Chem Soc 119:2060
34. Tomioka K, Fujieda H, Hayashi S, Hussein MA, Kambara T, Nomura Y, Kanai M, Koga K (1999) Chem Commun 715
35. Benaglia M, Cinquini M, Cozzi F (2000) Eur J Org Chem 563
36. Mames A, Stecko S, Mikozajczyk P, Soluch M, Furman B, Chmielewski M (2010) J Org Chem 75:7580
37. Lo MMC, Fu GC (2002) J Am Chem Soc 124:4572
38. Basak A, Ghosh SC, Bhowmick T, Das AK, Bertolasic V (2002) Tetrahedron Lett 43:5499
39. Shintani R, Fu GC (2003) Angew Chem 115:4216
40. Corey EJ, Guzman-Perez A (1998) Angew Chem Int Ed 37:388
41. Christoffers J, Mann A (2001) Angew Chem Int Ed 40:4591
42. Basak A, Rudra KR, Bdour HMM (2003) Indian J Chem Sect B: Org Chem Including Med Chem 42B(6):1508
43. Ye MC, Zhou J, Huang ZZ, Tang Y (2003) Chem Commun 2554
44. Evans DA, Kleinbeck F, Rueping M (2007) Asymm Synth 72
45. Whisler MC, Beak P (2003) J Org Chem 68:1207

46. Brandi A, Garro S, Guarna A, Goti A, Cordero F, De Sarlo F (1988) J Org Chem 53:2430
47. Goti A, Cordero FM, Brandi A (1996) Top Curr Chem 178:1
48. Cordero FM, Pisaneschi F, Goti A, Ollivier J, Salaün J, Brandi A (2000) J Am Chem Soc 122:8075
49. Paschetta V, Cordero FM, Paugam R, Ollivier J, Brandi A, Salaün J (2001) Synlett 1233
50. Cordero FM, Pisaneschi F, Salvati M, Paschetta V, Ollivier J, Salaün J, Brandi A (2003) J Org Chem 68:3271
51. Basak A, Ghosh SC (2004) Synlett 9:1637
52. Basak A, Pal R (2005) Bioorg Med Chem Lett 15:2015
53. Basak A, Ghosh SC, Das AK, Bertolasi V (2005) Org Biomol Chem 3:4050
54. Pal R, Basak A (2006) Chem Commun 2992
55. Langley DR, Doyle TW, Beveridge DL (1991) J Am Chem Soc 113:4395
56. Kawata S, Ashizawa S, Hirama S (1997) J Am Chem Soc 119:12012
57. Nicolaou KC, Sorensen EJ, Discordia R, Hwang C-K, Minto RE, Bharucha KN, Bergman RG (1992) Angew Chem Int Ed 104:1094
58. Basak A, Chandra K, Pal R, Ghosh SC (2007) Synlett 10:1585
59. Coyne AG, Müller-Bunz H, Guiry PJ (2007) Tetrahedron Asymmetry 18:199
60. Stecko S, Mames A, Furman B, Chmielewski M (2008) J Org Chem 73:7402
61. Stecko S, Mames A, Furman B, Chmielewski M (2009) J Org Chem 74:3094
62. Zhang X, Hsung RP, Li H, Zhang Y, Johnson WL, Figueroa R (2008) Org Lett 10:3477
63. Saito T, Kikuchi T, Tanabe H, Yahiro J, Otani T (2009) Tetrahedron Lett 50:4969
64. McKay CS, Kennedy DC, Pezacki JP (2009) Tetrahedron Lett 50:1893
65. Khangarot RK, Kaliappan KP (2011) Eur J Org Chem 6117
66. Masamune S, Choy W, Petersen JS, Sita LR (1985) Angew Chem Int Ed Engl 24:1
67. Alcaide B, Almendros P (2001) Chem Soc Rev 30:226
68. Evans DA, Kleinbeck F, Rueping M (2008) Asymm Synth 77
69. Alcaide B, Pardo C, Sáez E (2002) Synlett 1:85
70. Alcaide B, Almendros P, Alonso JM, Aly MF, Pardo C, Sáez E, Torres MR (2002) J Org Chem 67:7004
71. Alcaide B, Sáez E (2005) Eur J Org Chem 1680

Top Heterocycl Chem (2013) 30: 111–146
DOI: 10.1007/7081_2012_89

Published online: 7 August 2012

Preparation of Bis-β-Lactams by Ketene–Imine Cycloadditions

Thomas T. Tidwell

Abstract Bis-β-lactams are potential substrates for new drugs, but apparently no naturally occurring examples are known. Synthetically these have been prepared by dimerization of naturally occurring penicillin derivatives, by coupling of β-lactams, from ketene reactions with imino(β-lactams), from ketene reactions with bis (imines), and from bis(ketene) reactions with imines. Therapeutic applications of bis-β-lactams are of increasing interest, and further studies of these potentially valuable materials may be anticipated.

Keywords β-lactams · Amide formation · Bis(imines) · Bis(ketenes) · Bis-β-lactams · Cycloaddition · Grubbs catalyst · Imines · Ketenes · Macrocycles · Mukaiyama reagent · Olefin metathesis · Penicillin · Schiff bases · Stereoselectivity

Contents

T.T. Tidwell (✉)
Department of Chemistry, University of Toronto, Toronto, ON, Canada M5S 3H6
e-mail: ttidwell@chem.utoronto.ca

1 Introduction

Penicillin (**1**), the original β-lactam antibiotic, has been the object of intense investigation for many years because of the continuing need for new and better drugs, especially as resistance to earlier forms develops. Bis-β-lactams are one possibility for such new drugs, but there appear to be no known naturally occurring examples. This is perhaps surprising in view of an early serendipitous but simple bis-β-lactam preparation in 1983 by treatment of β-lactam ester **2**, which is the benzyl ester of the β-lactamase inhibitor clavulanic acid, with triethylamine in methylene chloride giving the isomerized ester **3**, which was unstable and on standing spontaneously formed the stereosiomeric bis-β-lactams **4a** and **4b**, in a process which can be depicted as shown in Scheme 1 [1]. Whether any naturally occurring examples of bis-β-lactams will be found remains a question for the future.

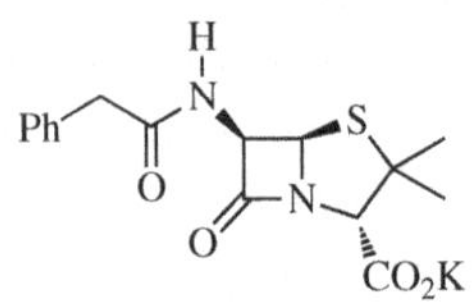

Penicillin G **1**

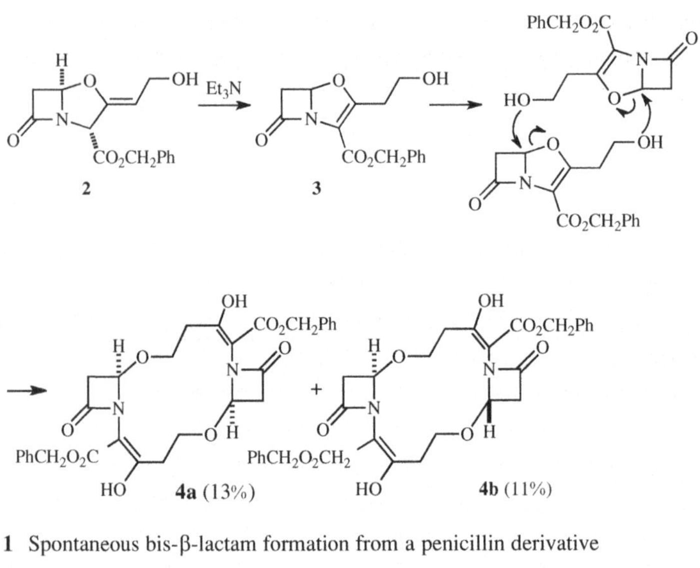

Scheme 1 Spontaneous bis-β-lactam formation from a penicillin derivative

An earlier synthesis of bis-β-lactams had already appeared in 1981, including among other examples azidoketene addition to imino-β-lactam **5** giving the two *cis*-stereoisomeric products in 41% and 39% yields [Eq. (1)] [2], as described in more detail below. This was followed by the discovery of unusual fused bis-β-lactams

(2,6-diazabicyclo[2.2.0]hexane-3,5-diones) by photochemical cyclization of 3,6-dihydro-6-oxo-1,3-diphenyl-l-pyrimidinium-4-olates [Eqs. (2) and (3)] [3].

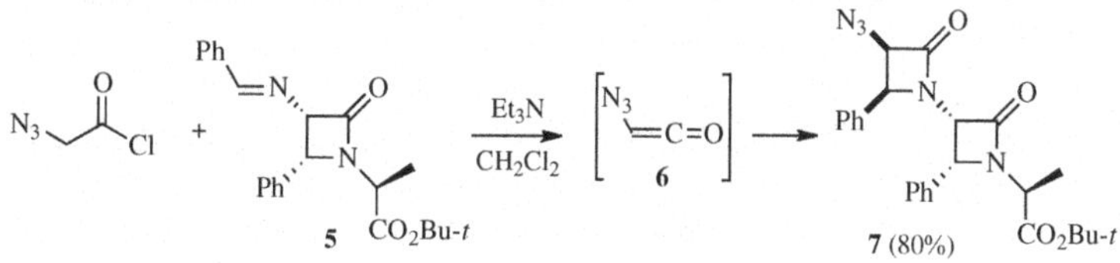

(1)

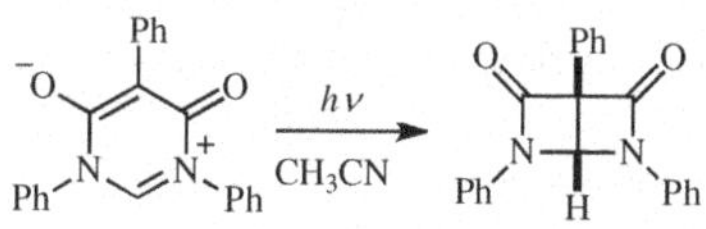

8 (96%)

(2)

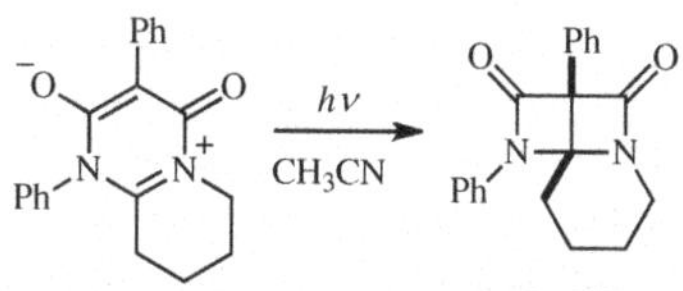

9 (64%)

(3)

Research directed toward the preparation and evaluation of such bis-β-lactams was initially rather modest, but has increased in intensity in recent years, and is the subject of this review.

2 Bis-β-Lactams from Coupling of β-Lactams

Another early approach to bis(lactam) formation utilized coupling of diacids with the methyl ester of penicillanic acid using DCC, followed by reduction of the aldehyde group with tritiated sodium borohydride to give tritium labeled bis-β-lactam **10** [Eq. (4)] [4]. Other β-lactams **11–14** were prepared by similar procedures. These act as bifunctional specific cross-linking reagents for the penicillin-binding proteins of Escherichia coli, and bind more strongly than the corresponding mono-β-lactams.

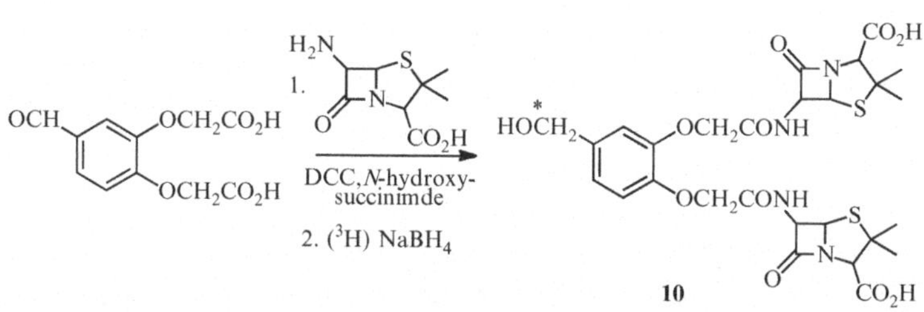

(4)

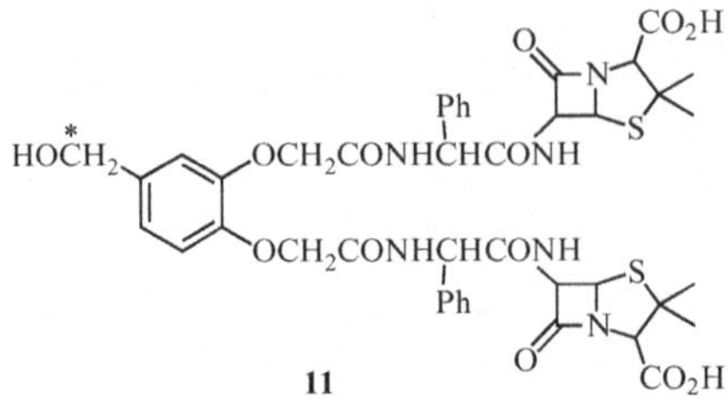

11

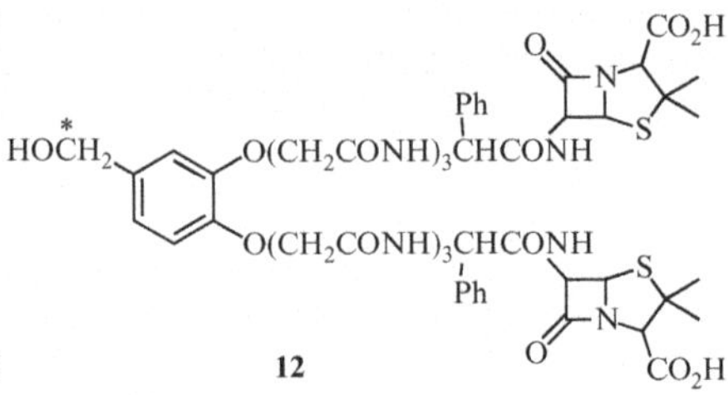

12

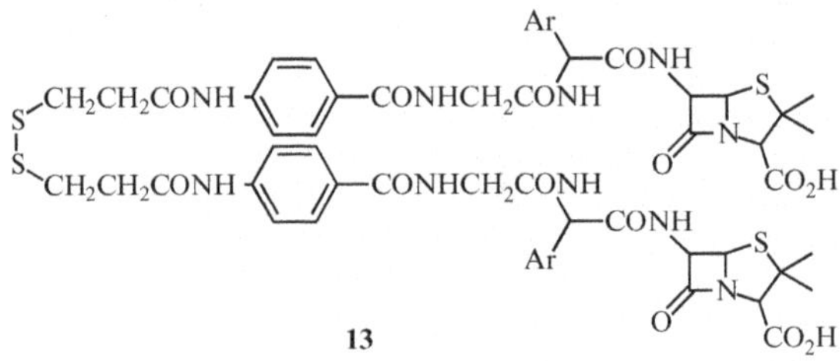

13

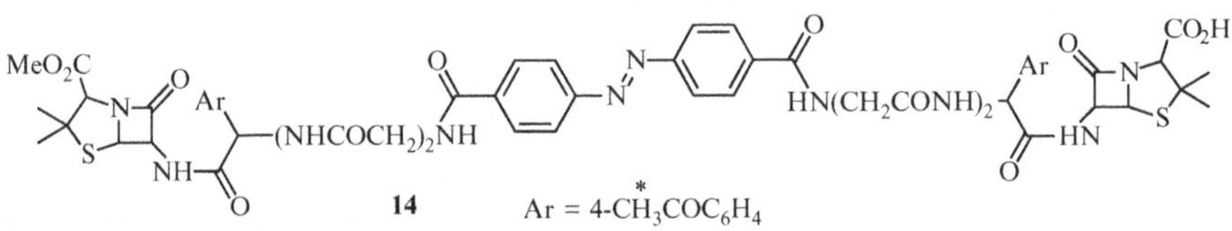

14 Ar = 4-$\overset{*}{C}H_3COC_6H_4$

A similar approach was reported by Ojima et al. [5], in which bis(lactam) formation was achieved by DCC coupling of chiral lactam carboxylic acids and amino lactams [Eq. (5)].

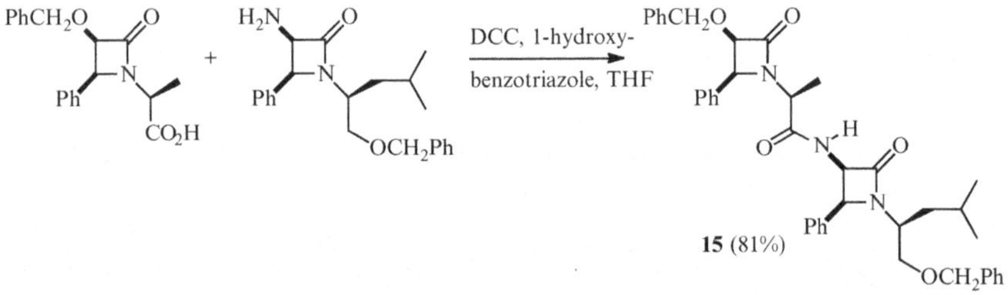

(5)

A different version of this approach of coupling two β-lactam units was the synthesis of the bis-penicillin derivative **16**, which utilized the incorporation of two molecules of an amino-β-lactam forming amide linkages with a bis(acyl chloride). It was envisaged that **16** would be an ester prodrug of carbenicillin (α-carbethoxy benzylpenicillin), a broad-spectrum antimicrobial agent, and would have enhanced bioavailability. The synthesis was achieved by reaction of the polyethylene glycol-based bis(α-carboxy–α-phenylacetyl) chloride and 6-aminopenicillanic acid [Eq. (6)] [6].

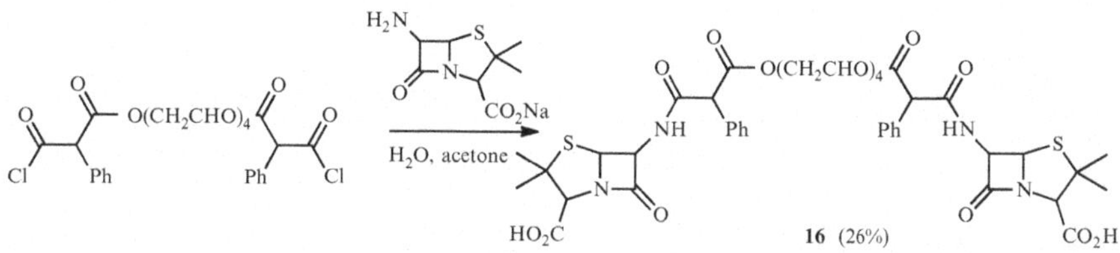

(6)

Copper promoted dimerization of enantiopure β-lactam substituted allenyl alkynes with copper(II) acetate gave the corresponding bis(β-lactams) **17**, which upon thermolysis gave double allene–alkyne [2+2] cycloaddition forming new bis (β-lactams) **18** (Scheme 2) [7]. Analogous unsymmetrical enantiopure bis (β-lactams) were prepared by similar means [7].

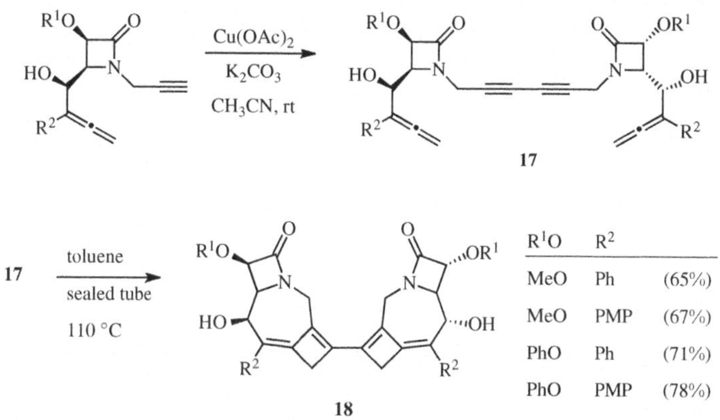

Scheme 2 Bis(β-lactams) from bis(alkynes)

Formation of bis-β-lactam **19** was achieved by dimerization of a selenium substituted mono-β-lactam, and the bis-β-lactam was converted to desired modified mono-β-lactams [Eq. (7)] [8].

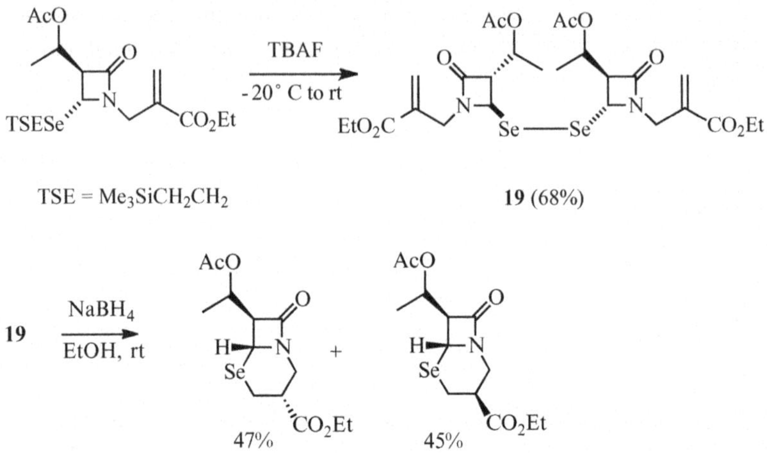

(7)

Another route to bis-β-lactam formation involved the multistep conversion of **20** to the biphenyl structure **21** [Eq. (8)] [9].

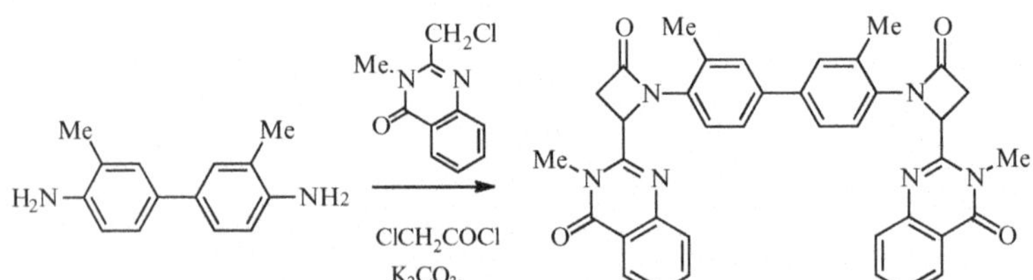

20 **21**

(8)

3 Bis-β-Lactams from Ketene Reactions with Imino (β-Lactams)

As noted above [Eq. (1)] the earliest reported example as well as the first synthesis of a bis-β-lactam was in 1981 [2] and utilized the Staudinger reaction. This procedure was introduced in 1907 [10], in the first synthesis of a β-lactam, utilizing the reaction of diphenylketene with benzylideneaniline [Eq. (9)] [10], long before there was any known use for these compounds.

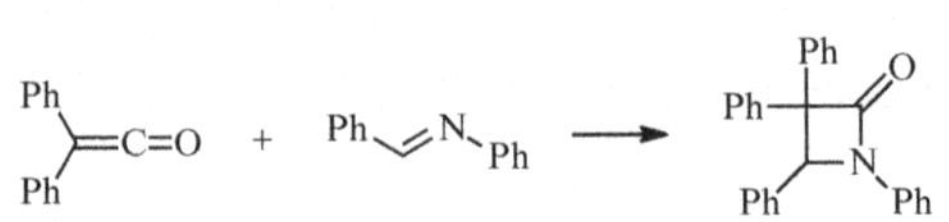

(9)

With the elucidation in the 1940s of the β-lactam structure of penicillin [11] and the subsequent discovery of the cephalosporins, the synthesis of members of this family quickly assumed a position of importance. Ketene–imine [2+2] cycloaddition provided in 1950 the first synthetic penicillin **22** [12] [Eq. (10)], while the potassium salt of optically active natural penicillin G1 was first synthesized in 1957 [13], and the synthesis of β-lactams is of continuing interest [14].

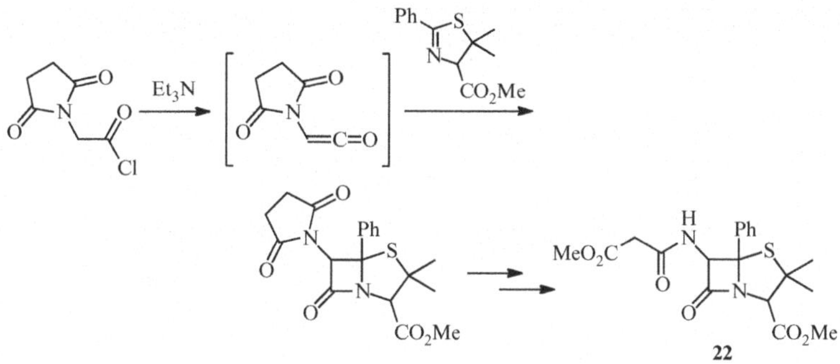

(10)

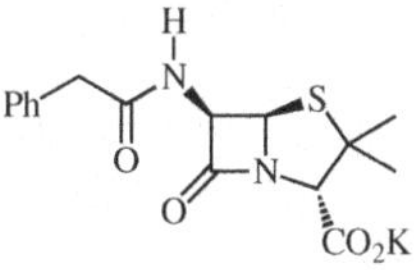

Penicillin G **1**

Imines, or Schiff bases, were first characterized by Schiff in 1864 [15, 16] and are one necessary component for synthesis of β-lactams by combination with ketenes [17–19] by [2+2] cycloaddition, and remain as an essential part of current synthetic and mechanistic chemistry.

As described below bis(β-lactams) have been prepared most readily in one step from ketene reactions with readily available bis(imines). Bisketenes have not been as extensively explored, and their use in bis(β-lactam) formation by reaction with imines is less common. The possible route of bis(β-lactam) formation from a bisketene reaction with a bis(imine) has not yet been definitively demonstrated, but the generation of poly(β-lactams) has been reported in some examples.

The original syntheses of bis(β-lactams) utilized imino(β-lactams), and in pioneering studies chiral imine-substituted β-lactams were prepared in a two-step process by first using the [2+2] cycloaddition of in situ generated azidoketene (**6**) with the chiral imine **23**, forming two diastereomeric azido-substituted *cis*-β-lactams **24** and **25** in 80% yield as a 1:1 mixture [Eq. (11)] [20]. These were separated and converted to the imine-substituted β-lactams **5** [Eq. (12)] and **26** [Eq. (13)], respectively, and these upon similar cycloaddition with azidoketene **6** of the separated diastereomers gave *cis*/*cis*-bis(β-lactams) **7** [Eq. (12)] and **27** [Eq. (13)] in 78% and 48% yields, respectively, and 99.5% stereoselectivity in each case, and with the opposite stereoselectivity from the first cycloaddition. The remote chiral *N*-substituent did not affect the stereoselectivity [Eq. (13)]. Similar reaction of the imino-β-lactam **28** gave the bis(β-lactam) **29** [Eq. (14)] and confirmed that the stereoselectivity of ketene–imine cycloaddition was unaffected by the *N*-substituent in these examples [20].

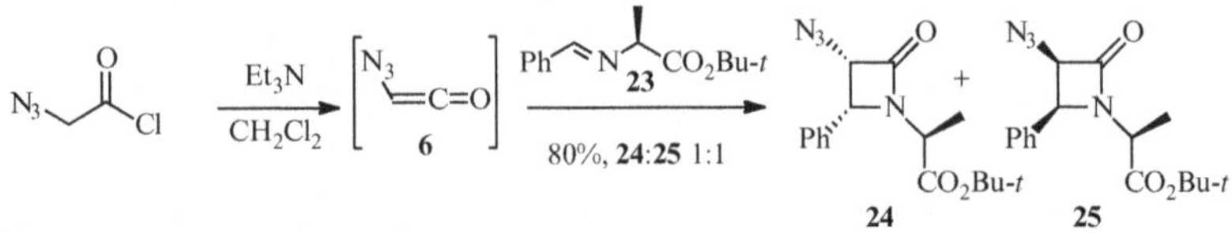

(11)

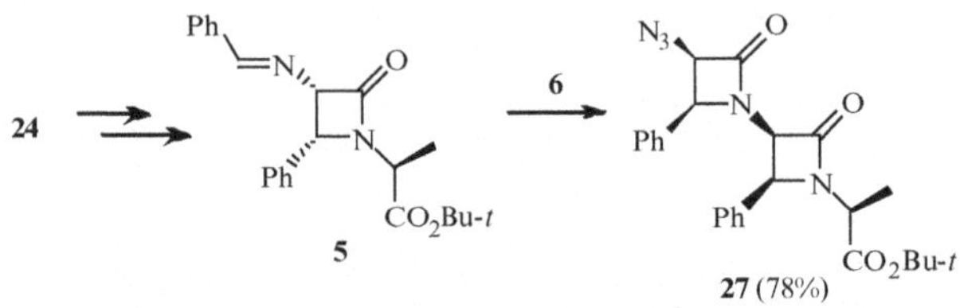

(12)

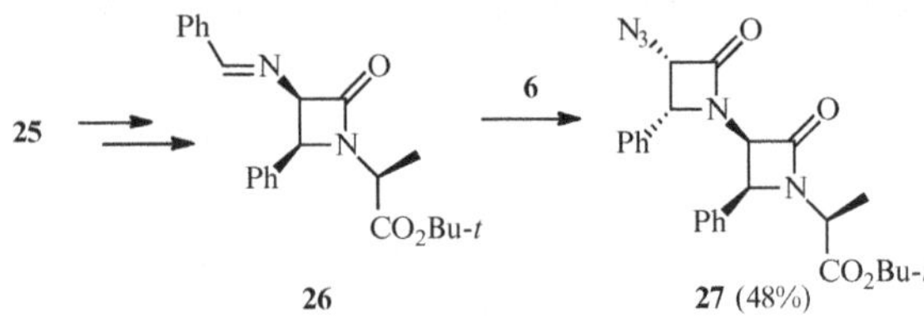

(13)

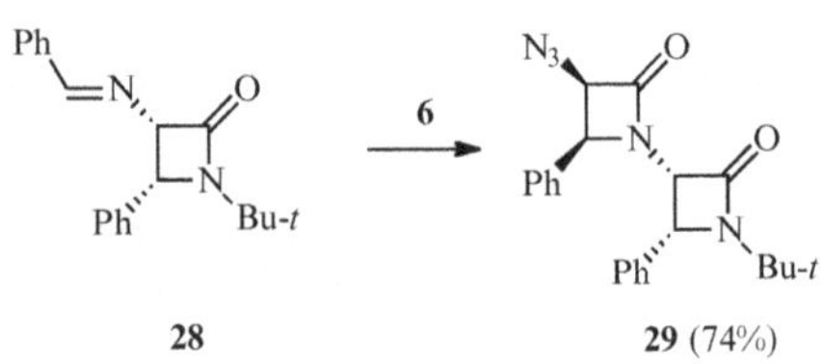

(14)

The origin of the high stereoselectivity in these reactions was further elucidated from the study of cycloadditions with azidoketene **6** of the imino-β-lactams **30**–**32** forming bis(β-lactams) **33**–**36** [Eqs. (15)–(17)] [20]. Both **30** and **31** gave highly selective formation of single bis(β-lactams) **33** and **34**, respectively, whereas **32** gave a mixture of products **35** and **36** [Eq. (17)]. Analysis of X-ray and calculated structures indicated that the product distributions were determined by the conformation of the β-lactams group in the reactants, and particularly the orientation of the carbonyl group which favored formation of a single product from **30**, whereas **32** was suggested to involve competitive formation of two products.

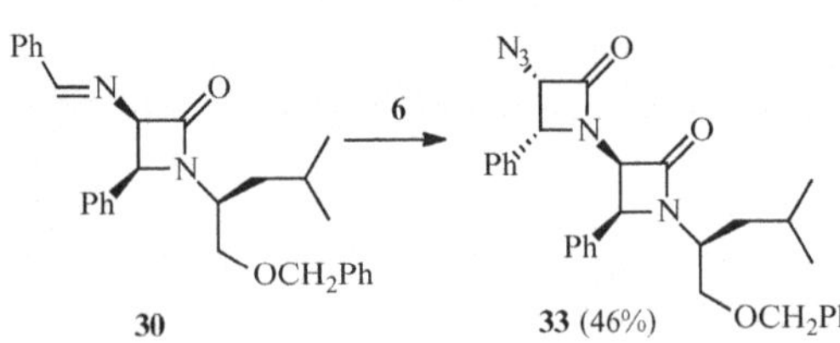

(15)

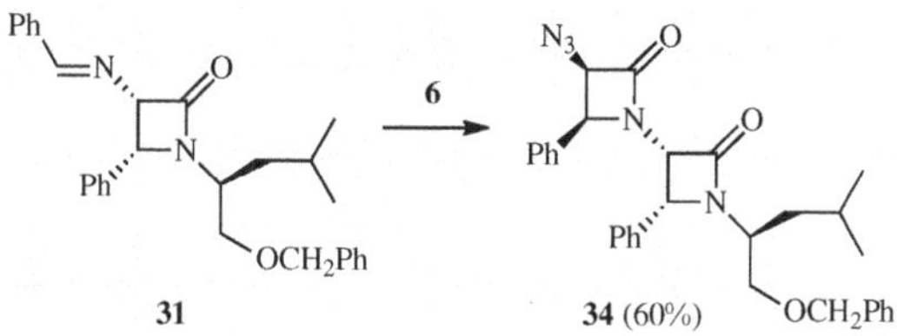
(16)

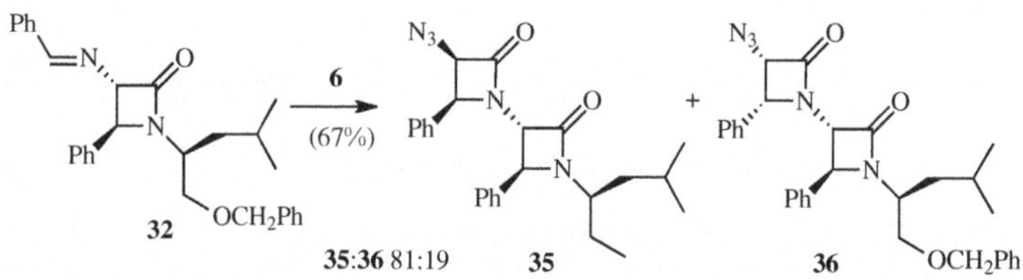
(17)

Ojima et al. [5] also examined [2+2] cycloaddition of preformed *cis*-imino-β-lactams **26** with oxygen substituted ketenes generated in situ by dehydrochlorination of acyl chlorides forming diastereomeric products **38** and **39** of *cis*-addition [Eqs. (18)–(20)]. The *cis*-imino(β-lactam) **40** similarly gave the diastereoisomeric bis(β-lactams) **41** and **42** [Eq. (20)] (Table 1).

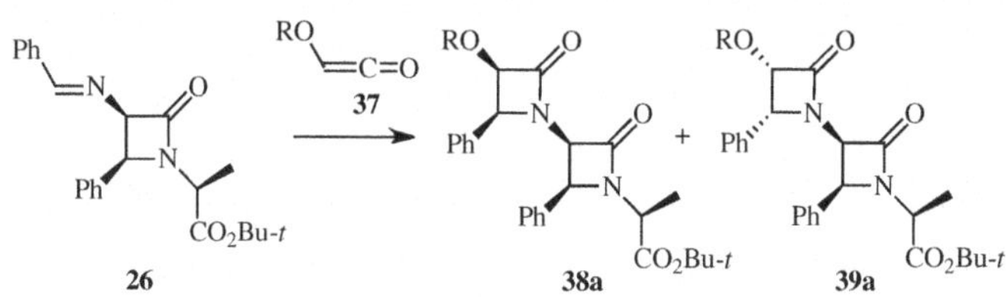
(18)

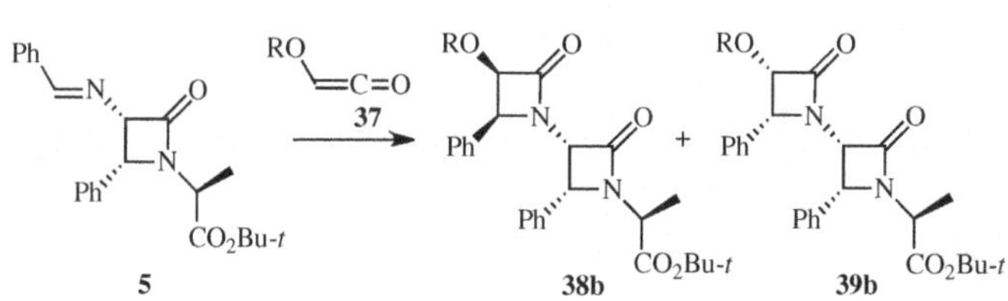
(19)

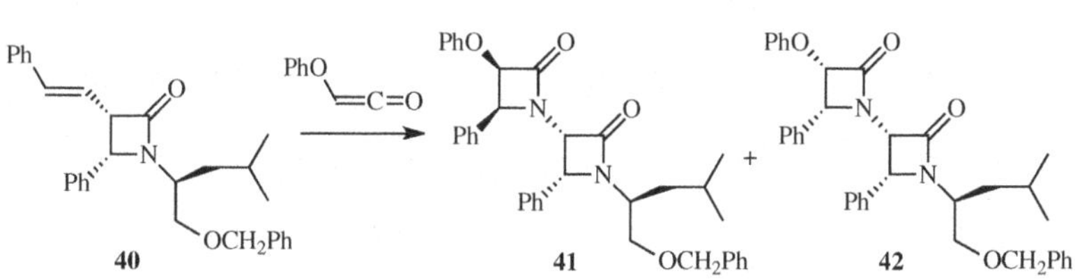
(20)

Table 1 Bis(β-lactams) from Imino(β-lactams)

Imine	RO	Yield (%)	**38a**:**39a**
26	PhO	80	33:67
26	$PhCH_2O$	58	70:30
26	AcO	72	17:83
Imine	RO	Yield (%)	**38b**:**39b**
5	PhO	71	36:64
5	$PhCH_2O$	83	52:48
5	AcO	77	23:77
Imine	RO	Yield (%)	**41**:**42**
40	PhO	87	31:59

Racemic β-lactams **43** with 4-imino-substituents react in situ with oxygen and phthalimido-substituted ketenes generated by dehydrochlorination giving *cis,cis*-bis(β-lactams) **44**, and for R = PhO these were separated using a chiral column into the enantiomers in a 1:1 ratio, showing the high stereoselectivity of the reaction [Eq. (21)] [21].

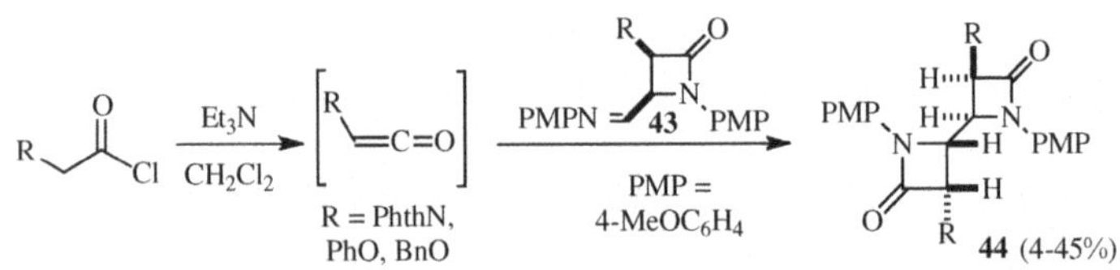

(21)

Ketenes generated in situ reacted with optically active β-lactam substituted azadienes **45** and gave diastereomeric *cis,cis*-bis(β-lactams) **46a** and **46b** with chiral methylene linkers between N1,C4′ (Scheme 3) [22].

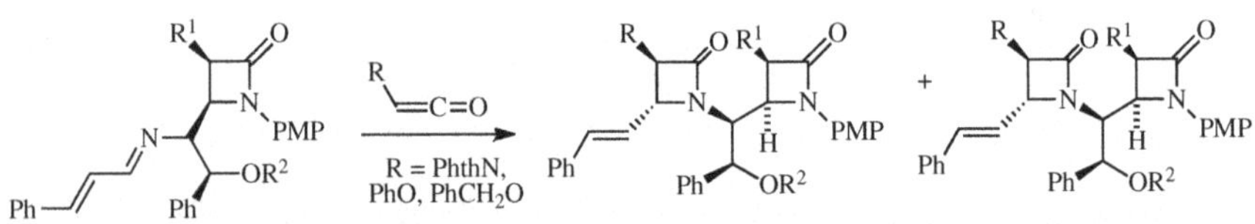

45 **46a** **46b**

R	R^1	R^2	Yield (%)	**46a**:**46b**	R	R^1	R^2	Yield (%)	**46a**:**46b**
BnO	TMS	PhO	82	77:23	PhthN	TBDMS	PhO	70	66:34
PhO	TMS	PhO	78	65:35	PhthN	TBDMS	BnO	96	70:30
PhthN	TMS	PhO	86	78:22	PhthN	TBDMS	PhthN	69	>95:5
PhthN	TMS	BnO	87	75:25					

Scheme 3 Methylene N1,C4′ linked bis(β-lactams) **46** by [2+2] cycloaddition

Ketene reactions with 4-imino-β-lactams **47** gave *cis,cis*-bis-β-lactams **48** with C4,C4′-bonding between the β-lactam groups (Scheme 4) [23, 24].

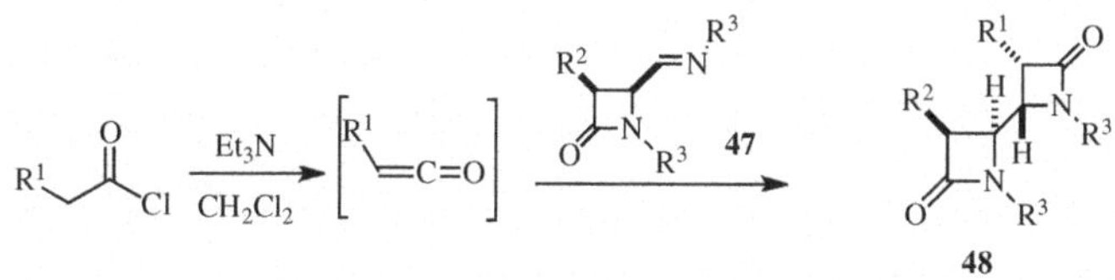

R^1	R^2	R^3	Yield (%)	R^1	R^3	Ar	Yield (%)
PhO	Me	PMP	90	Cl	PhO	PMP	85
PhO	Me	DAM	40	Phth	PhS	PMP	62
PhO	Ph	PMP	76	PhS	PhO	PMP	70
Md	PhO	PMP	70				

Scheme 4 C4,C4′ linked bis(β-lactams) **48** by [2+2] cycloaddition. *PMP* 4-$MeOC_6H_4$, *DAM* (4-$MeOC_6H_4$)CH, *Phth* phthalimido, *Md* malimidyl

N,N-dialkenyl-bis-β-lactams were prepared in a two-step procedure from *N*-alkenyllactam (+)-**49** and cyclized using Grubbs catalyst in refluxing toluene to tricyclic bis-β-lactams **50** [Eq. (22)] [25].

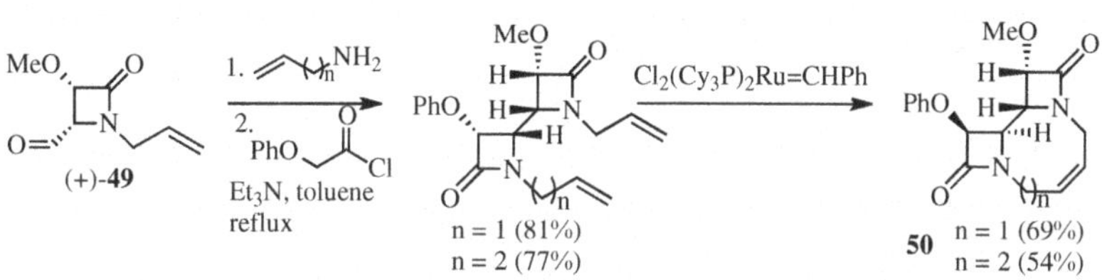
(22)

4 Bis-β-Lactams from Ketene Reactions with Bis(Imines)

Bis(imines) have been prepared by a variety of methods and provide direct access to bis-β-lactams by ketene cycloaddition. For example, cycloaddition of bis(imine) **51** with methoxyketene generated in situ gave a mixture of mono- and bis-*cis*-β-lactams, with a 2:3 *syn:anti* ratio of the bis(adducts) [Eq. (23)]. Further studies included the reactions of other ketenes, and of **51** with methyl substitution on the dimethylene bridge [26].

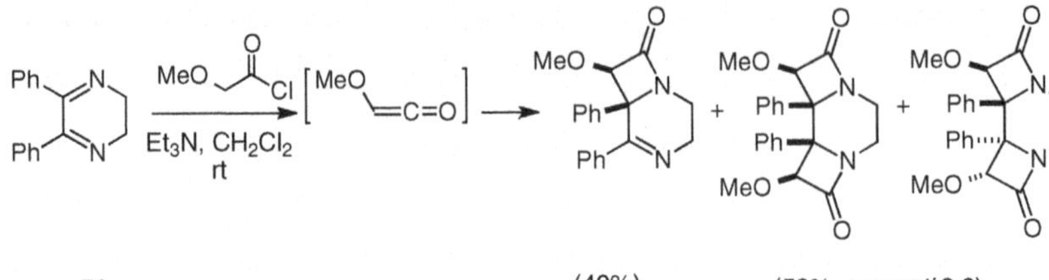

51 (40%) (53%, *syn:anti* 2:3)

(23)

One-step formation of *cis,cis*-bis-β-lactams **53** with C4,C4′-bonding between the β-lactam groups was achieved by reaction of bis(imines) **52** with ketenes generated in situ by dehydrochlorination of acyl chlorides (Scheme 5) [23]. Base-induced ring-expansion of the bis(β-lactams) **53** gave fused *cis,cis*-bis(γ-lactams).

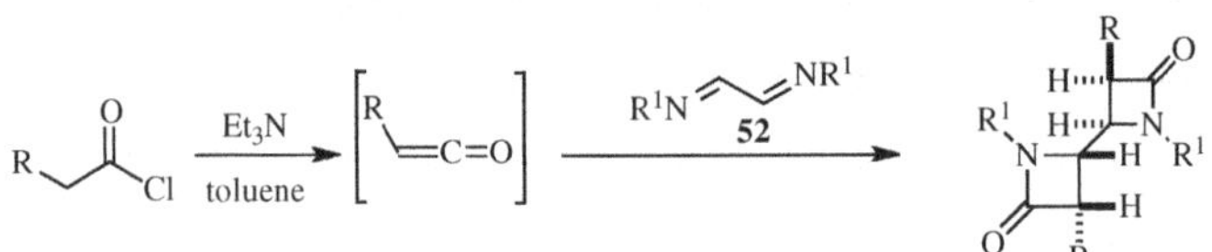

53

R	R^1	Yield (%)	R	R^1	Yield (%)	R	R^1	Yield (%)
PhO	PMP	90	MeO	PMP	85	Phth	DAM	44
Md	PMP	82	BnO	PMP	80	BnO	DAM	56
Phth	PMP	89	PhO	DAM	64			

Scheme 5 Bis-β-lactams from bis(imines). *PMP* 4-$MeOC_6H_4$, *DAM* (4-$MeOC_6H_4$)CH, *Phth* phthalimido, *Md* malimidyl

A single diastereomer of bis(β-lactam) **55** with a C4,C4′ linkage was obtained by bis(imine) reaction with the in situ generated chiral ketene **54** [Eq. (24)] [24]. Reaction of ketenes with chiral bis(imine) **56** yielded the stereoisomeric bis (β-lactams) **57** and **58** [Eq. (25)] [24]. Other optically pure bis(β-lactams) were obtained by two-step procedures with initial preparation of imino substituted monocyclic 2-azetidinones [24].

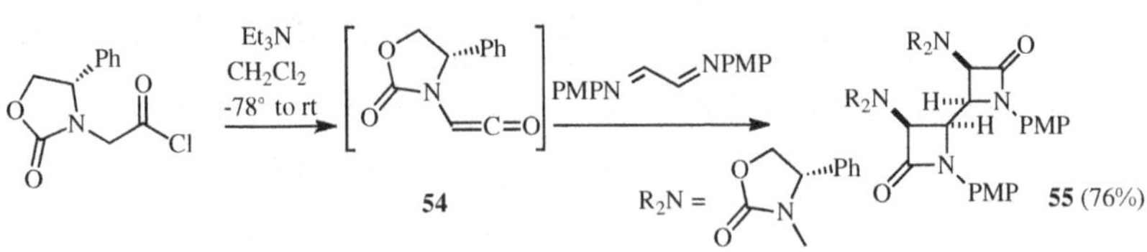

(24)

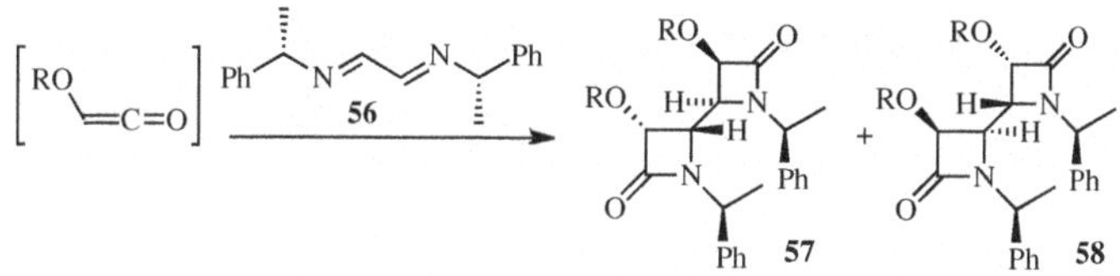

RO	% 57	% 58
AcO	35	30
MeO	40	36
BnO	33	29

(25)

Further studies of the selectivity in ethoxyketene reactions with vicinal dimines include the regio- and stereoselective formation of bis(β-lactams) **59** [Eq. (26)] [27]. *cis*-stereochemistry of the initial *cis*-4-iminyl mono β-lactams formed in these reactions was observed and was attributed to the electron withdrawing character of the imino group in the vicinal dimines.

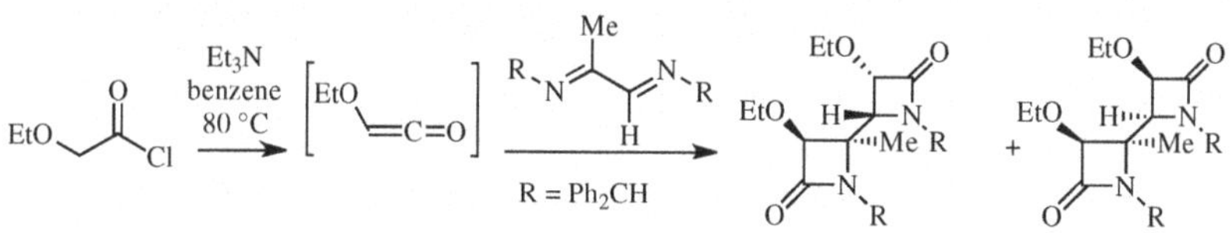

(26)

59a (65%) **59b** (14%)

Two equivalents of ketenes generated in situ by dehydrochlorination in the presence of methylene bridged bis(imines) with reaction from 0°C to 70°C gave two diastereomeric methylene N1,N1′ linked *cis,cis*-bis(β-lactams) **60** (Scheme 6) [28]. Successive reaction with R = MeO and BnO gave the mixed bis (lactam).

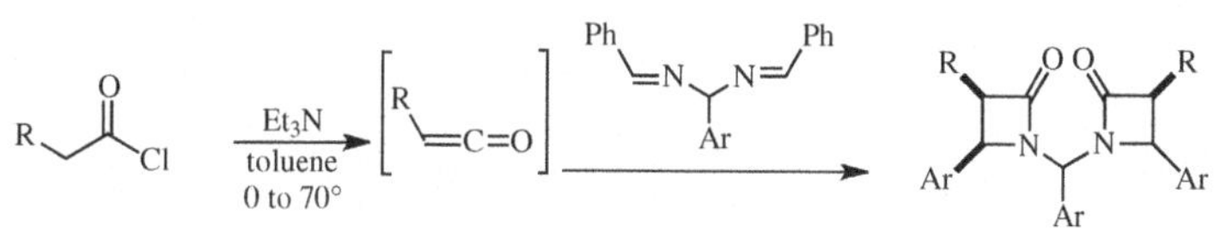

60

	Ar	Yield %	*dr*	R	Ar	Yield %	*dr*
MeO	Ph	66	70:30	MeO	4-MeOC$_6$H$_4$	66	70:30
BnO	Ph	84	75:25	BnO	4-MeOC$_6$H$_4$	73	80:20
				MeO	3-BrC$_6$H$_4$	74	70:30

Scheme 6 Methylene N1,N1′ linked *cis,cis*-bis(β-lactams) **60**

Successive reaction of bis(imine) **61** with two different ketenes generated in situ from one equivalent of each ketene precursor and raising the temperature from 0°C to 70°C was reported to give the unsymmetrical bis(β-lactam) **62** as a mixture of diastereomers [Eq. (27)] [28]. The products were also prepared containing ^{15}N label as possible Taxol analogue synthons.

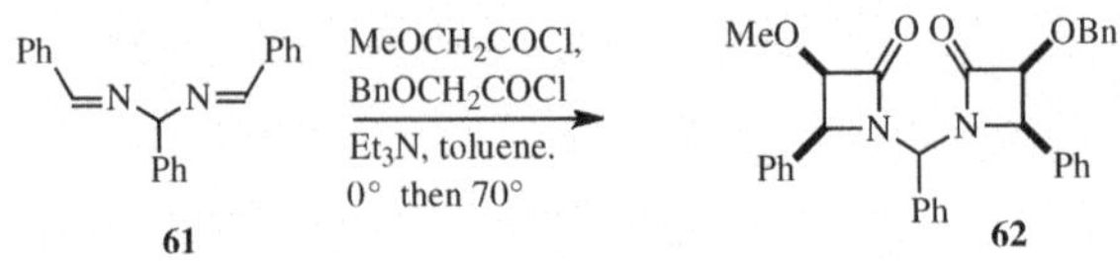

61 62 (27)

Reaction of chiral tartaric acid-derived bis(imines) with ketenes generated in situ gave bis(β-lactams) **63** linked by a two-carbon C4,C4′ chain in 52–73% yields, with only minor amounts of other diastereomers (Scheme 7) [29]. For R = MeO the product was, however, **64**. The bis(β-lactams) were hydrolyzed to diols, which were cleaved to mono β-lactams.

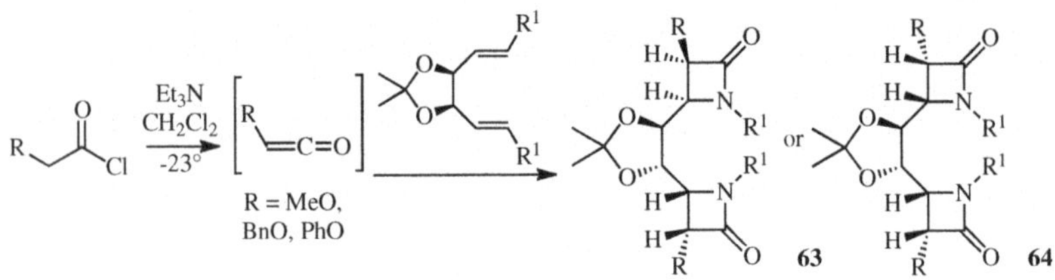

R	R^1	Yield (%)	R	R^1	Yield (%)	R	R^1	Yield (%)
PhO	Bn	61	PhO	PMP	73	PhO	2-Furyl	63
BnO	Bn	53	BnO	PMP	52	BnO	2-Furyl	58

Scheme 7 Bis-β-lactams with two-carbon linkers

Ketenes generated in situ by dehydrochlorination of acyl chlorides reacted with dimethylene bridged bis(imines) forming N1,N1′ linked bis(β-lactams) **65** and **66** as mixtures of *cis/cis meso* and *dl* stereoisomers (Scheme 8) [30].

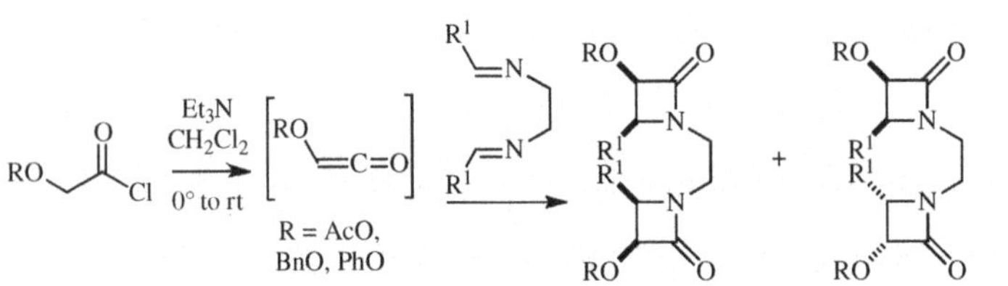

65 **66**

RO	R^1	Yield %	**65:66**	RO	R^1	Yield %	**65:66**
PhO	4-$MeOC_6H_4$	80	58:42	BnO	Ph	85	70:30
BnO	4-$MeOC_6H_4$	86	45:55	BnO	PhCH=CH	60	42:58
AcO	4-$MeOC_6H_4$	75	49:51	PhO	PhCH=CH	66	39:61
PhO	Ph	79	62:38				

Scheme 8 N1,N1′ linked bis(β-lactams) from bis(imines)

Chiral ketene **67** generated by in situ activation of the acid reacted with the bis(imine) forming the bis(β-lactam) in 93% yield as a single diastereomer [Eq. (28)] [30].

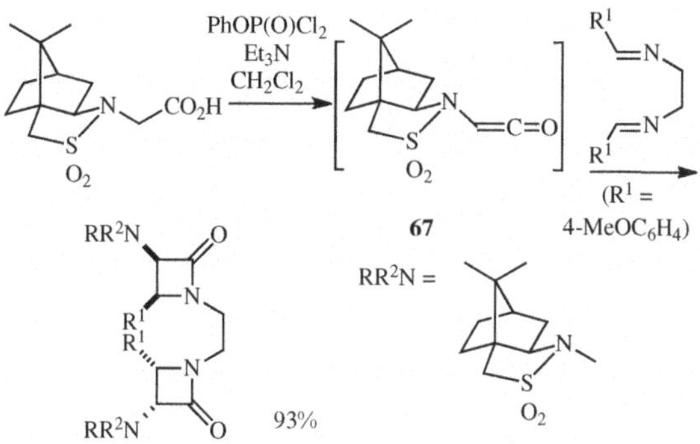

(28)

1,2-*trans*-Bis(iminyl)cyclohexanes **68** reacted with an excess of ketenes generated in situ to form *meso* and *dl* substituted bis(β-lactams) **69a** and **69b**, with a preference for the *meso*-isomer **69a** (Scheme 9) [31]. The use of enantiomerically pure bis(imine) (R^1 = Ph) with phenoxylketene led to the pure diastereomers isolated by chromatography.

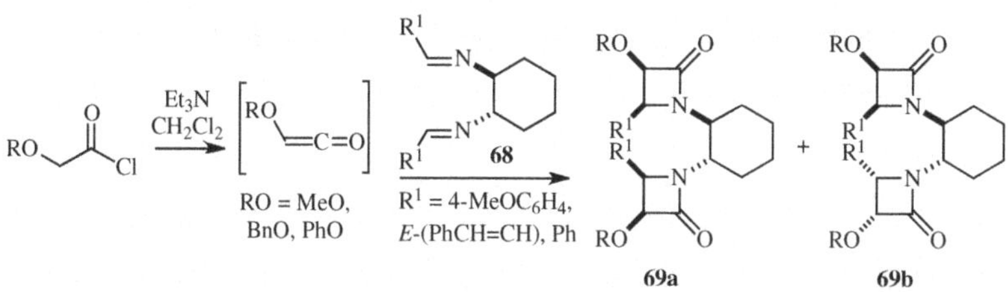

RO, R^1	*meso:dl*	Yield (%)	R, R^1	*meso:dl*	Yield (%)
PhO, Ph	66:34	90	MeO, PMP	76:24	73
BnO, Ph	66:34	87	PhO, styryl	64:36	88
MeO, Ph	77:23	86	BnO, styryl	54:46	81
PhO, PMP	78:22	93	MeO, styryl	56:42	86
BnO, PMP	72:28	80			

Scheme 9 N1,N1′ cyclohexyl linked bis(β-lactams) from bis(imines)

Stereoisomeric bis(β-lactams) **70** were formed in a similar manner from acylic 1,2-bis(imines) by reaction with ketenes [Eq. (29)] [31]. The use of enantiomerically pure 1,2-bis(imines) ($R^1 = R^2 = Ph$) with phenoxyketene led to the pure diastereomers of **70** isolated by chromatography.

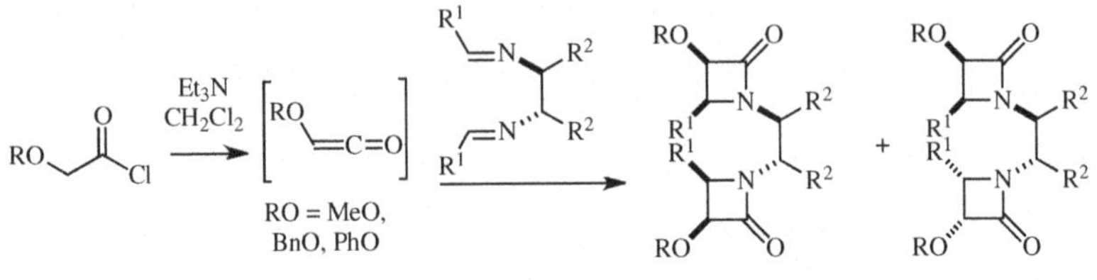

70a **70b**

R, R^1, R^2	*meso*:*dl*	Yield (%)	R, R^1, R^2	*meso*:*dl*	Yield (%)
PhO, Ph, Ph	58:42	81	MeO, Ph, Ph	60:40	88
BnO, Ph, Ph	56:44	84	PhO, Ph, Me	70:30	91

Azidoketene generated by reaction of potassium azidoacetate with triphosgene and triethylamine reacted in situ by stereoselective *cis*,*cis*-[2+2] bis(imine) cycloaddition forming the stereoisomeric bis(β-lactams) **71** [Eq. (30)], and these underwent copper catalyzed [3+2] cycloaddition with acetylenic esters or amides **72** giving bile acid linked bis-β-lactams as diastereomeric mixtures in yields of 92–95% [Eq. (31)] [32]. These products were designed to combine the therapeutic activity of the β-lactam and bile acid moieties and were evaluated for their antifungal and antibacterial activity.

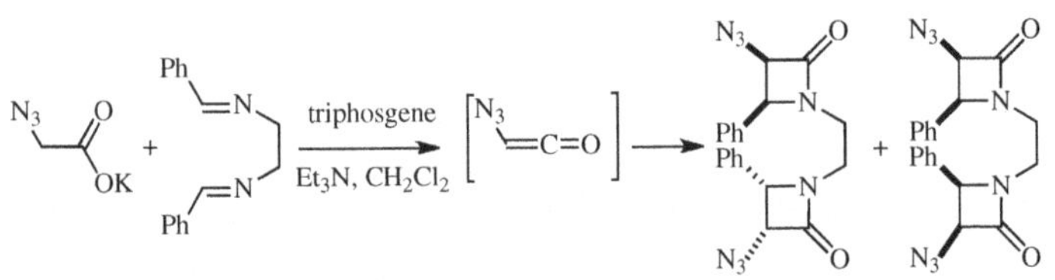

71a (35%) **71b** (41%)

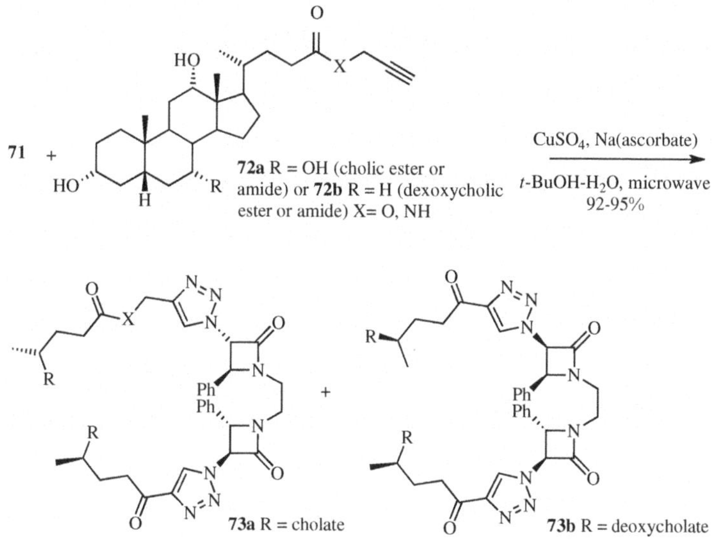

(31)

Different separation of the β-lactam units has been achieved in various architectures, for example ketene generation using separately formed Vilsmeier reagent **74** [Eq. (29)] added to a solution of phenoxyacetic acid and the bis(imine) **75a** with a 1,2-phenylene bridge gives [2+2] cycloaddition and formation of bis-β-lactam **76a** [Eq. (32)] [33]. A similar procedure was carried out with a bis(imine) **75b** with a 1,3-phenylene bridge forming bis(β-lactam) **76b** [Eq. (33)] [34]. The latter reaction was suggested to take place by successive reactions beginning at the less crowded imine functionality.

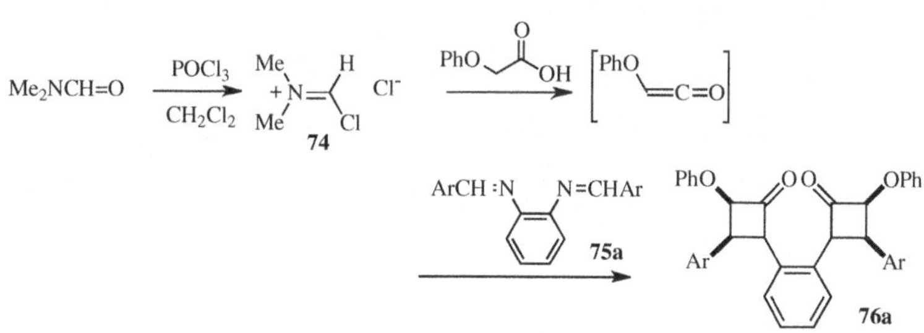

(32)

Ar	Yield (%)	Ar	Yield (%)	Ar	Yield (%)
Ph	78	2-HOC_6H_4	80	4-HOC_6H_4	82
2-$O_2NC_6H_4$	77	3-$O_2NC_6H_4$	72	4-$Me_2NC_6H_4$	69
4-$MeOC_6H_4$	71	4-FC_6H_4	76		

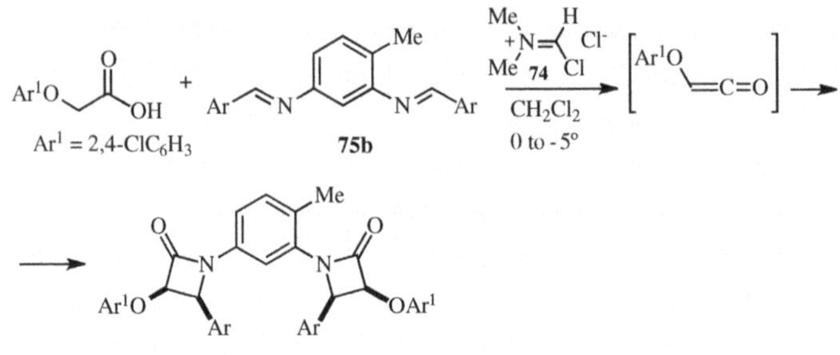

76b

(33)

Ar	Yield (%)	Ar	Yield (%)
Ph (a)	81	3-$O_2NC_6H_4$	72
2-HOC_6H_4	79	4-$MeOC_6H_4$	69
4-HOC_6H_4	82	2-ClC_6H_4	71
2-$O_2NC_6H_4$	75	4-ClC_6H_4	76

Bis-β-lactams are also obtained from bis(imines) which are derived from anthryl-substituted 1,4-diaminobenzene. Thus, formation of ketene **77** as an unobserved reactive intermediate by dehydration of 9*H*-xanthene-9-carboxylic acid using tosyl chloride and triethylamine in the presence of the bis(imine) **78a** gave in situ double [2+2] cycloaddition forming the bis(spiro-β-lactam) **79a** [Eq. (34)] [35]. Ketene **77** also reacted with dibenzo cyclopentylidene bisimines **78b** to form the corresponding tetrakis(spiro-β-lactams) **79b** [Eq. (35)] [36].

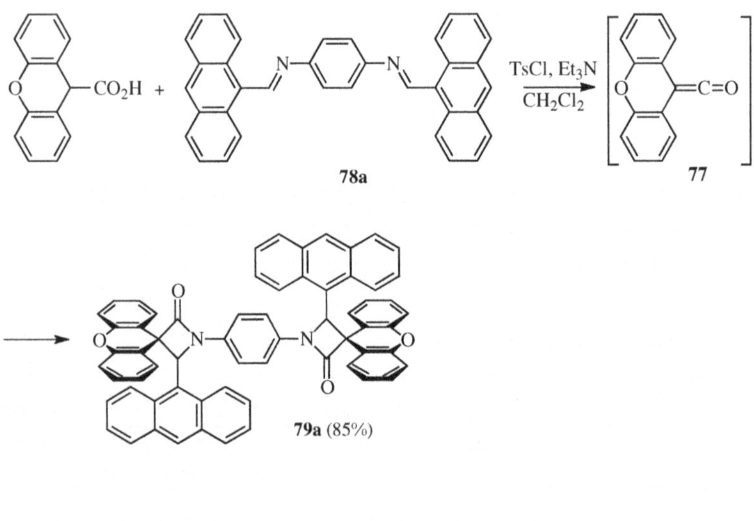

79a (85%)

(34)

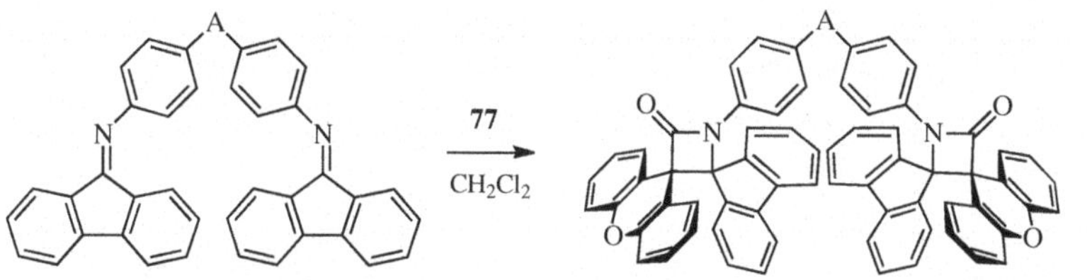

(35)

78b A = CH_2, O **79b** A = CH_2 (45%), O (48%)

The *meta* bridged bis(imines) **78c** and the *para* bridged isomers **78d** react similarly with ketene **77** to form spiro-bis(β-lactams) **79c** [Eq. (36)] [35] and **79d** [Eq. (37)] [35, 36].

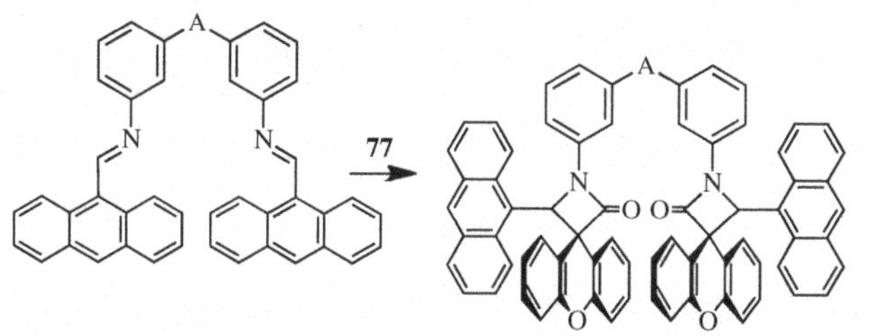

(36)

78c **79c** (A = CH_2, 92%; A = O, 96%)

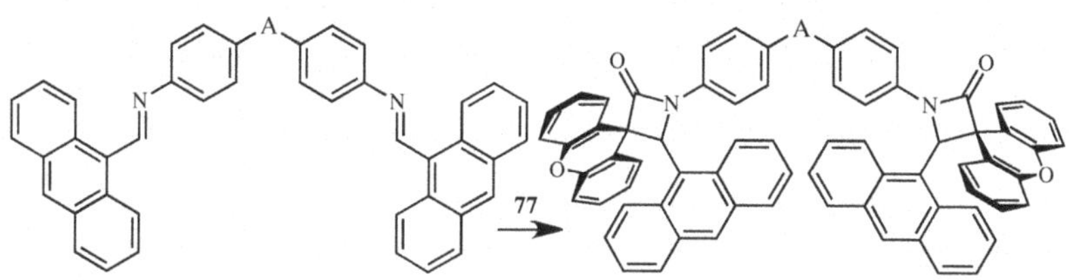

(37)

78d **79d** (A = CH_2, 92%; A = O, 96%)

Phenoxyketene generated by dehydration of phenoxyacetic acid with tosyl chloride and triethylamine reacted with imine **78e** forming the corresponding bis (spiro)lactam in 45% yield [Eq. (38)] [36]. Other examples of bis(β-lactams) **80–85** from phenoxyketene reaction with bis(imines) are given below (Table 2) [35, 36].

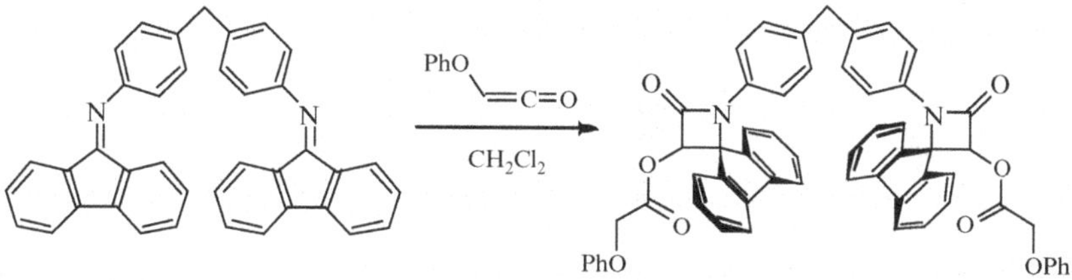

78e (45%)

(38)

Table 2 Bis(β-lactams) from phenoxyketene reactions with bis(imines)

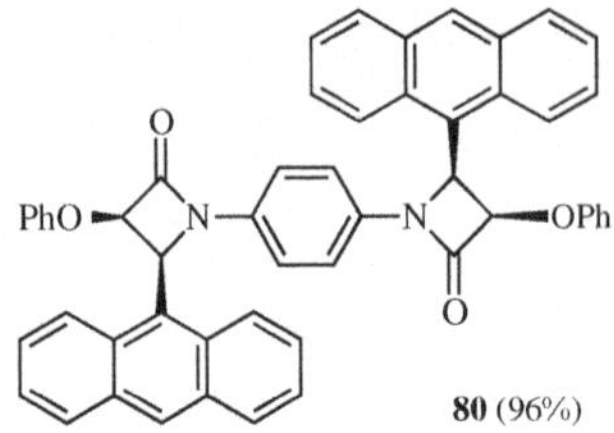

80 (96%)

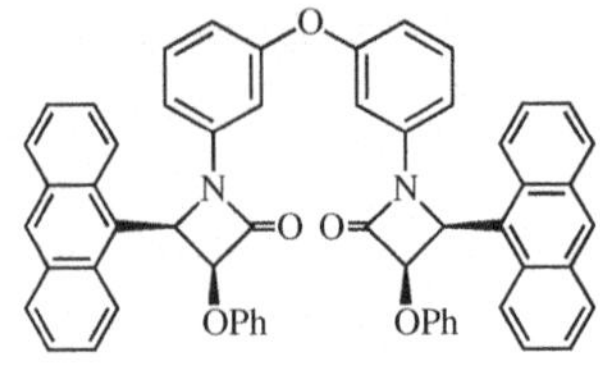

81 (90%)

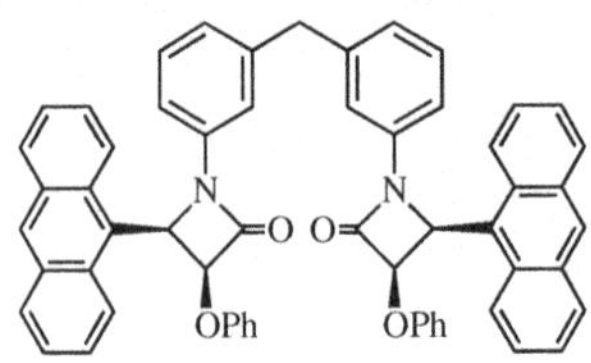

82 (87%)

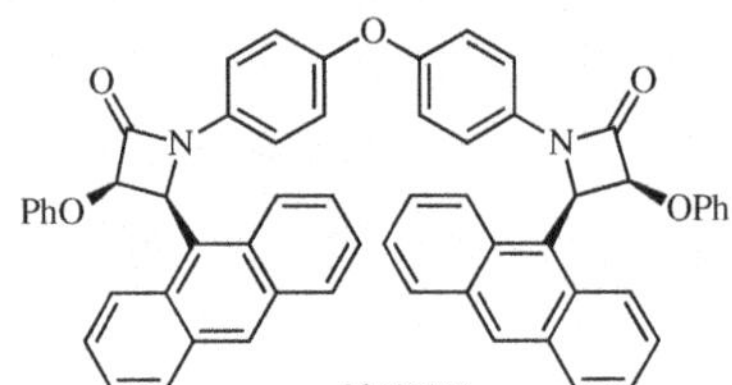

83 (90%)

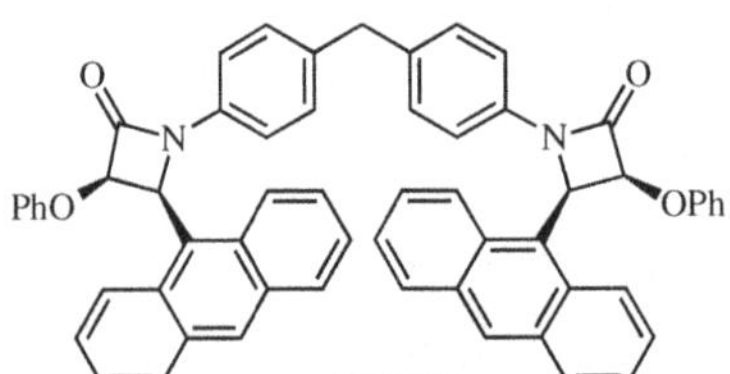

84 (80%)

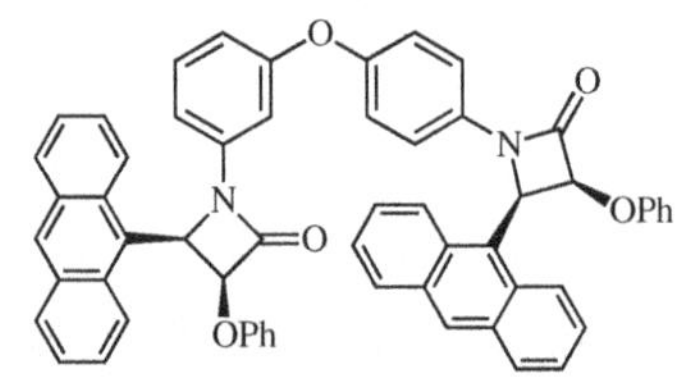

85 (95%)

Phenoxyketene generated by dehydrochlorination reacted with bis(alkynylphenyl) imines with three different linker groups connecting the imino groups by [2+2] cycloaddition forming bis-β-lactams with either 3-alkynyl or 4-alkynyl substituents. The bis-β-lactams from **86** to **90** were obtained as diastereomeric mixtures of *cis,cis* products [37] (Table 3).

Desilylated bis(β-lactams) from **86** to **90** were reacted with bisazides **91–94**, as in the reaction of **94** with **95** forming **96** [Eq. (39)] [37].

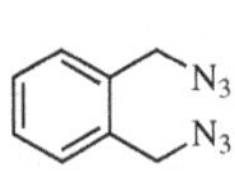

91

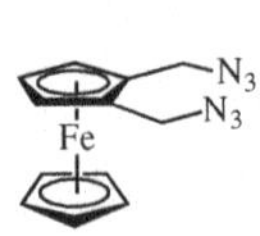

92

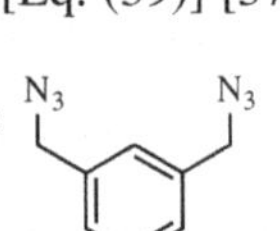

93

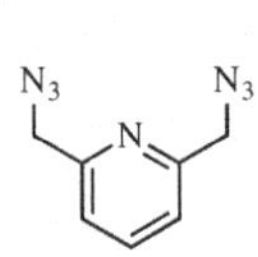

94

Table 3 Bis(β-lactams) from phenoxyketene reactions with alkynyl substituted bis(imines)

Imine reactants	Bis-β-lactam products
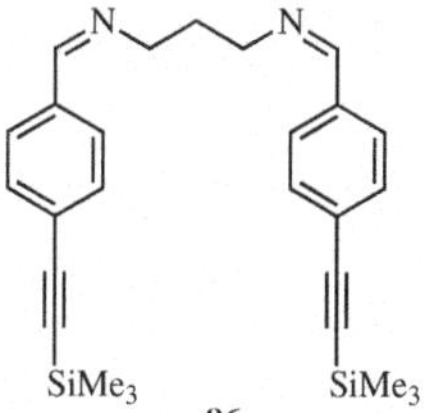 **86**	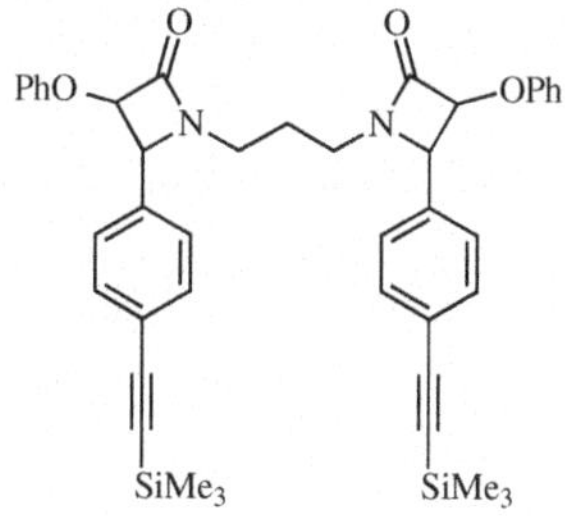 (30% *meso*, 32% *dl*)
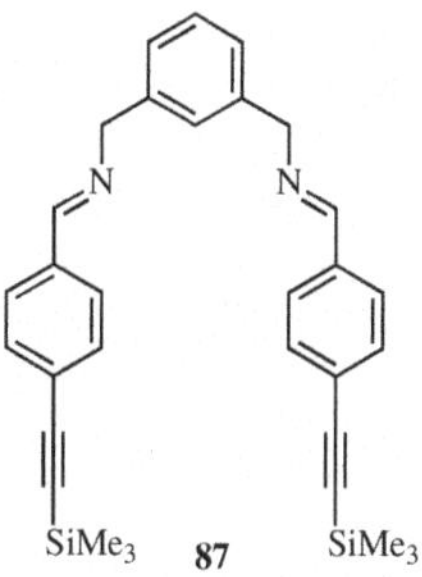 **87**	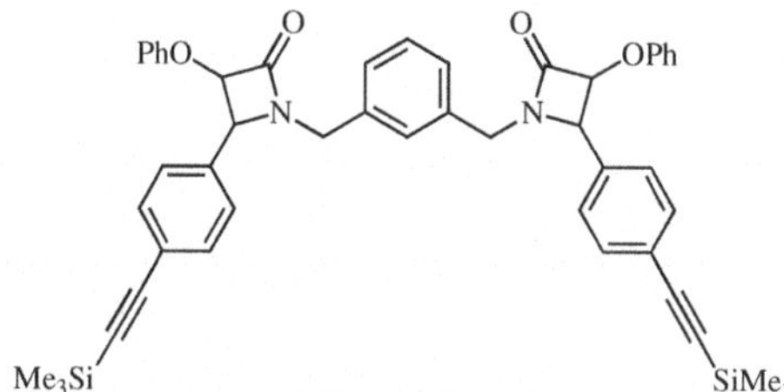 (31% *meso*, 33% *dl*)
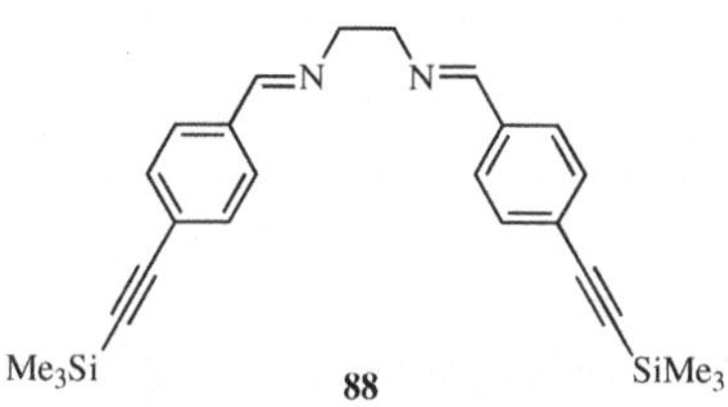 **88**	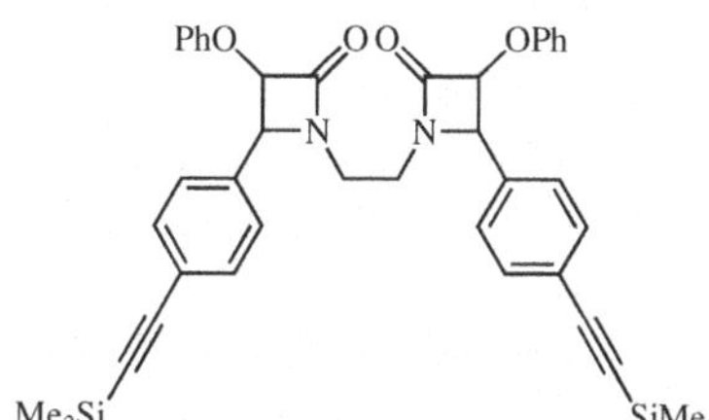 (26% *meso*, 31% *dl*)
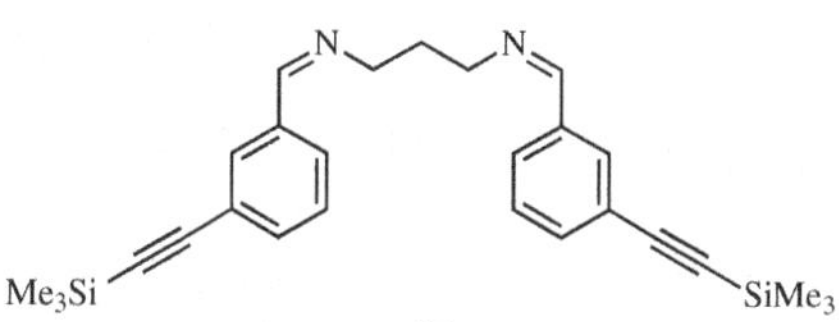 **89**	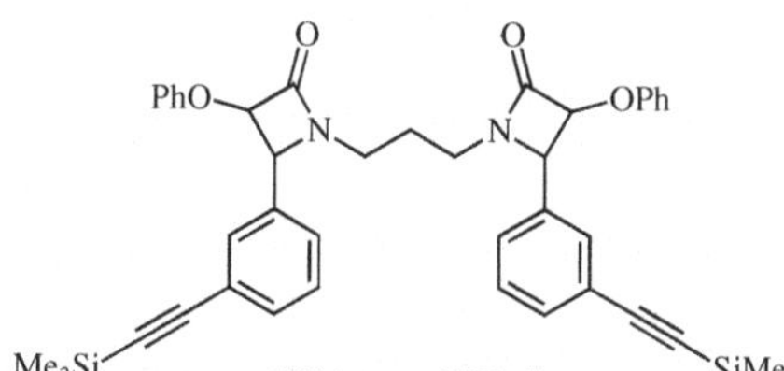 (29% *meso*, 33% *dl*)

(continued)

Table 3 (continued)

Imine reactants	Bis-β-lactam products
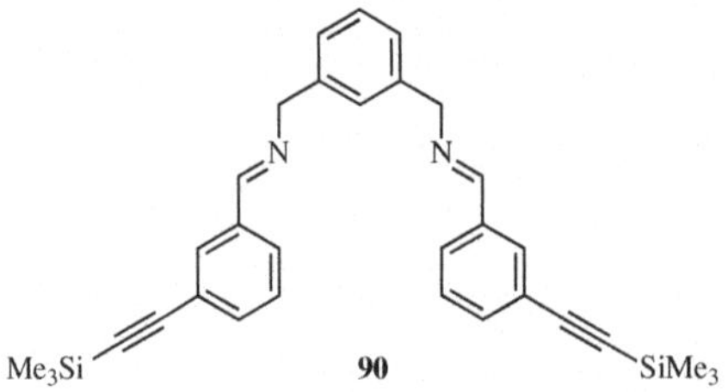 **90**	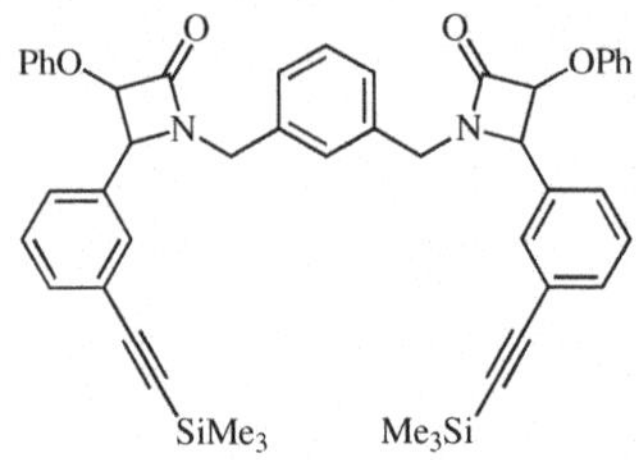 (35% *meso*, 33% *dl*)

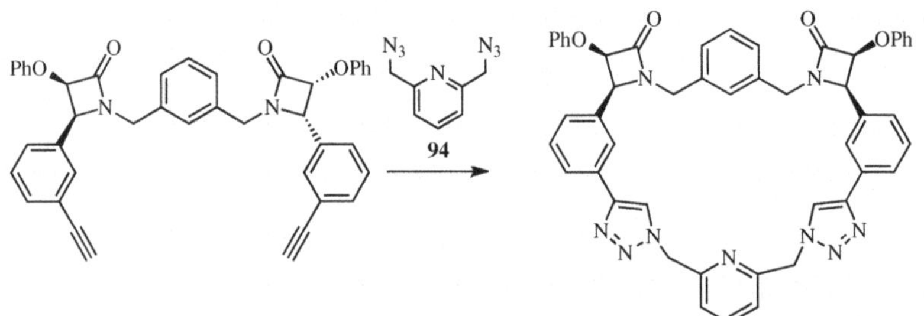

95 **96** (51%)

(39)

Generation of the chiral ketene **54** in the presence of bis(imine) **86** gave the enantiomerically pure bis(β-lactam) **97** in 95% yield [Eq. (40)] [37]. The alkynyl groups were desilylated with TBAF at 0° and reacted with azides **91–94** to give the corresponding macrocycles analogous to **96** [37]. Isomerization of the *cis*-bis(β) lactam groups to *trans* could also be accomplished by carrying out the desilylation at room temperature, and the corresponding macrocycles were obtained upon reaction with the azides.

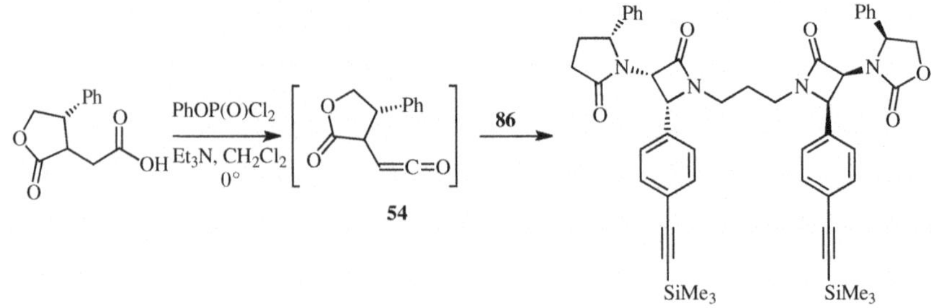

97 (95%)

(40)

Phenoxyketene generated by dehydrochlorination reacted in situ with bis(imine) **86** forming a 1:1 mixture of the bis(lactam) diastereomers in 75% yield, which after desilylation and separation of the stereoisomers followed by Sonogashira coupling with 4-iodopyridine gave **98** and **99** in 82% and 76% yields, respectively [Eq. (41)]. These were reacted with *cis*-[Pd(dppe)(OTf)$_2$] and with *cis*-[Pt(dppe)(OTf)$_2$] to give the respective cyclic complexes **100** and **101** [Eq. (42)], each in 98% yields, although the platinum complex **101** could not be separated from oligomeric by-product [38].

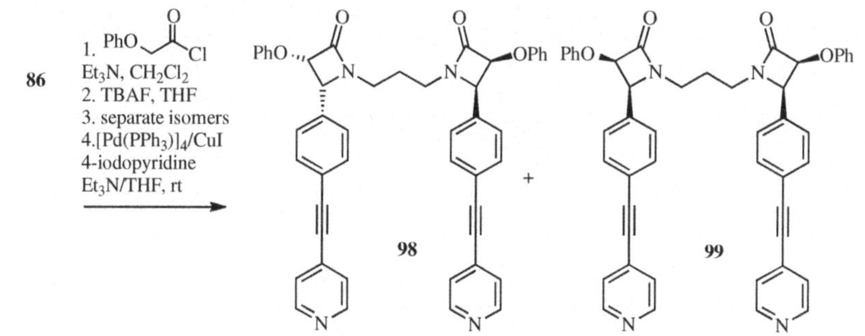

(41)

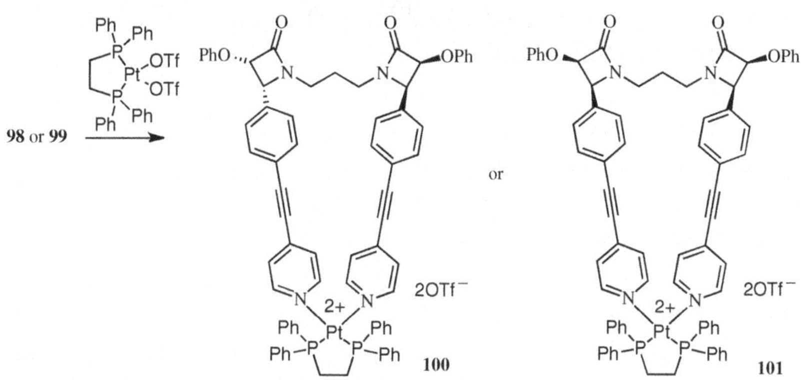

(42)

The alkynes **102** and **103** were similarly converted to the corresponding platinum complexes [Eq. (43)] [38].

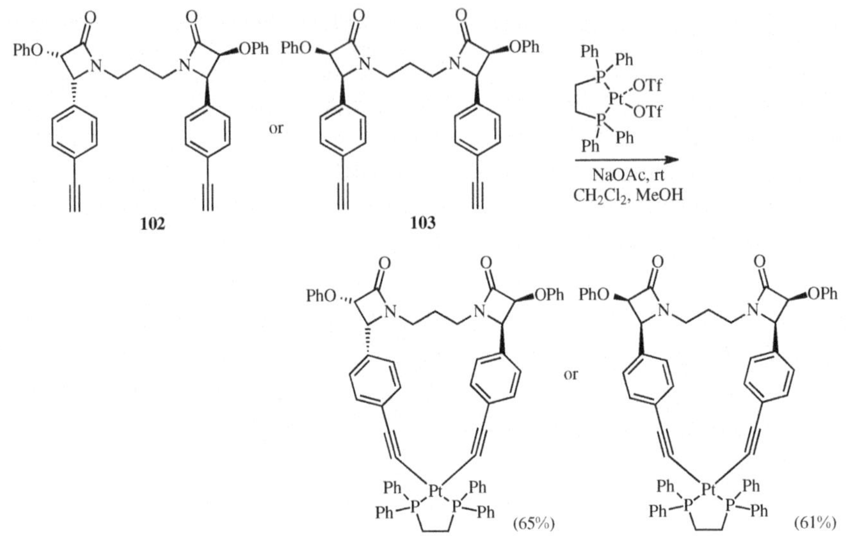

(43)

Coupling of **102** with *cis*-[Pt(dppe)Cl_2] and triethylamine with 10% CuI as a catalyst gave tetra-β-lactams **104** and **105** in 68% yield as an inseparable 1:1 mixture of diastereomeric *syn*/*anti* forms [Eq. (44)] [38].

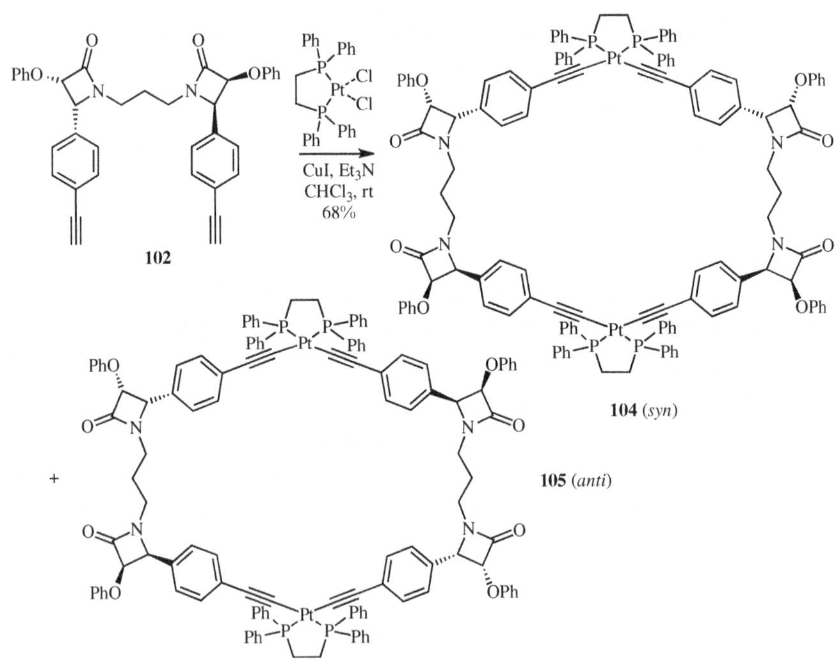

(44)

Phenoxyketene generated in situ by dehydrochlorination reacted with the bis (imine) **104** forming the bis(β-lactam) with N1,N1′ and C4,C4′ linkages in 78% yield [Eq. (45)] [39]. Reactions of **105–107** with phenoxyketene similarly formed the respective bis(β-lactams), Table 4 [39], while **108** and **109** gave the respective bis(β-lactams) as mixtures of *cis/cis* diastereomers, Table 4 [40].

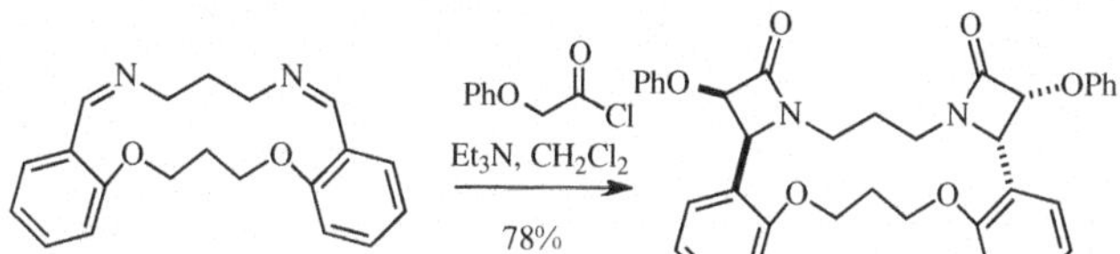

104 (45)

Table 4 Bis(β-lactams) from phenoxyketene reactions with bis(imines) **105–109**

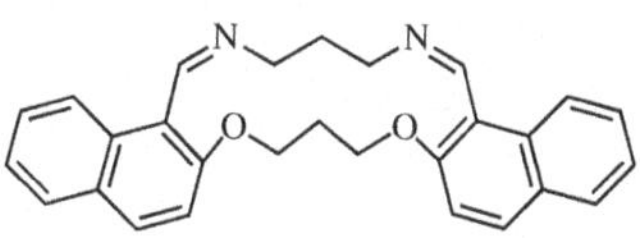

105

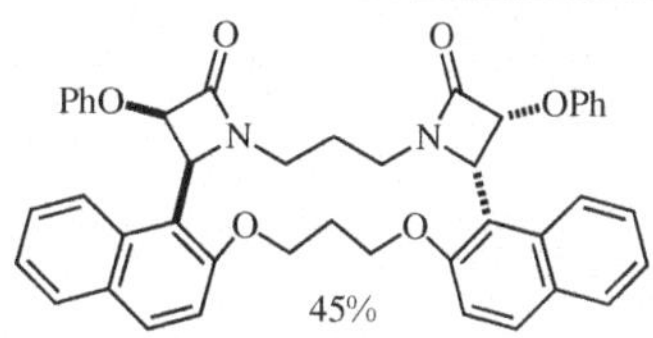

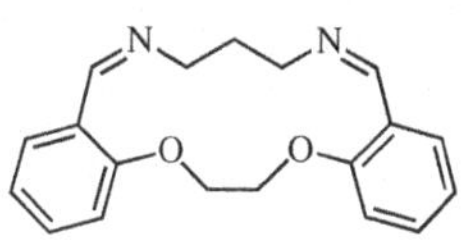

106

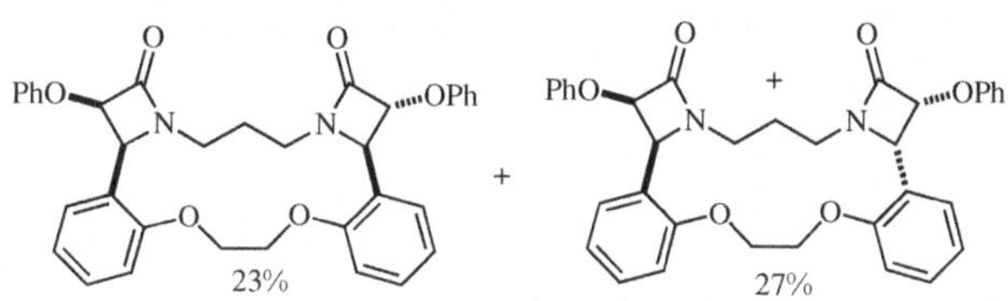

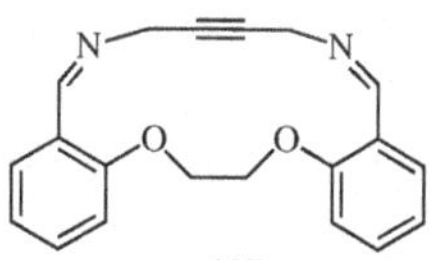

107

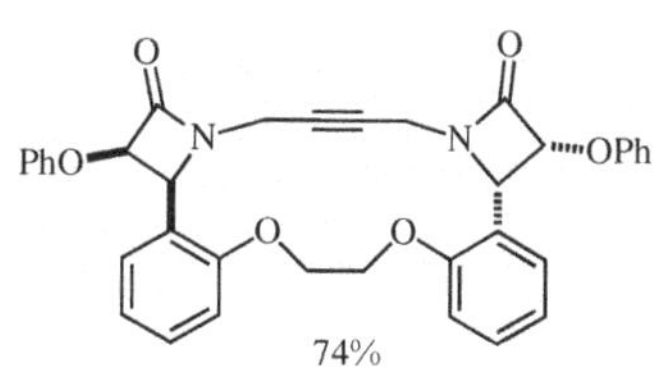

(continued)

Table 4 (continued)

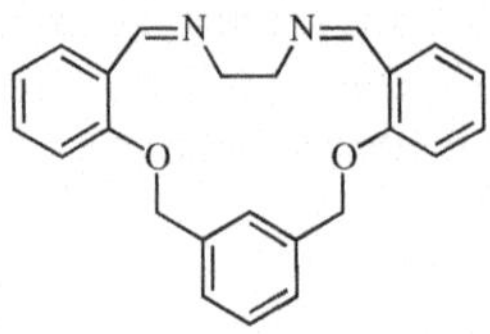

108

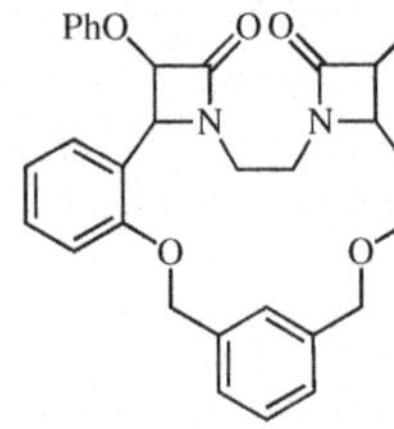

(50%)

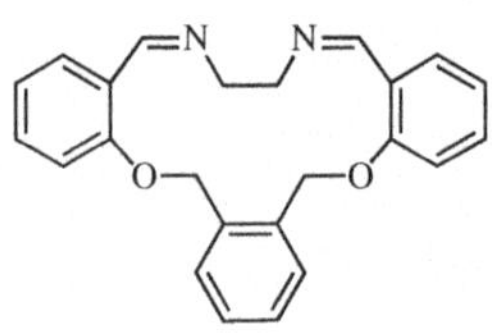

109

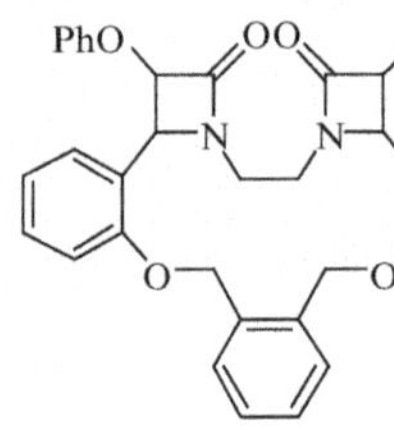

45%

Bis(imines) derived from benzylisatin with either 1,4- or 1,3-aryl linkages reacted with ketenes generated in situ by Et_3N dehydrochlorination of acyl chlorides in CH_2Cl_2 from $-10°$ to room temperature forming bis(spiro-β-lactams) [Eqs. (46) and (47)] [41].

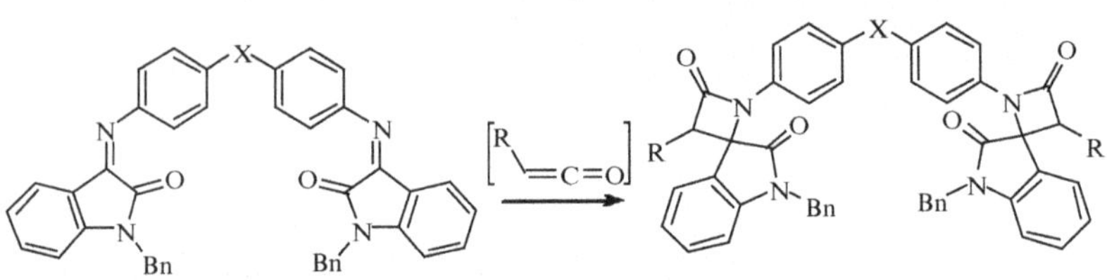

(46)

X	R	Yield (%)	X	R	Yield (%)
CH_2	PhO	67	O	PhO	66
CH_2	PhthN	63	O	PhthN	66

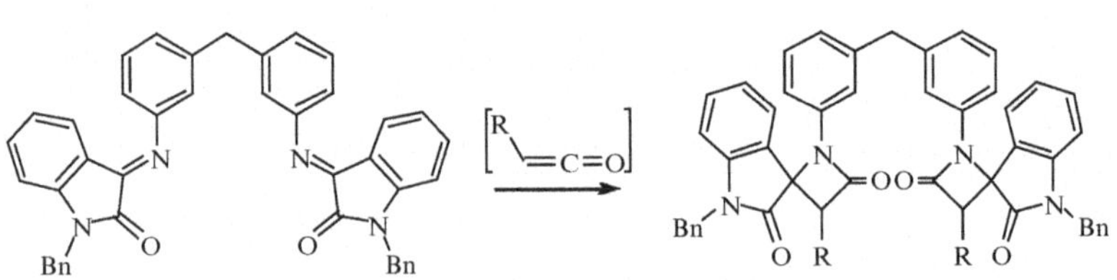

(47)

70% (R = PhO); 58% (R = PhthN)

Phenoxyketene reacts with the 1,1′-bis(iminyl)ferrocene **110** at 0° in methylene chloride forming *cis*/*cis*-bis(β-lactams) **111** in a 76% yield with a 2:1 *syn*:*anti* ratio [Eq. (48)]. When the reaction is carried out in toluene at reflux the *trans*/*trans* products are formed, in 34% yield and a 1:1 ratio of diastereomers [Eq. (49)]. The formation of the *trans*-β-lactams at the higher temperature is consistent with the formation of an intermediate zwitterion which undergoes bond-rotation favoring formation of the more stable product, as opposed to the usual course of ketene cycloaddition favoring the less stable *cis*-product [42].

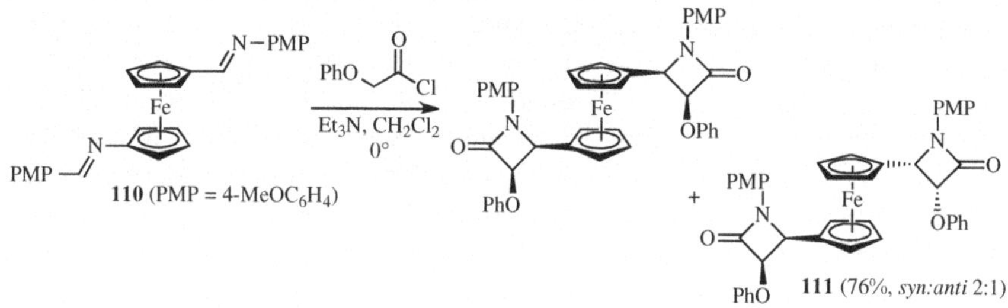

(48)

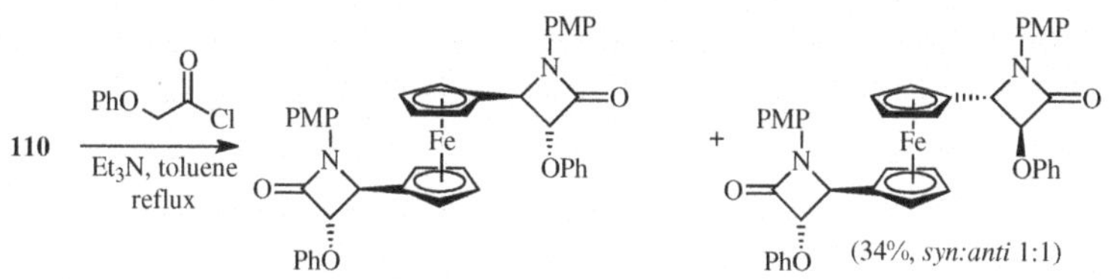

(49)

Bis(3-butenyl) substituted ferrocenyl bis(β-lactam) **112** was prepared as a *cis*,*cis* diastereomeric mixture by phenoxyketene cycloaddition to the *N*,*N*′-di-3-butenyl bis(imine) in 93% yield, and in a subsequent step underwent ring closing olefin metathesis with a second generation Grubbs' catalyst in refluxing dichloromethane giving after chromatographic separation the *syn*- and *anti*-diastereomeric bridged ferrocenes [Eq. (50)] [42].

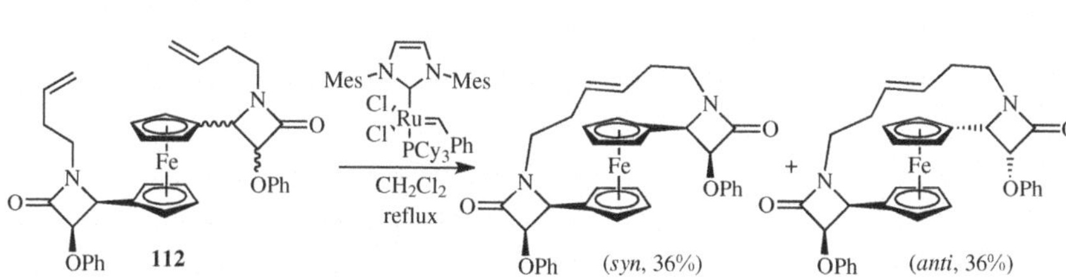

(50)

Dialkynyl bis(β-lactam) **113** prepared by cycloaddition to the bis(imine) underwent ring closure by copper(II) acetate catalyzed oxidative coupling yielding the ring closed product as an inseparable 1:1 mixture of *syn*/*anti*-diastereomeric bridged ferrocenes in 30% yield [Eq. (51)] [42].

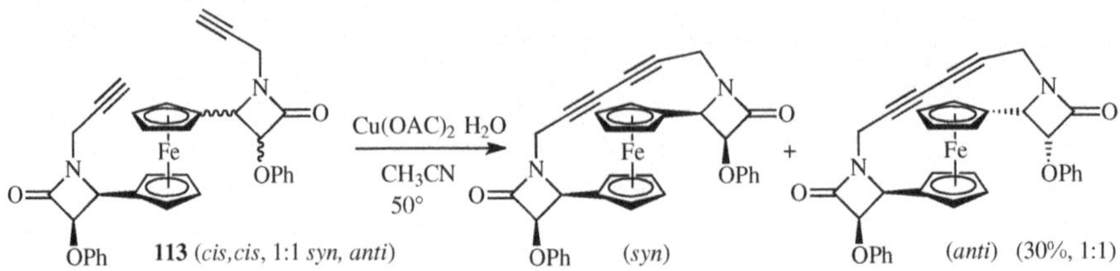

(51)

Ketene imine [2+2] cyclizations for formation of macrocycles have been implemented as three-component reactions in which diamines and dialdehydes are reacted together in methylene chloride and after sufficient time for formation of macrocyclic oligoimines triethylamine and methoxyacetyl chloride are added at −50°, and the solution is stirred overnight at room temperature. Macrocycles **114–116** were obtained in this way as inseparable mixtures of diastereomers, with *cis*-stereochemistry in the β-lactams [Eqs. (52)–(54)] [43].

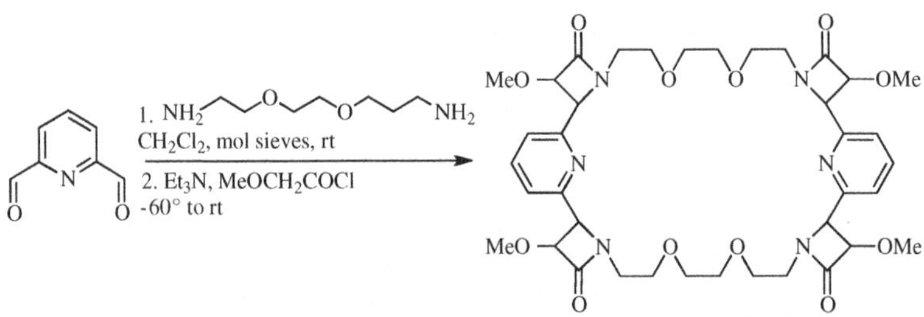

(52)

114 (44%)

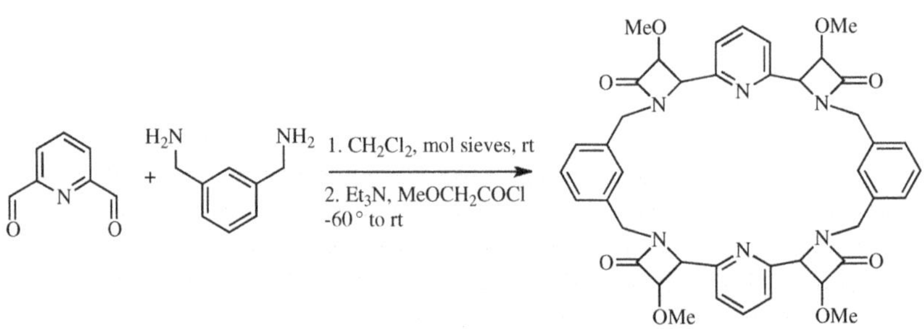

115 (33%)

(53)

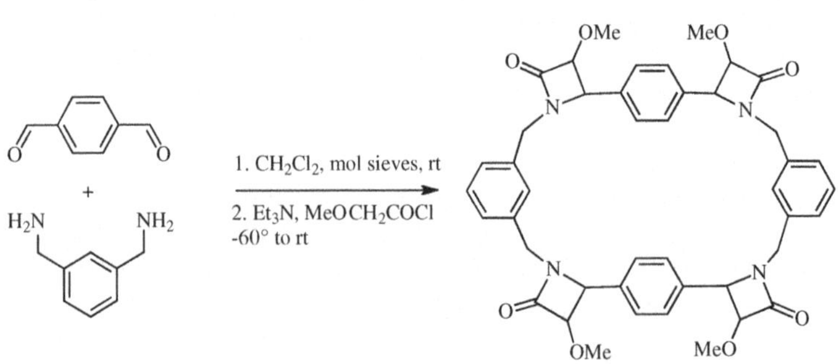

(54)

116 (82%)

Bis(β-lactams) **117** are formed by aryloxyketene additions to bis(imines), and for phenoxylketene there is an 85% yield of the *cis,cis* diastereomers with a *syn/anti* ratio of 47:53 [44]. Ring closing metathesis of the *anti* diastereomer with the Grubbs' catalyst forms the macrocyclic alkene in 98% yield with a *Z/E* ratio of 17:83 for the alkene [Eq. (55)]. The bis(imine) **119** gave similar results [Eq. (56)]. Other aryloxyketenes gave similar results, as collected in Table 5 [44].

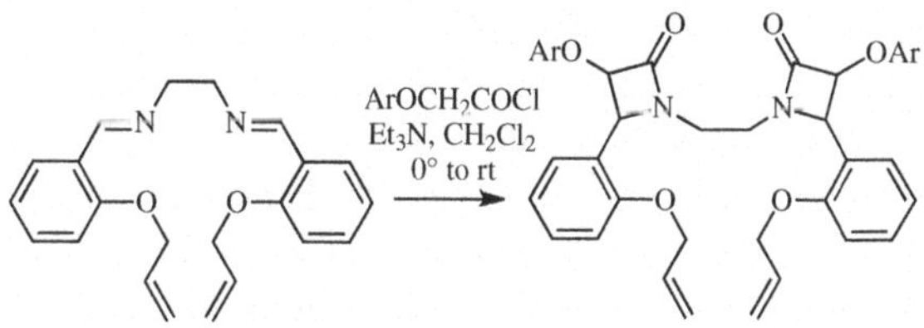

117 (Ar = Ph) 85%
cis, cis; *syn/anti* 47:53

(55)

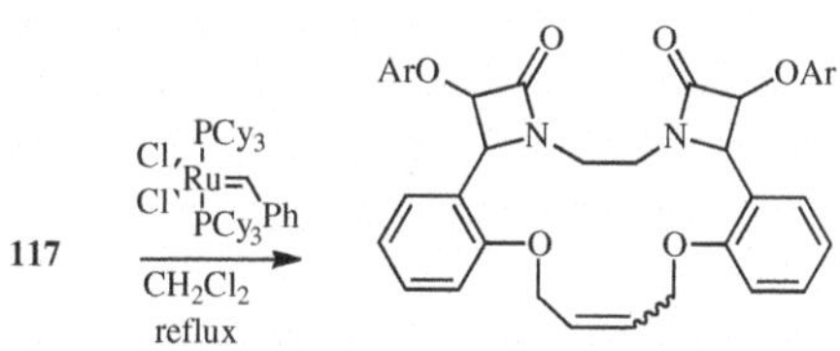

118 (Ar = Ph) 98%, *Z/E* 17:83)

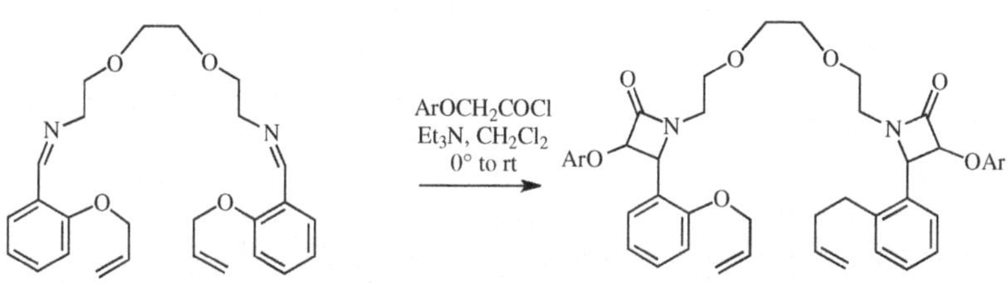

119 (Ar = Ph) 75%
cis, cis; *syn/anti* 48:52

(56)

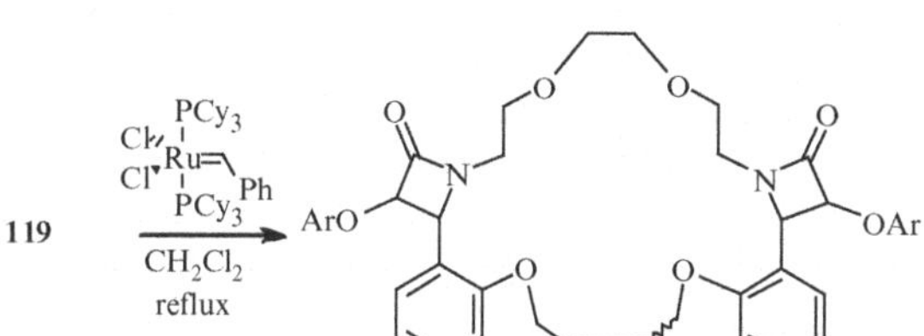

120 (Ar = Ph) 82%, *Z/E* 1:1.7)

Table 5 Bis(β-lactams) from aryloxyketene additions to imines

Aryl	Yield **117**, *syn*/*anti*	Yield **118** from *syn*, *Z*/*E*	Yield **118** from *anti*, *Z*/*E*
Ph	85, 47:53	98, 1:5	98, 1:5
1-Naphthyl	91, 49:51	95, 1:4	98, 1:4
2-Naphthyl	93, 49:51	57, 1:3	60, 1:1.6
Aryl	Yield **119**, *syn*/*anti*	Yield **120** from *syn* **119**, *Z*/*E*	Yield **120** from *anti* **119**, *Z*/*E*
Ph	75, 48:52	82, 1:1.7	82, 1.9:1
1-Naphthyl	90, 50:50	80, 1.2:1	80, 1.5:1
2-Naphthyl	80. 50:50	75, 1.8:1	75, 1.1:1

For the purposes of confirming the *E*/*Z* stereochemistry of bis(lactams) **120** the *E*-isomers **122** were also prepared by reactions of the aryloxyketenes generated by dehydrochlorination with the separated cyclic bis(imines) *E*-**121**, Eq. (57). Isomerization of the *E*- to the *Z*-products was carried out using the Grubbs catalyst [44].

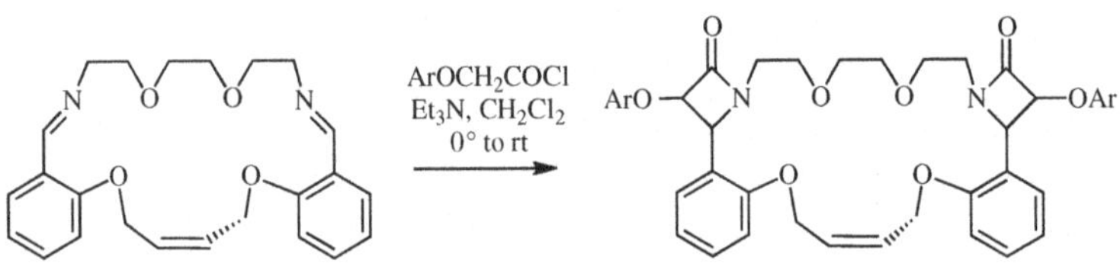

(57)

121

E-**122**.Ar = Ph 78%, *syn/anti* 36:64)
Ar = 1-Naphthyl 85%, *syn/anti* 45:55
Ar = 2-Naphthyl 84%, *syn/anti* 45:55

5 Bis-β-Lactams from Imines and Bisketenes

The use of bisketenes in bis-β-lactam synthesis has been examined in only a few examples. Although bisketenes were discovered very early in the history of ketenes [45], their chemistry was developed more recently [17, 19, 46].

The use of bisketenes in the formation of bis-β-lactams has been examined in only a few examples, although many more possibilities appear to be feasible for this approach. Reaction of adipyl chloride in CH_2Cl_2 with Et_3N in the presence of imines leads to the formation of bis-β-lactams **124** [Eq. (58)] [47]. It is known that bisketene **123** can be formed in this reaction and directly observed by IR at 2114 cm^{-1} [46], but the extent to which the product is formed from **123** as opposed to a stepwise process with successive generation of a mono-ketene followed by conversion to a β-lactam is not determined.

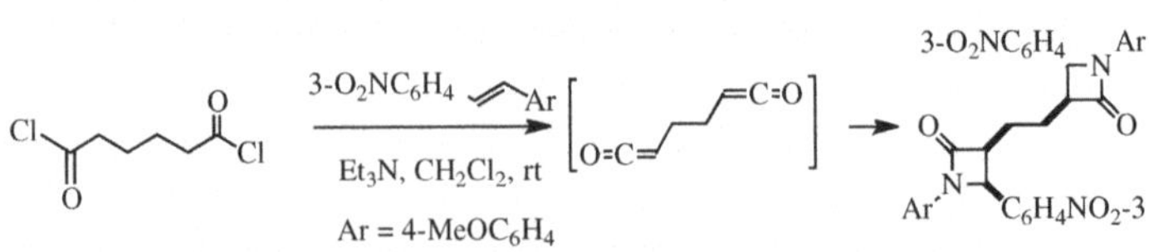

(58)

123 **124** (75%)

Diacid treatment with Mukaiyama's reagent in the presence of imines in a similar process gives formal generation of bisketene **125** that undergoes cycloaddition leading to bis(lactam) formation [Eq. (59)] [48]. The ketenyl groups are not formed simultaneously and may be generated and react in a stepwise fashion. Similar generation of **125** together with a bis(imine) formed a polymer with β-lactams in the polymer chain [48], and a number average molecular weight of 15,400, in 88% yield. The possibility of formation of a cyclic poly(lactam) by this process has not been elucidated.

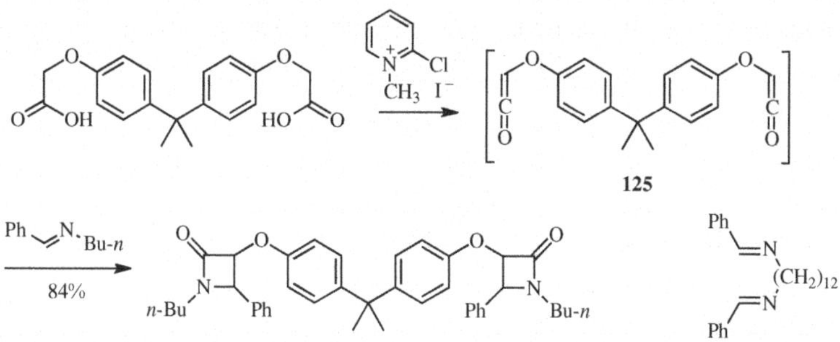

(59)

Bisketene **126** generated by photochemical Wolff rearrangement was directly observed by IR, and on reaction with benzanilide give *trans,trans*-bis(β-lactam) **127** as a mixture of *meso* and *dl* isomers [Eq. (60)] [49]. 1,2- and 1,4-Bisketenylbenzenes also observed by IR were generated by dehydrochlorination, and reacted with benzanilide giving mixtures of *meso* and *dl-trans,trans*-bis (β-lactams) [Eq. (61)] [49]. Trimethylsilyl-1,2-bisketenes **128**, however, did not give bis(β-lactams) with imines, but instead formed aziridines [Eq. (62)] [49].

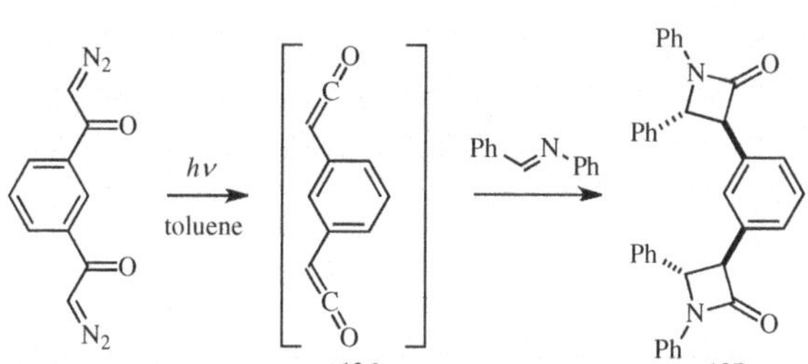

(60)

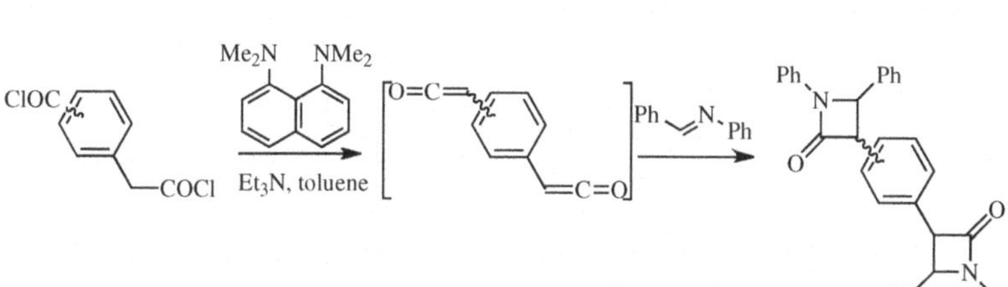

(61)

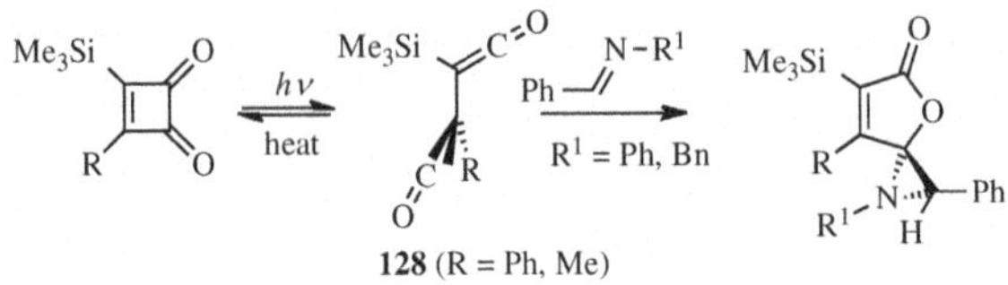

(62)

An alternative approach to bis(β-lactam) synthesis utilizes photolysis of bis (carbene) complexes in the presence of imidazoline **129** to give bis(β-lactams) as 1:1 mixtures of racemic diastereomers in 55–68% yields by a reaction which could formally have proceeded through bisketene complexes **130** which reacted with the imine functionality in a double [2+2] cycloaddition leading to the observed product [Eq. (63)] [50].

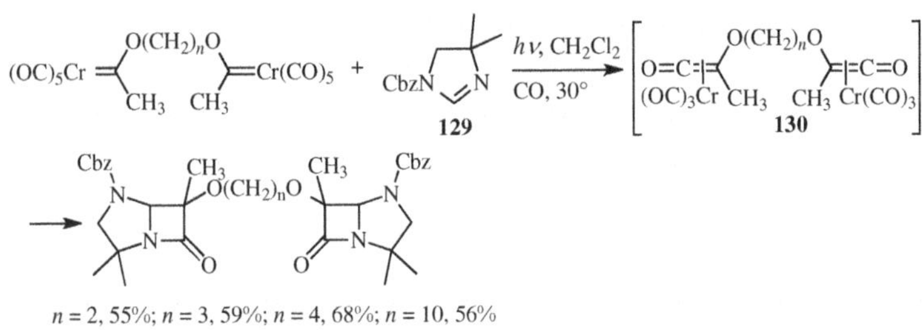

(63)

These bis(lactams) were subjected to a variety of structural modifications producing a large number of dimers and other structures. With photochemical activation the generation of bis(ketene) equivalents is enhanced, but stepwise processes are not disproven [50]. Bis(carbene) complex photolysis in the presence of the optically active imidazoline **131** gave the optically active bis(β-lactam) as a single stereoisomer [Eq. (64)]) [51].

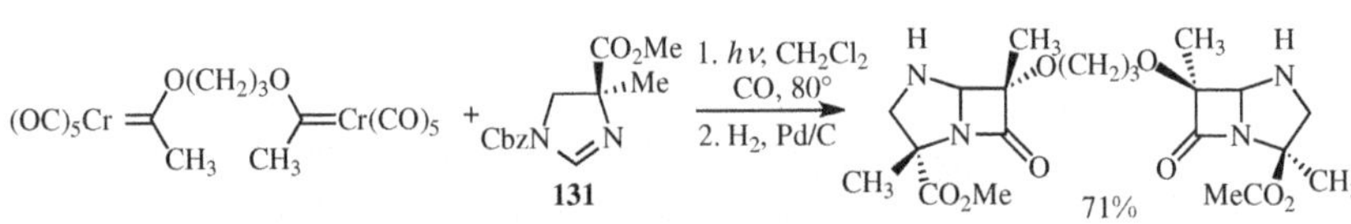

(64)

Bis(carbene) complexes **132** with hydrocarbon linkers reacted with imidazoline **129** forming bis(lactams) as mixtures of stereoisomers [Eq. (65)], and reacted with the optically active imidazoline **131** giving a single product [Eq. (66)] [52]. Poly (ethylene glycol) linked bis(lactams) were formed from the corresponding bis (carbene) complexes by the same methodology [Eq. (67)], and displayed effective complexing ability with alkali metal ions [53].

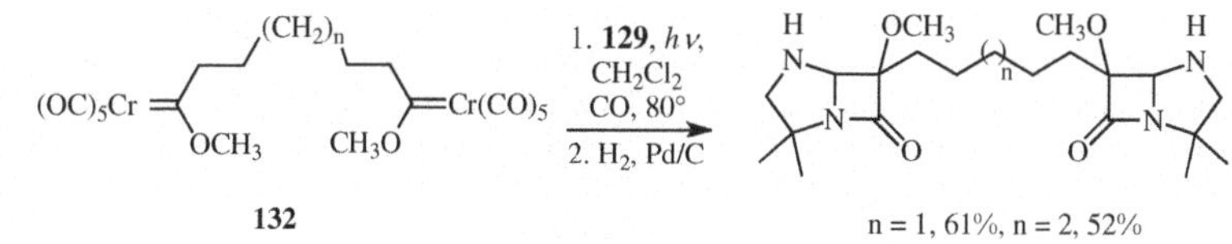

(65)

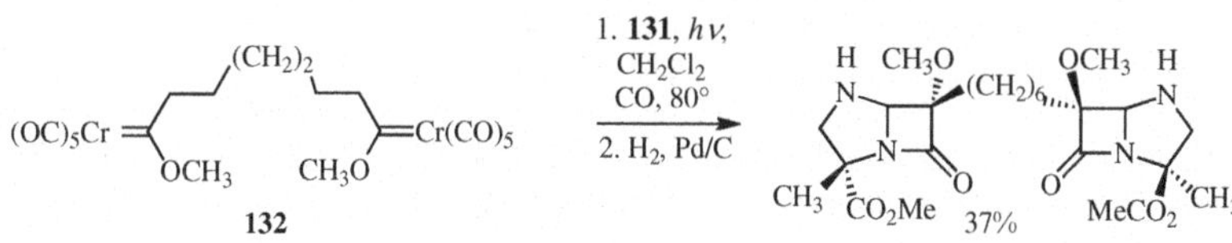

(66)

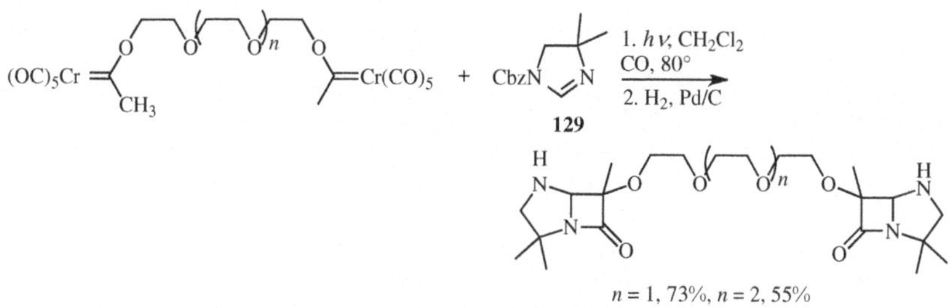

(67)

6 Conclusion

Bis-β-lactams have been known for more than 30 years and recently have gained increasing attention. While these have not yet found significant therapeutic applications further study of these compounds may be expected, and the variety of methods already available for their preparation can lead to many new examples of these fascinating materials.

References

1. Brooks G, Hunt E (1983) J Chem Soc Perkin Trans 1:115–120
2. Hatanaka N, Ojima I (1981) J Chem Soc Chem Commun 344–346
3. Gotthardt H, Schenk K-H (1986) J Chem Soc Chem Commun 687–688
4. Rodríguez-Tébar A, Arán V, Vázquez D (1984) Eur J Biochem 139:287–293
5. Ojima I, Zhao M, Yamato T, Nakahashi K, Yamashita M, Abe R (1991) J Org Chem 56:5263–5277
6. Kim YT, Jung YJ, Kim YM (1999) Drug Dev Ind Pharm 25:523–528
7. Alcaide B, Almendros P, Aragoncillo C, Gomez-Campillos G (2011) Eur J Org Chem 364–370
8. Garud DR, Makimura M, Koketsu M (2011) New J Chem 35:581–586
9. Reddy PSN, Mittapelli V, Reddy VD (2010) Rasayan J Chem 3:635–640
10. Staudinger H (1907) Justus Liebigs Ann Chem 356:51–123

11. Woodward RB, Johnson JR, Robinson R (1949) In: Clarke HT, Johnson JR, Robinson R (eds) The chemistry of penicillin. Princeton University Press, Princeton, pp. 440–454 (Chap. 15)
12. Sheehan JC, Buhle EL, Corey EJ, Laubach GD, Ryan JJ (1950) J Am Chem Soc 72:3828–3829
13. Sheehan JC, Henery-Logan KR (1957) J Am Chem Soc 79:1262–1263
14. Fu N, Tidwell TT (2008) Tetrahedron 64:10465–10496
15. Schiff H (1864) Justus Liebigs Ann Chem 131:118–119
16. Schiff H (1866) Justus Liebigs Ann Chem 140:92–137
17. Tidwell TT (2006) Ketenes, 2nd edn. Wiley, Hoboken
18. Allen AD, Tidwell, TT (2012) Eur J Org Chem 1081–1096
19. Danheiser RL (ed) (2006) Science of synthesis, vol. 23. Georg Thieme Verlag, Stuttgart
20. Ojima I, Nakahashi K, Brandstadter SM, Hatanaka N (1987) J Am Chem Soc 109:1798–1805
21. Bose AK, Womelsdorf JF, Krishnan L, Urbanczyk-Lipkowska Z, Shelly DC, Manhas MS (1991) Tetrahedron 47:5379–5390
22. Jayaraman M, Puranik VG, Bhawal BM (1996) Tetrahedron 52:9005–9016
23. Alcaide B, Martín-Cantalejo Y, Pérez-Castells J, Rodríquez-López J, Sierra MA, Monge A, Pérez-Garcia V (1992) J Org Chem 57:5921–5931
24. Alcaide B, Martín-Cantalejo Y, Pérez-Castells J, Sierra MA, Monge A (1996) J Org Chem 61:9156–9163
25. Alcaide B, Almendros P, Alonso JM, Redondo MC (2003) J Org Chem 68:1426–1432
26. Nakahara K, Yamaguchi K, Yoshitake Y, Yamaguchi T, Harano K (2009) Chem Pharm Bull 57:846–852
27. Wang Z, Chen N, Xu J (2011) Tetrahedron 67:9690–9699
28. Park SH, Lee SY, Bose AK (2001) Bull Kor Chem Soc 22:493–498
29. Jayaraman M, Deshmukh ARAS, Bhawal BM (1994) J Org Chem 59:932–934
30. Karupaiyan K, Puranik VG, Deshmukh ARAS, Bhawal BM (2000) Tetrahedron 56:8555–8560
31. Shaikh AL, Puranik VG, Deshmukh ARAS (2005) Tetrahedron 61:2441–2451
32. Vatmurge NS, Hazra BG, Pore VS, Shirazi F, Deshpande MV, Kadreppa S, Chattopadhyay S, Gonnade RG (2008) Org Biomol Chem 6:3823–3830
33. Ali P, Meshram J, Tiwari V, Dongre R, Sheikh J, Ahemad M (2010) Der Pharma Chemica 2:138–147
34. Meshram J, Ali P, Tiwari V (2010) J Heterocycl Chem 47:1454–1458
35. Jarrahpour A, Ebrahimi E (2010) Molecules 15:515–531
36. Jarrahpour A, Ebrahimi E, De Clercq E, Sinou V, Latour CD, Bouktab L, Brunel JM (2011) Tetrahedron 67:8699–8704
37. Pellico D, Gómez-Gallego M, Ramírez-López P, Mancheño MJ, Sierra MA, Torres MR (2010) Chem Eur J 16:1592–1600
38. Pellico D, Gómez-Gallego M, Ramírez-López P, Mancheño MJ, Sierra MA, Torres MR (2009) Chem Eur J 15:6940–6952
39. Sierra MA, Pellico D, Gómez-Gallego M, Macheño MJ, Torres R (2006) J Org Chem 71:8787–8793
40. Arumugam N, Ragunathan R (2006) Tetrahedron Lett 47:8855–8857
41. Jarrahpour A, Khalili D (2007) Tetrahedron Lett 48:7140–7143
42. Sierra MA, Rodríguez-Fernández M, Casarrubios L, Gómez-Gallego M, Allen CP, Mancheño MJ (2009) Dalton Trans 8399–8405
43. Leon F, Rivera DG, Wessjohann LA (2008) J Org Chem 73:1762–1767
44. Ibrahim YA, Al-Azemi T, Abd El-Halim MD, John E (2009) J Org Chem 74:4305–4310
45. Diels O, Wolf B (1906) Chem Ber 39:689–697
46. Allen AD, Fenwick MH, Jabri A, Rangwala H, Saidi K, Tidwell TT (2001) Org Lett 3:4095–4098
47. Pagadala R, Meshram JS, Chopde HN, Jetti V (2010) Int J Pharm Res 2:1581–1585
48. Sato A, Endo T (1998) Macromolecules 31:7996–7998

49. Allen AD, Godoy J, Fu N, Nagy M, Spadaro S, Tidwell TT, Vukovic S (2008) J Am Chem Soc 130:2386–2387
50. Dumas S, Lastra E, Hegedus LS (1995) J Am Chem Soc 117:3368–3379
51. Hsiao Y, Hegedus LS (1997) J Org Chem 62:3586–3591
52. Kuester E, Hegedus LS (1999) Organometallics 18:5318–5321
53. Puntener K, Hellman MD, Kuester E, Hegedus LS (2000) J Org Chem 65:8301–8306

Top Heterocycl Chem (2013) 30: 147–182
DOI: 10.1007/7081_2012_90

Published online: 8 August 2012

The Chemistry and Biology of *N*-Thiolated β-Lactams

Edward Turos

Abstract *N*-thiolated β-lactams are a recently discovered family of bioactive agents having antibacterial, antifungal, and anticancer properties. Although our initial interest in these compounds was for their utility in organic synthesis, our laboratory began investigating their biological properties about a decade ago. The selectivity these β-lactams show as bacteriostatic agents towards certain pathogenic bacteria, including *Staphylococcus aureus* and *Bacillus anthracis*, and their structure–activity profiles and mode of action, are highly unique. This chapter describes in roughly chronological sequence results from our laboratory in the discovery and ongoing development of these biologically active compounds.

Keywords Antibacterials · Antibiotics · Fatty acid biosynthesis · MRSA · *N*-thiolated β-lactams

Contents

E. Turos (✉)
Department of Chemistry, Center for Molecular Diversity in Drug Design, Discovery, and Delivery, University of South Florida, 4202 East Fowler Avenue, CHE 205, Tampa, FL 33620, USA
e-mail: eturos@usf.edu

1 Introduction to β-Lactam Antibiotics

For the last 80 years, β-lactam antibiotics have been used ubiquitously as therapeutics for a wide range of bacterial infections and diseases [1]. The first β-lactams were actually synthesized by Staudinger over 100 years ago, utilizing a bimolecular coupling reaction between an acid chloride and an aldimine (Fig. 1) [2].

It was not until several decades later that the biological implications of β-lactam compounds became known. The entry point in this occurred in 1928, when Alexander Fleming, a British microbiologist, accidentally discovered a fungal metabolite produced from a *Penicillium* mold that was able to kill bacteria [3]. He referred to this microcidal substance as penicillin. It came as a surprise to many when Dorothy Hodgkin proved by X-ray crystallography that penicillin contained within its chemical structure a β-lactam ring [4]. From that point on, a number of different classes of bicyclic β-lactam ring systems have been investigated in terms of both chemical and biological applications (Fig. 2). The chemical and biological literature is rich with reports documenting just how valuable β-lactams are as bioactive molecules as well as building blocks for organic synthesis.

The now well-known antimicrobial activity of bicyclic β-lactam compounds arises from the amido nitrogen occupying the site of ring fusion, which enables the nitrogen center to be sufficiently pyramidalized to perturb amide planarity and resonance stabilization of the four-membered amide (2-azetidinone) ring. Long ago, Nature established that N-fused bicyclic β-lactams have elevated electrophilicity and reactivity toward nucleophilic ring opening, which makes them effective biological acylating agents and antibiotics. Derived biochemically from two molecules of L-cysteine, penicillin mimics the D-alanine–D-alanine termini of bacterial peptidoglycans, making itself readily recognizable to bacterial transpeptidases [3]. As a consequence, penicillin can block the enzymatic crosslinking of peptidoglycan strands in the bacterial cell wall (murein) by irreversibly acylating the catalytic serine within the enzyme active site. The ability of fungi to biochemically produce penicillin as a secondary metabolite provides a remarkably effective means for its defense against bacteria.

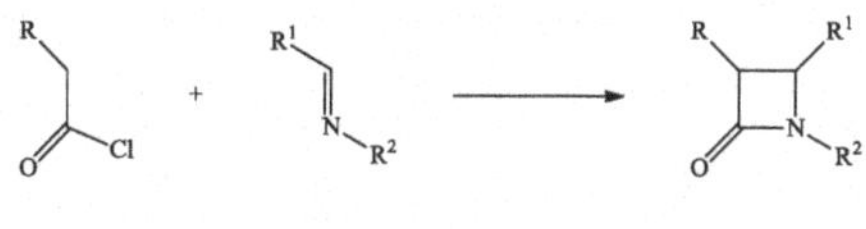

Fig. 1 Staudinger's synthesis of β-lactams from an acid chloride and an aldimine

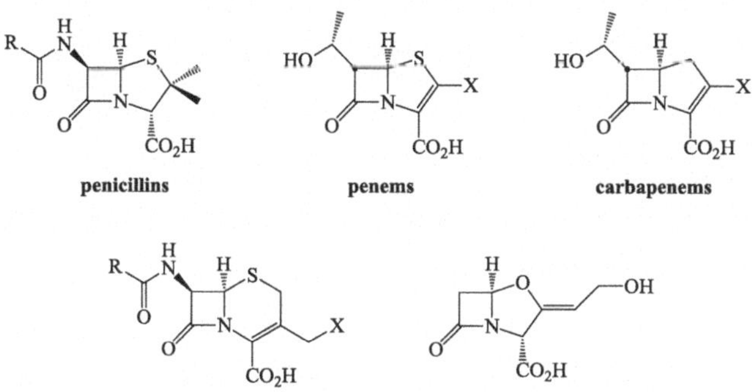

Fig. 2 Common classes of bicyclic β-lactams

2 Discovery of the Synthetic Utility of *N*-Thiolated β-Lactams

Shah and Cama provided the first report of *N*-sulfenylated β-lactams in the mid-1980s, showing that an *N*-methylthio-substituent could protect the β-lactam during synthetic procedures in a manner that enables ready removal by a thiolate anion [5] (Fig. 3).

We took advantage of this unusual chemistry when we began our studies on *N*-thiolated β-lactams in the mid-1990s [6–8]. We became interested in determining whether the β-lactam ring could conformationally influence regioselectivity of ring closing during electrophile-promoted cyclizations of unsaturated sulfides [9, 10], such as shown in Fig. 4. Treatment of an alkenyl (or alkynyl) *N*-methylthio β-lactam with iodine or bromine can generate an electrophile-activated intermediate, which provokes a nucleophilic addition of the sulfur center to give either a six-membered ring through a 6-endo ring closure process or a five-membered ring via a 5-exo cyclization. Our experimental results indicated that the smaller, entropically favored five-membered ring cycloadduct was formed exclusively in these reactions [6–8]. The structural similarity of these bicyclic β-lactam ring products to those of the classical cephalosporins and penicillins caused us to refer to them more commonly as "isocephams" and "isopenams".

We used this procedure to synthesize bicyclic structures in which the β-lactam ring is placed at unusual positions within the framework. In considering the penam system, as an example, one can conceive of three "isopenam" structures as shown in Fig. 5.

Semiempirical calculations on the three isomeric ring structures suggested that the N–S fused ("inversely fused") bicyclic ring III has about the same thermodynamic

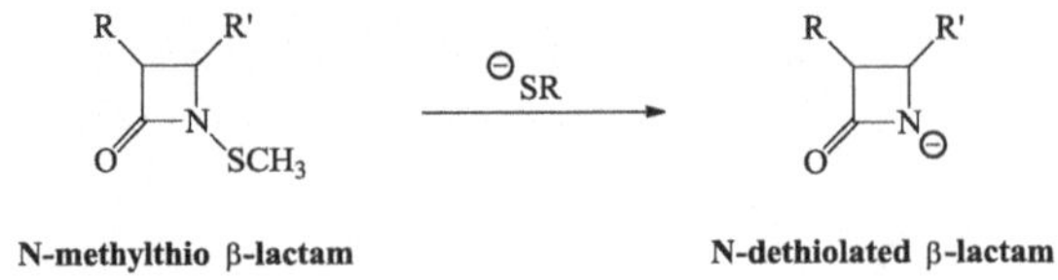

Fig. 3 Thiophilic removal of an *N*-methylthio group from a β-lactam

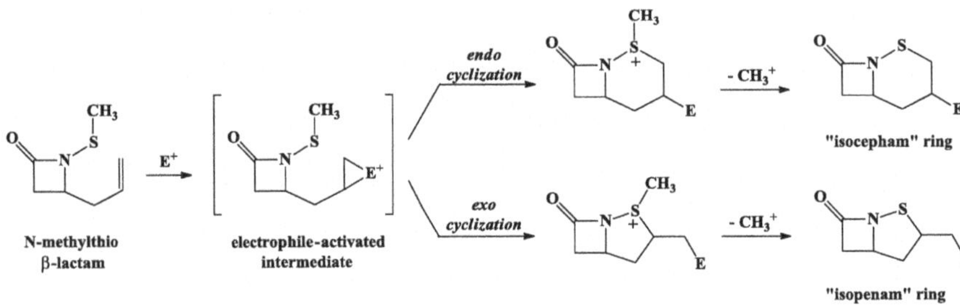

Fig. 4 Electrophile-promoted cyclization can proceed through either an endo or exo cyclization process

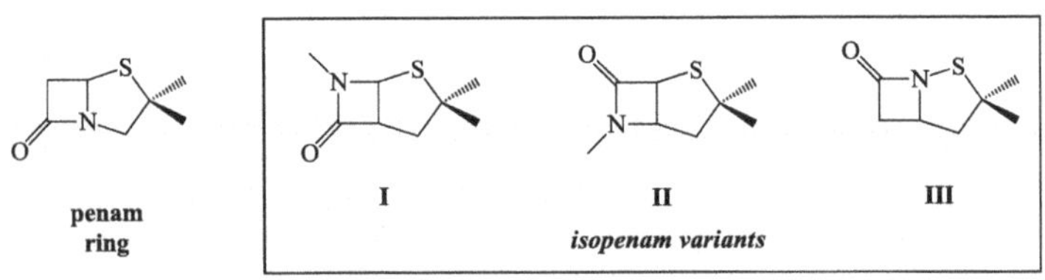

Fig. 5 Three *isopenam* ring variants based on the classical bicyclic penam ring system

stability as the classical penam ring found in penicillin, although with somewhat less pyramidalization at nitrogen and a reduced bend angle between the fused rings. Ab initio calculations indicated that the N–S fused ring is also somewhat more twisted than the classical penam ring, thus lowering the energy of its LUMO. In addition to these electronic effects, the sulfur center can withdraw electron density from the lactam nitrogen, and thereby enhance the electrophilicity of the β-lactam carbonyl.

The N–S-fused bicyclic β-lactams of type III could be prepared synthetically for further studies, as shown in Scheme 1, starting from *N*-methylthio azetidinones **3**.

Staudinger coupling of the *N*-*p*-methoxyphenyl imine, prepared fresh from phenylpropynal, with commercially available methoxyacetyl chloride provided exclusively the *cis*-disubstituted β-lactam **1** (as a racemate). Oxidative removal of the para-methoxyphenyl protecting group of lactam **1** using aqueous ceric ammonium nitrate gave *N*-protio β-lactam **2**, which was then converted to *N*-methylthio β-lactam **3** by deprotonation with *n*-BuLi at −78°C and subsequent thiolation of the *N*-lithio species with methyl methanethiosulfonate. The cycloaddition reaction of lactam **3** to the bicyclic isopenem **4** took place in high yield with molecular iodine

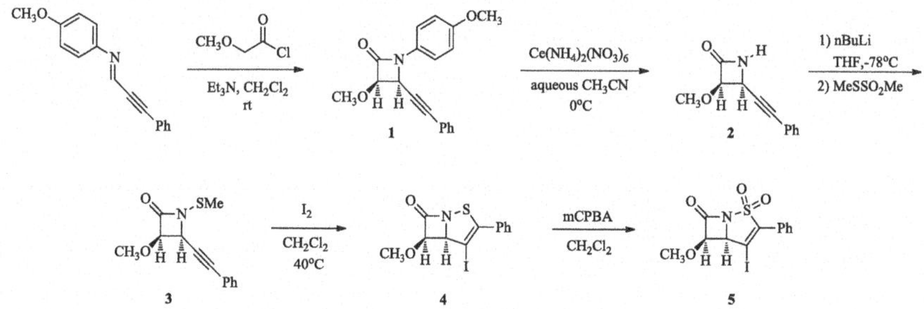

Scheme 1 Synthesis of *isopenems* **4** and **5** by halocyclization

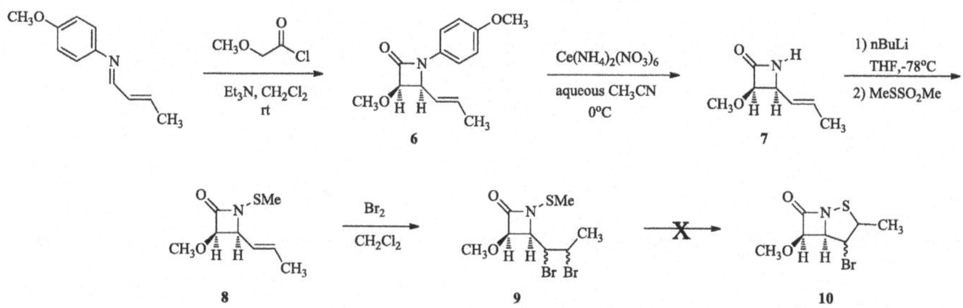

Scheme 2 Attempted synthesis of *isopenam* **10** by halocyclization of β-lactams **8** or **9**

in refluxing methylene chloride solution. The structure of this bicycloadduct was determined by proton NMR and infrared spectroscopy, which showed a strong C=O stretching signal close to 1,790 cm^{-1}. This ultra-high carbonyl stretching frequency confirmed the strong electron-withdrawal by sulfur on the amide. S-oxidation of sulfenamide **4** to the sulfonamide **5** further increased the C=O stretching frequency to ~1,800 cm^{-1}. Although the iodocyclization of lactam **3** took place efficiently, it noticeably required more forcing conditions (40°C) than those of our prior halocyclizations of unsaturated sulfides, owing to the electron-withdrawing effect of the β-lactam nitrogen on sulfur's nucleophilicity.

Attempting to use this same synthetic approach on the corresponding alkenyl β-lactam **8**, prepared as shown in Scheme 2, we were not able to reach the bicyclo β-lactam (*isopenam*) adduct **10**. Instead, halogenation of **8** gave only the dibromination products **9** (in the case of bromination), or alkene **8** was recovered after workup (in the case of iodination). Efforts to cyclize dibromo adduct **9** to the fused 4,5-bicyclic ring **10** were similarly unsuccessful. Thus, it appears that the conformational constraints of the alkenyl or alkyl substrates (versus alkynyl β-lactam **3**) were sufficiently perturbed to prevent cyclization to the *isopenam* ring system **10**.

In stark contrast, the homologous 5-exo ring closures of alkynyl β-lactam **14** could be easily carried out to construct clavulanic acid-type ring **15** (Scheme 3). Substrate **14** was prepared in several steps from alkenyl β-lactam **6**, starting with

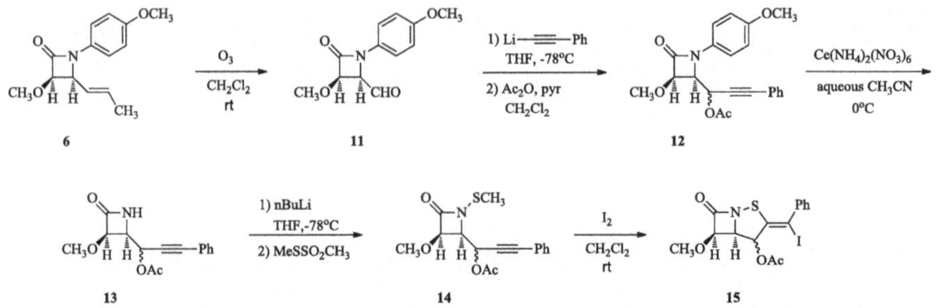

Scheme 3 Synthesis of *isoclavulanate* **15** by way of an efficient 5-*exo-dig* halocyclization of alkynyl β-lactam **14**

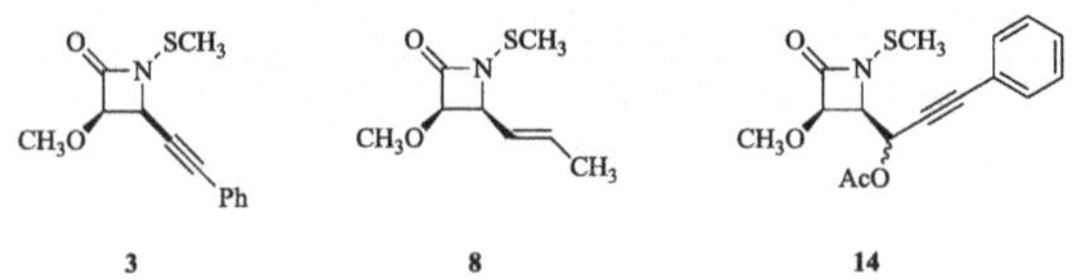

Fig. 6 *N*-methylthio β-lactams **3**, **8**, and **14**

ozonolysis of the alkenyl side chain in **6** to afford aldehyde **11**, followed by stereoselective addition of lithium phenylacetylide and *O*-acetylation of the resulting propargylic alcohol to arrive at propargylic acetates **12**. Oxidative dearylation of *N*-aryl group from lactam **12** and subsequent introduction of the *N*-methylthio substituent provided **14**. We were pleased to find that the iodocyclization proceeded as intended to allow us to prepare *isoclavulanate* adduct **15**. Infrared spectroscopy indicated again that the β-lactam carbonyl in **15** is electrophilically enhanced, with a stretching frequency of ~1,780 cm^{-1}. The formation of the 4,5-bicyclo β-lactam fused ring of **15** occurs with much more facility than that leading to the 4,5-bicycloadduct **4** (Scheme 1), which we believed was reflective of different stereoelectronic arrangements achievable during the ring closures. In spite of their highly electrophilic carbonyl center, each of the N–S fused bicyclic β-lactams we prepared was very stable and highly resistant to hydrolysis over a wide pH range (pH 1–10). However, we found that they were susceptible to rapid cleavage of the N–S bonds by reductive processes. As an example, our efforts to perform palladium-catalyzed Stille couplings and carbonylation cross-couplings on vinyl iodide **15** led only to the ring-opened disulfide, which we attributed to PPh_3 in the reaction media. Disappointingly, upon in vitro testing of all these compounds, not one of the bicyclic β-lactam compounds showed antibacterial behavior against a selection of laboratory strains of common Gram-positive or Gram-negative bacteria. During the microbiological testing, however, we were dumbfounded by data indicating that all of the *monocyclic* *N*-methylthio β-lactams **3**, **8**, and **14** had really strong activity against *Staphylococcus aureus*, including methicillin-resistant *S. aureus* (or MRSA) (Fig. 6). Thus, we saw this as an opportunity to look further at other monocyclic *N*-thiolated β-lactams to evaluate antibacterial capabilities.

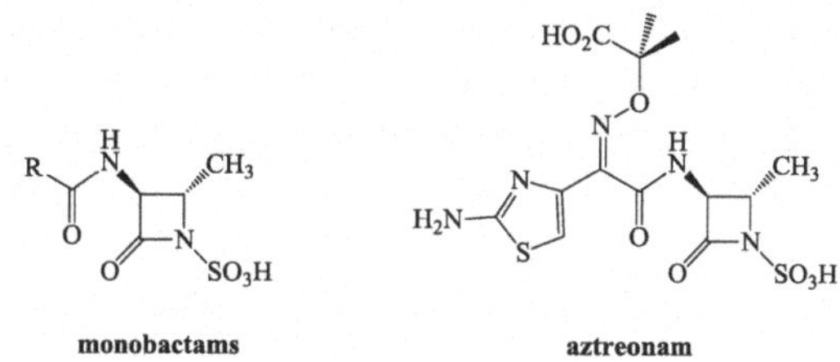

Fig. 7 Monobactams and aztreonam

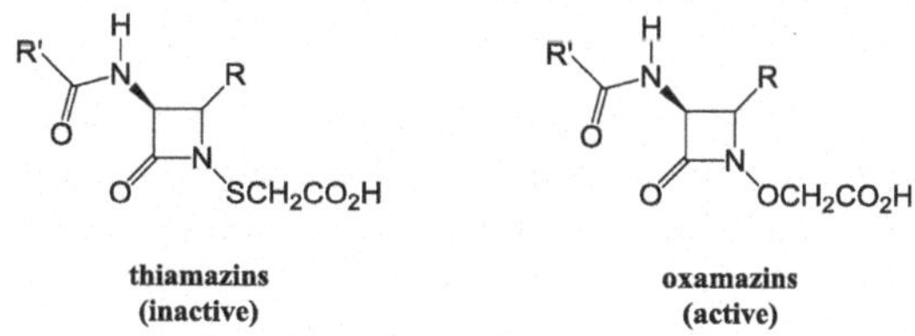

Fig. 8 Thiamazins versus the oxamazin antimicrobials

3 Discovery of the Biological Properties of *N*-Thiolated β-Lactams

In 1981, two research laboratories independently reported new monocyclic β-lactams isolated from *Pseudomonas* bacteria that bear an *N*-sulfonic acid moiety on the nitrogen center (Fig. 7) [11, 12]. They referred to these monocyclic β-lactams as "*monobactams*," of which the synthetic analogue Aztreonam became the first example developed commercially. Interestingly, these monobactams show complementary activity to that of the traditional bicyclic β-lactams (which are mostly effective against Gram-positive microbes), having broad activity against various aerobic Gram-negative bacilli such as *P. aeruginosa*. These discoveries showed convincingly for the first time that β-lactams do not have to have a conformationally restricted bicyclic ring to have antibacterial capabilities, marking the start of a new chapter in the history of β-lactam antibiotics.

Also in the 1980s, Marvin Miller's laboratory at University of Notre Dame reported an interesting study comparing the bioactivity of another type of antibacterially active monobactam, the oxamazins [13], with their inactive sulfur counterparts, the thiamazins (Fig. 8) [14–17].

The observed difference in microbiological activity of these two monocyclic β-lactams was attributed to the somewhat longer N–S bond of the thiamazins compared to the N–O bond of the oxamazins, which appears to be sufficient to prevent proper alignment of the thiamazin β-lactam ring within the active site cavity of the target enzyme. It is well documented that antibacterial activity in β-lactams requires an ionizable group such as a carboxylic acid be present on the ring system and within 3.6 Å of the β-lactam carbonyl to be able to bind to the transpeptidase protein. Based on Miller's conclusions and the observation that the *N*-thio-substituted thiamazins are totally devoid of antibacterial activity, we were perplexed over why our *N*-methylthio β-lactams possess potent anti-MRSA capabilities.

Similar to the thiamazins, *N*-methylthio β-lactams **3** and **8** are stable from pH 1 ± 10 and completely resistant to hydrolysis by penicillinases under conditions in which penicillin G is rapidly hydrolyzed. It was becoming more and more obvious to us that *N*-methylthio β-lactams do not behave in the same manner, or even on the same bacterial targets, as all previously known β-lactam drugs.

4 Structure–Activity Studies of *N*-Thiolated β-Lactams

We quickly began looking in more detail at a possible structure–activity relationship for these β-lactams, by examining systematically the effects of substituents on the azetidinone ring. Beginning with the lead *N*-methylthio β-lactam **3**, we began changing the lipophilic and steric nature of the C_3 and C_4 substituents. Going into this, we considered the possibility that the molecule may need to undergo bioactivation by an electrophilic species in the bacterial cell in order to exhibit its antibacterial effects. Our prior studies had shown that electrophilic reagents such as molecular iodine react with the alkynyl moiety of β-lactam **3** to induce a halocyclization to the *isopenem* **16** (Scheme 4).

In the reaction itself, we postulated the initial formation of an electrophilic complex **I**, with subsequent generation of a sulfonium intermediate **II**, which we deemed to be a powerful alkylating agent. We wondered whether this electrophile-promoted cyclization pathway could also take place to account for the antimicrobial activity of these β-lactams. Countering this possibility, we thought, was our earlier observation that C_4-alkenyl analogue **8** had strong bioactivity even though it could not undergo bromine-promoted cyclization to the *isopenem* **10** (Scheme 2). We reexamined this by preparing alkenyl lactams **17** and **18** in their geometrically pure E and Z forms (Fig. 9) and determined that both had strong antimicrobial bioactivity [14–17].

Of course, we could not totally exclude the possibility that the electrophile-induced cyclization depicted in Scheme 4 could occur in vitro, even though there was no tendency for this to take place using Br_2 in organic media. We excluded this likelihood, however, by synthesizing C_4-phenylethyl-substituted and C_4-phenyl-substituted β-lactams **19** and **20** and showed they both possessed similar bioactivity to lactam **3**. Thus, quite clearly π-unsaturation within the C_4 side chain was not a requirement for antibacterial activity of these compounds, and we therefore ruled out the involvement of an electrophile-promoted cyclization mechanism leading to the generation of a sulfonium alkylating species. Following this, we decided to examine additional lactams to see whether anti-MRSA activity could be enhanced by appropriate aryl ring substitution. Our first effort was to prepare C_4 aryl-substituted analogues **21–29** (Fig. 10). What we observed was that the effect of electron-donating or -withdrawing groups on the aryl ring diminished the activity to some degree, but all were active towards MRSA [18–21].

We also examined halogenated aryl derivatves **30–41** (Fig. 11). *ortho*-Substituted variants seemed to display the best activities, regardless of the halogen

Scheme 4 Electrophile-promoted activation and cyclization of **3** to bicyclolactam **16**

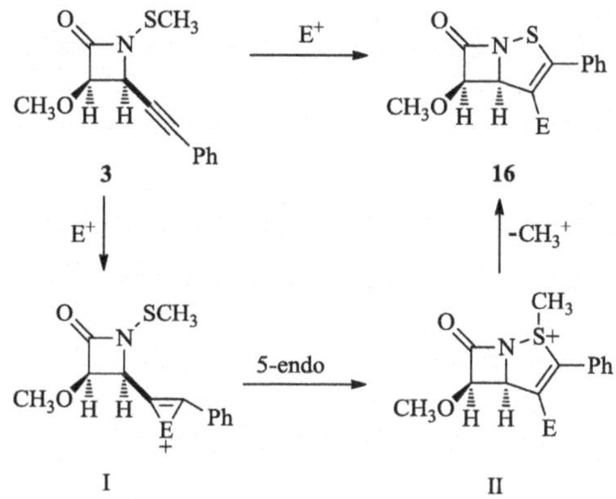

Fig. 9 *N*-methylthio β-lactams **17–20**

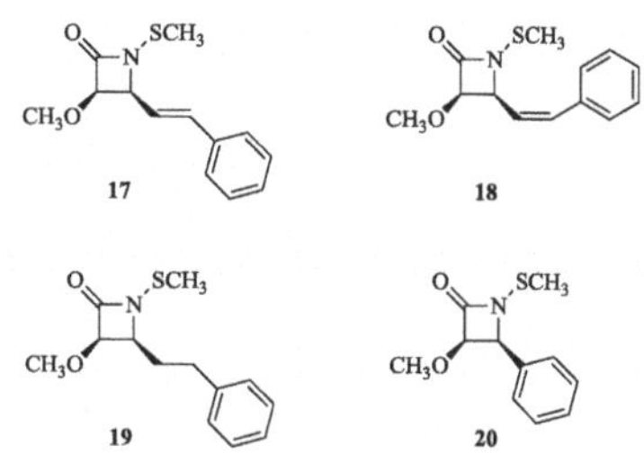

Fig. 10 *N*-methylthio β-lactam variants **21–29**

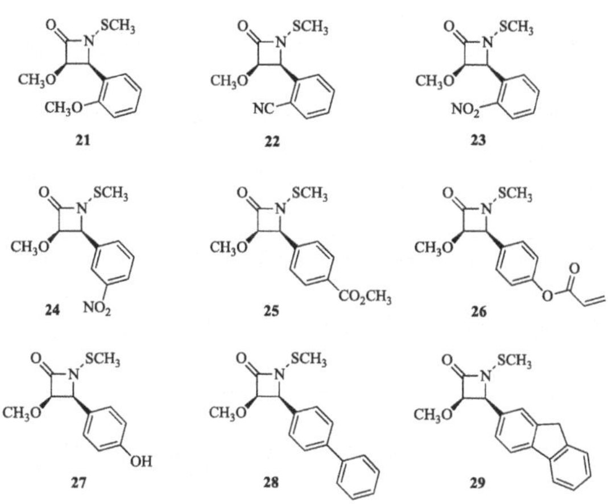

or the number of halogens resided on the aryl ring. Our most active analogue was the *ortho*-chlorophenyl lactam **30**.

We moved on next to evaluating the C_3 ring substituent, by replacing the methoxy group of β-lactam **3**. The phthalimidyl derivative **42** and the *trans*-disubstituted benzylthio lactam **43** were each significantly weaker in activity than **3**, while the C_3-ethyl β-lactam **44** had no activity at all (Fig. 12).

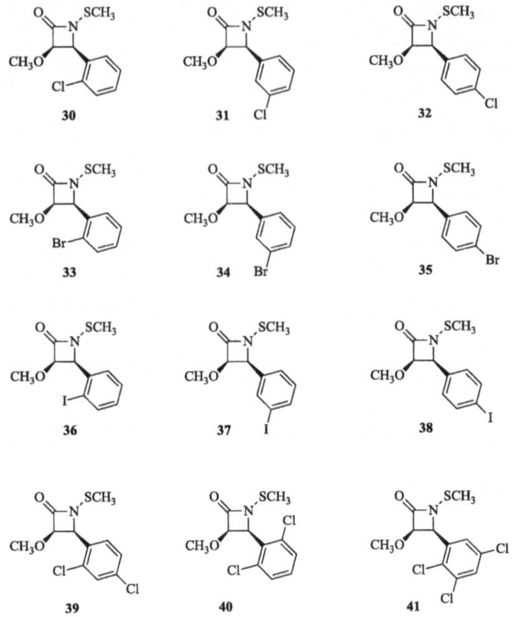

Fig. 11 C_4-halophenyl-substituted *N*-methylthio β-lactams **30–41**

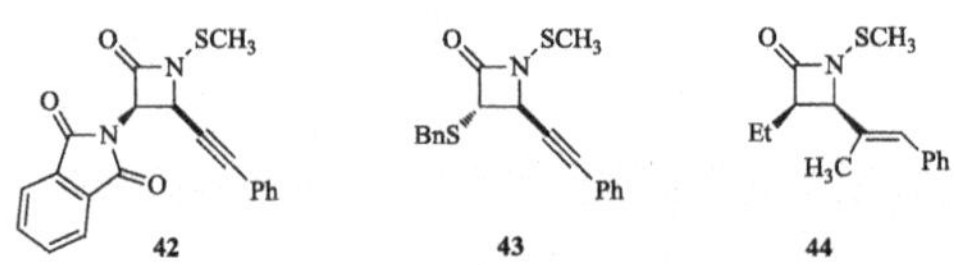

Fig. 12 C_3-heterosubstituted *N*-methylthio β-lactams **42–44**

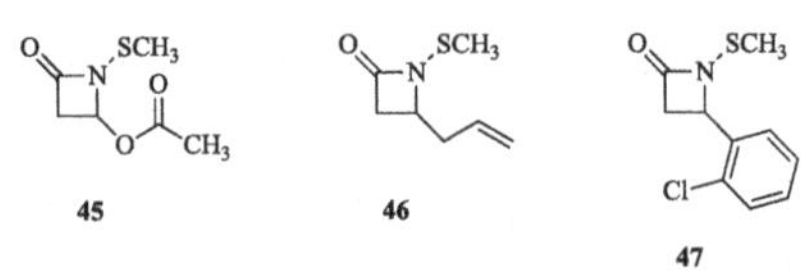

Fig. 13 C_3-unsubstituted *N*-methylthio β-lactams **45–47**

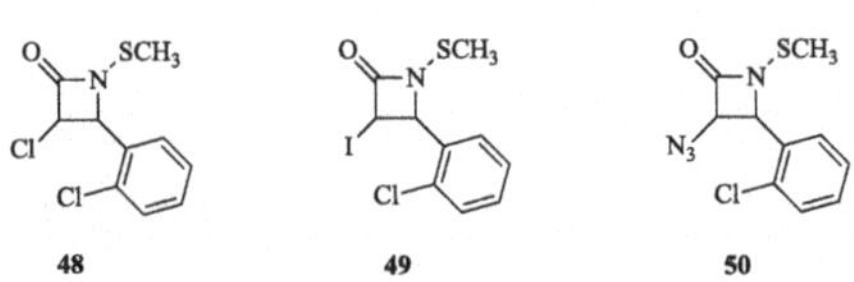

Fig. 14 C_3-heterosubstituted *N*-methylthio β-lactams **48–50**

We likewise saw that the C_3-unsubstituted β-lactams **45–47** had very weak anti-MRSA activity, indicating to us that perhaps a polar group positioned next to the carbonyl center may be required for bioactivity (Fig. 13).

We wanted to explore this more by comparing our most active compound, C_3-methoxy lactam **30**, to C_3-chloro, iodo, and azido β-lactam derivatives **48–50** (Fig. 14). Of these, chloro derivative **48** had the best in vitro activity, with zones of inhibition against MRSA being more than double that of penicillin and equivalent to our lead methoxy compound **30**. So far, we found for these that anti-MRSA activity follows the trend of **30** = **48** > **49** > **50** > **47** > **45** = **46**.

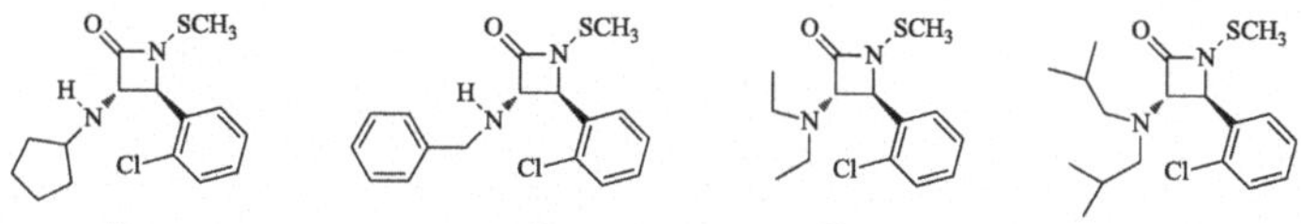

51 52 53 54

Fig. 15 C_3-amino-substituted *N*-methylthio β-lactams **51–54**

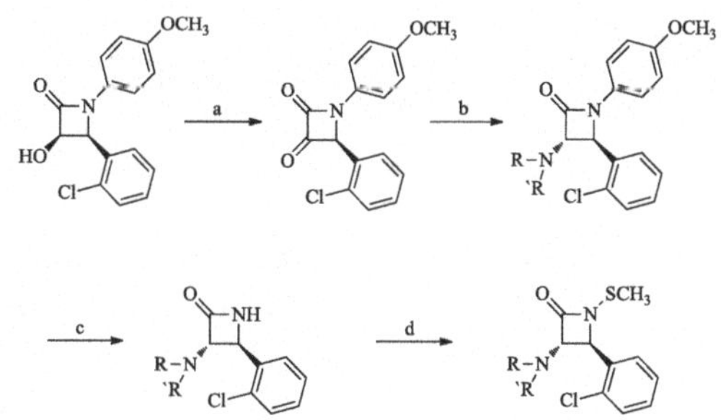

51-54

Conditions: a) P_2O_5, DMSO, rt; b) 1° or 2° amine, $NaBH(OAc)_3$, AcOH, $ClCH_2CH_2Cl$, rt; c) $Ce(NH_4)_2(NO_3)_6$, aq CH_3CN; d) BuLi, THF, -78°C; then $CH_3SO_2SSCH_3$

Scheme 5 Synthesis of C_3-amino-substituted β-lactams **51–54**

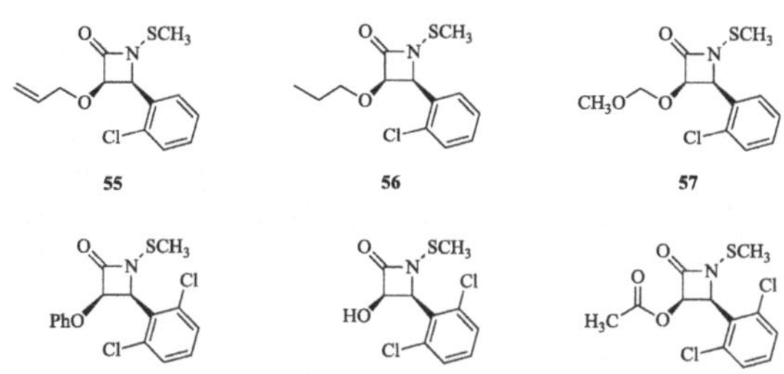

55 56 57

58 59 60

Fig. 16 C_3-oxygenated β-lactams **55–60**

Next, we set out to examine some C_3-amino-substituted derivatives **51–54**, shown in Fig. 15, which we synthesized stereospecifically by the route depicted in Scheme 5. Antimicrobial screening ascertained all of these C_3-amino β-lactams to be significantly *less* potent against MRSA compared to the C_3-methoxy compound **30** or to the halo or azido β-lactams above.

Our next series of analogues that we examined consisted of various C_3-oxygenated derivatives of lactam **30**, including C_3-alkoxy β-lactams **55–57**, C_3-phenoxy compound **58**, C_3-hydroxy lactam **59**, and C_3-acetoxy derivative **60** (Fig. 16). Every one of these compounds showed better anti-MRSA activity compared to penicillin, but still, none were as potent as C_3-methoxy β-lactam **30**.

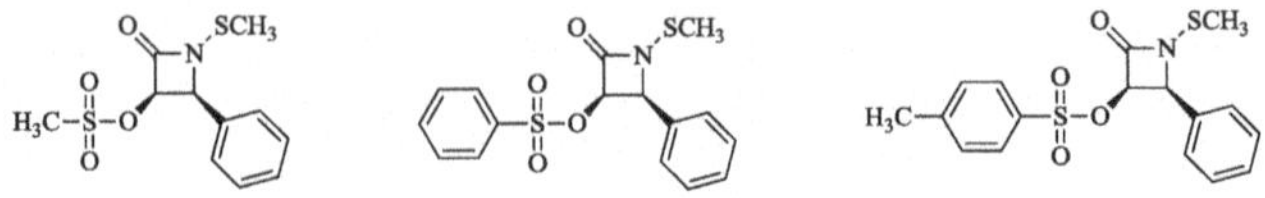

61 62 63

Fig. 17 C_3-sulfonated β-lactams **61–63**

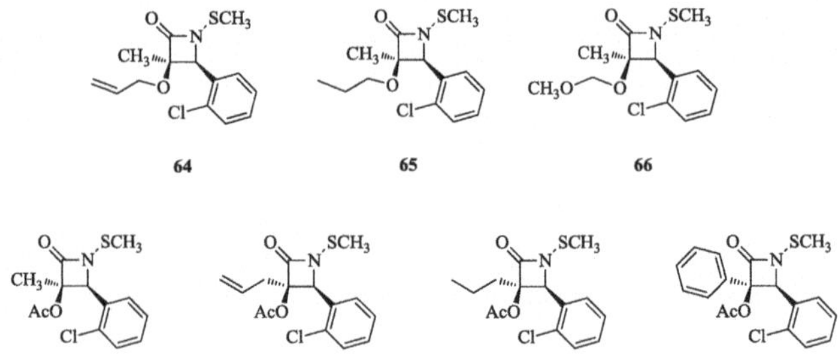

64 65 66

67 68 69 70

Fig. 18 C_3-disubstituted β-lactams **64–70**

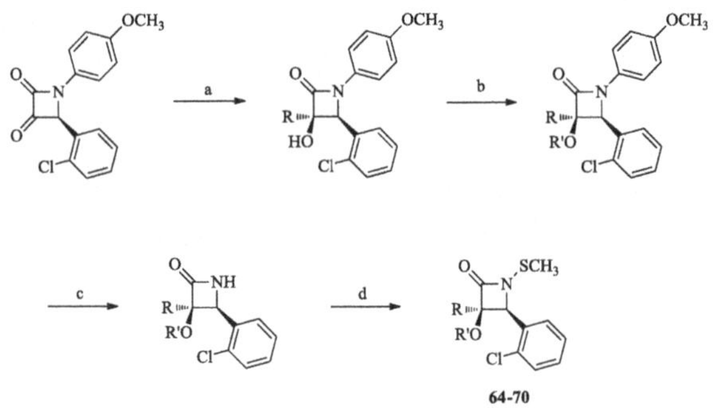

Conditions: a) RMgBr, THF, NH_4Cl, -78 °C; b) for **64-66**: NaH, R'-X, TBAI, CH_2Cl_2, rt; for **67-70**: NaH, AcCl, CH_2Cl_2, rt; c) $(NH_4)_2Ce(NO)_6$, MeCN-H_2O, 0 °C; then for **65** and **69** (from allyl precursor) H_2, Pd/C, EtOAc, rt; d) *N*-(methylthio)phthalimide, Et_3N, C_6H_6, 70 °C

Scheme 6 Synthesis of C_3-disubstituted β-lactams **64–70**

In addition to these ether derivatives, we prepared three C_3-sulfonate lactams **61–63** and found that anti-MRSA activity increases with lipophilicity of the sulfonate moiety: methyl < phenyl < *p*-tolyl (Fig. 17).

Several sterically congested C_3-disubstituted analogues **64–70** were also synthesized for antimicrobial testing (Fig. 18).

We prepared these compounds stereoselectively with the relative stereochemistry depicted, via Grignard addition to a keto β-lactam (Scheme 6) [22]. The resulting tertiary alcohol was then *O*-alkylated or *O*-acylated prior to being converted to the *N*-methythio derivative. Antibacterial activity among these derivatives steadily diminished as steric crowding on the ring increased. The allyl and propyl ethers

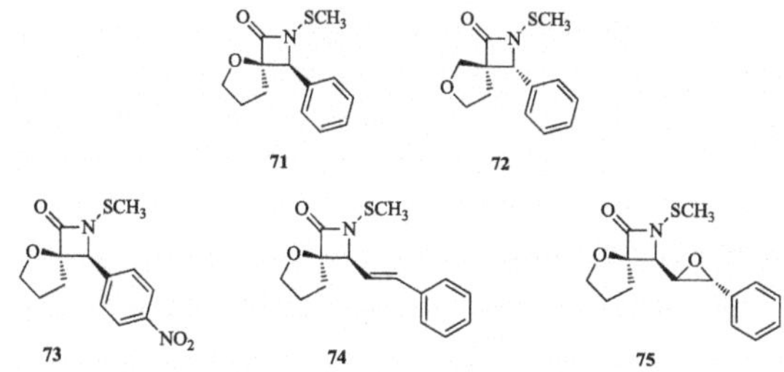

Fig. 19 C_3-spiro-substituted β-lactams **71–75**

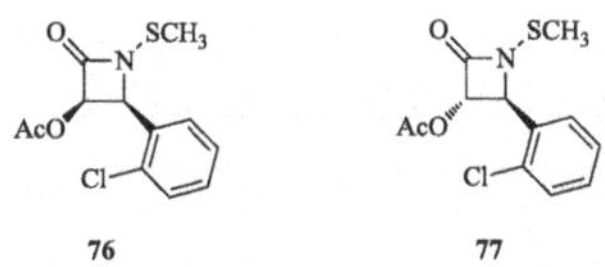

Fig. 20 Diastereomeric C_3-acetoxy-substituted β-lactams **76** and **77**

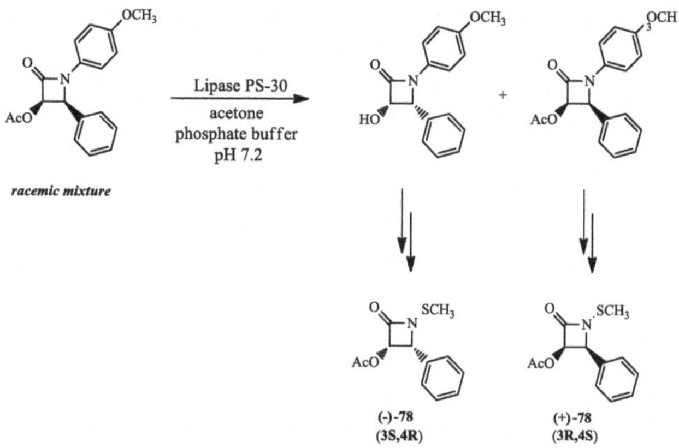

Fig. 21 Individual preparation of enantiomeric β-lactams **78**

64 and **65** were more active than methoxymethyl derivative **66**, while the methyl and propyl analogues **67** and **69** were more active than the allyl or phenyl-bearing β-lactams **68** and **70**.

The issue of steric congestion on the ring was further examined for spirocyclic β-lactams **71–75** (Fig. 19). Compound **71** was a bit more active than isomer **72**, and while all had reasonable anti-MRSA activity, they had 25% less activity than the open-ring analogue **65**.

In most cases, biologically active compounds are highly dependent on their stereochemistry, which can affect how they bind or interact with their targets. We therefore decided to examine whether ring stereochemistry could influence the activity of *N*-thiolated β-lactams. First, we evaluated this for the *cis* and *trans* stereoisomers of C_3-acetoxy *N*-methylthio β-lactams **76** and **77**. It turned out that the *trans* isomer **77** was about 10% more active towards MRSA than the *cis* β-lactam, based on in vitro testing on agar plates (Fig. 20).

Along with this, we explored the effect of *absolute* stereochemistry on bioactivity by independently synthesizing enantiomeric lactams (−)-**78** and (+)-**78** (Fig. 21). Both antipodes of lactam **78** produced identical zones of growth inhibition of MRSA on agar plates and minimum inhibitory concentration (MIC) values in broth microdilution experiments. Thus, we reached the conclusion that in vitro antibacterial activity of the lactams is not directly dependent on relative or absolute stereochemistry, and that this must be manifested by how the molecules act biochemically.

5 Biological Activity of *N*-Thiolated β-Lactams Depends on the Presence and Structure of the *N*-Organothio Substituent

Somewhat to our surprise, ab initio calculations on the *N*-methylthio β-lactams predicted that the sulfur and carbon centers in the *N*-methylthio group are electronically neutral. Also, the sulfur–carbon bond of the *N*-methylthio group prefers to be oriented orthogonally to the lactam ring, with the *cis* rotamer being about 1 kcal/mol more stable than the *trans*. The calculated barrier for rotamerization between the *cis* and *trans* forms is less than 8 kcal/mol, enabling the two rotamers to equilibrate rapidly in solution. We confirmed this by variable temperature NMR experiments. The computations also predicted that this stability difference in the *cis–trans* rotamers increases as the alkylthio moiety is enlarged.

These conformational studies agreed closely with the results reported by Marvin Miller's group for the thiamazins, but certainly failed to explain why the *N*-methylthio β-lactams are even antibacterially active given that the very closely analogous thiamazins are not. We further examined this by replacing the *N*-methylthio group of β-lactam **3** for other *N*-substituents (Fig. 22). Fortuitously, compounds **79** and **80** were precursors of **3**, and *N*-chloro and *N*-phenylseleno derivatives **81** and **82** were readily obtained by deprotonating **79** with *n*-butyllithium at −78°C, and capturing the *N*-lithiated species with Cl_2 or PhSeCl, respectively. None of these *N*-derivatives (except for *N*-methylthio β-lactam **3**) showed anti-MRSA activity, confirming the need for an *N*-organothio substituent for antibacterial activity[23].

With this clearly ascertained, we then investigated in much more detail how changing the *N*-methylthio group for other organothio residues could alter antibacterial activity (Fig. 23).

In vitro activity of *N*-alkylthio lactams **83**–**86** varied considerably, with the *N*-ethyl derivative **83** being more active than the *N*-methyl (**30**), but activity diminishing rapidly as the alkyl chain was elongated. *N*-octylthio β-lactam **86** had less than half of the anti-MRSA activity of **30**, as determined by both Kirby–Bauer agar diffusion studies and MIC determinations in broth.

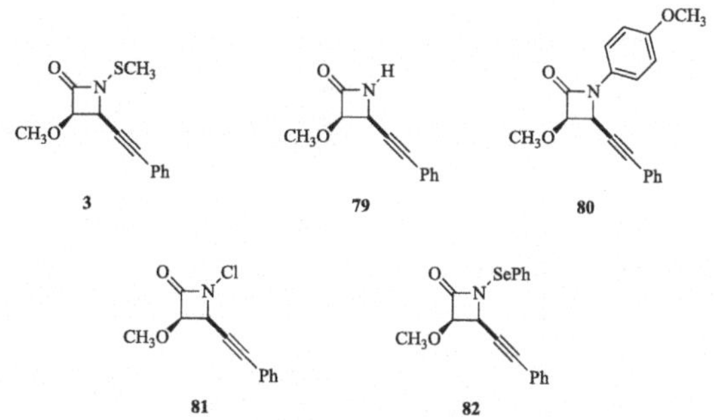

Fig. 22 Differentially *N*-derivatized β-lactams **3** and **79–82**

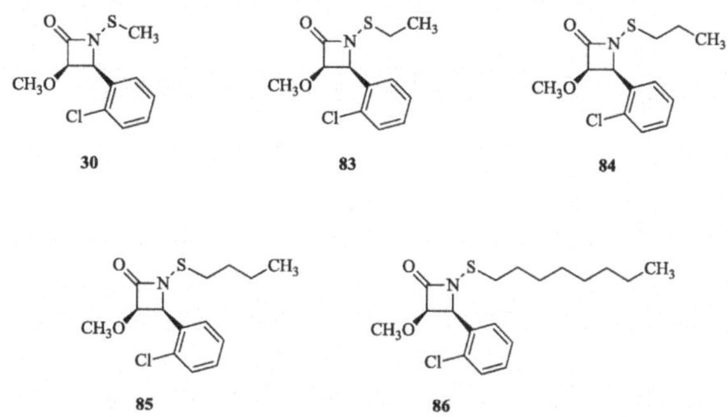

Fig. 23 Straight chain *N*-alkylthio β-lactams **30** and **83–86**

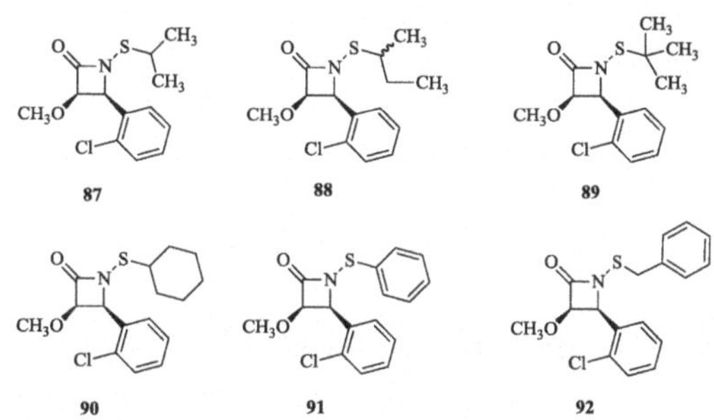

Fig. 24 Branched *N*-alkylthio β-lactams **87–92**

We also examined branching within the *N*-alkylthio chain for β-lactams **87–92** (Fig. 24).

N-isopropylthio and *N*-*sec*-butylthio analogues **87** and **88** displayed stronger bioactivity against MRSA compared to the *N*-methylthio or *N*-ethylthio β-lactams **30** and **83**, but *N*-cyclohexylthio, *N*-phenylthio, and *N*-benzylthio β-lactams **90–92** were far less effective than **87** or **88**. Of all the compounds tested in our structure--activity profiling, *sec*-butylthio β-lactam **88** had the strongest anti-MRSA bioactivity.

While it was clear now that *N*-thiolation of the lactam was required for bioactivity, we were not sure as to what oxidation state of the sulfur center would provide the best antibacterial activity. We examined this for *N*-sulfinyl β-lactam **93** and

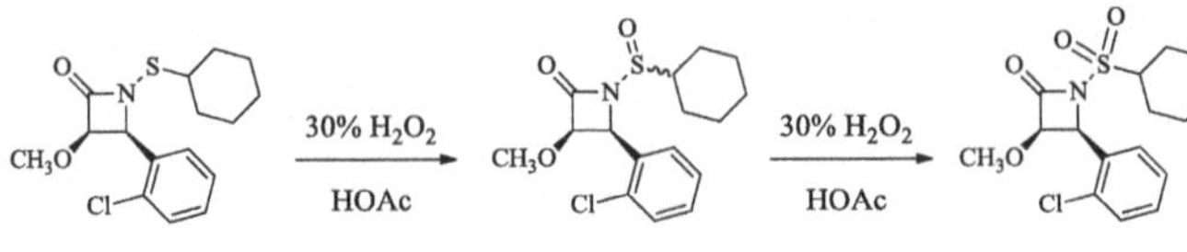

Scheme 7 Synthesis of *N*-sulfinyl and *N*-sulfonyl β-lactams **93** and **94**

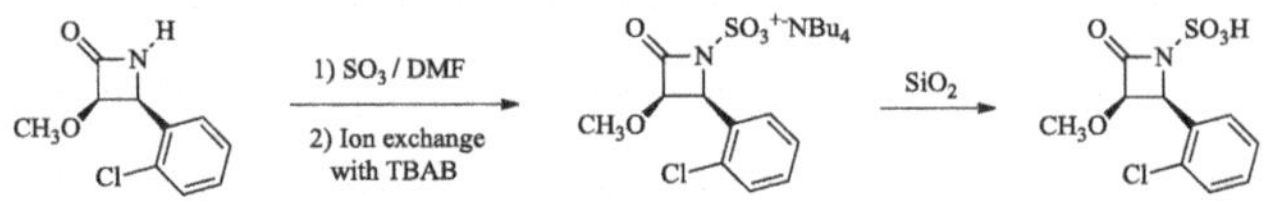

Scheme 8 Synthesis of *N*-sulfonated β-lactams **95** and **96**

N-sulfonyl analogue **94**, which we prepared by treating *N*-cyclohexylthio precursor **90** with 30% hydrogen peroxide in glacial acetic acid (Scheme 7). The *N*-sulfinyl product **93** was isolated as a 1:1 mixture of diastereomers, and adding excess hydrogen peroxide to this initial mixture gave the *N*-sulfonyl product **94**. The *N*-sulfinyl β-lactam **93** had similar in vitro activity to sulfenamide **90**, but the more highly oxidized *N*-sulfonyl analogue **94** had no activity at all.

Likewise, the *N*-sulfonate salt **95** and *N*-sulfonic acid **96** were each synthesized from the *N*-protio precursor and found to be inactive against MRSA (Scheme 8). We have not yet examined either **95** or **96** against Gram-negatives, which may show sensitivity given the structural relationship these compounds have with aztreonam and other *N*-sulfonic acid monobactams.

6 Studies on the Effect of Relative and Absolute Chirality of *N*-Thiolated β-Lactams on Anti-MRSA Activity

Earlier, we demonstrated that relative and absolute stereochemistry does not significantly affect bioactivity of *N*-methylthio β-lactams. Granted, we did find that *trans*-disubstituted analogues were slightly more active than the *cis* isomers, as for isomeric compounds **76** and **77**. We also found that the (+) and (−) enantiomers of 3-acetoxy β-lactam **78** had the same in vitro anti-MRSA bioactivity. With the eventual realization that the *N*-*sec*-butylthio β-lactam **88** provided better anti-MRSA bioactivity to that of the *N*-methylthio or *N*-ethylthio analogues, we recognized that this compound had originally been evaluated as a *mixture* of enantiomeric diastereomers **97**, **98**, *ent*-**97** and *ent*-**98** (Fig. 25). This did not tell us whether there could be differences in activity between the four stereoisomers, so we developed a stereospecific synthesis of each for individual testing.

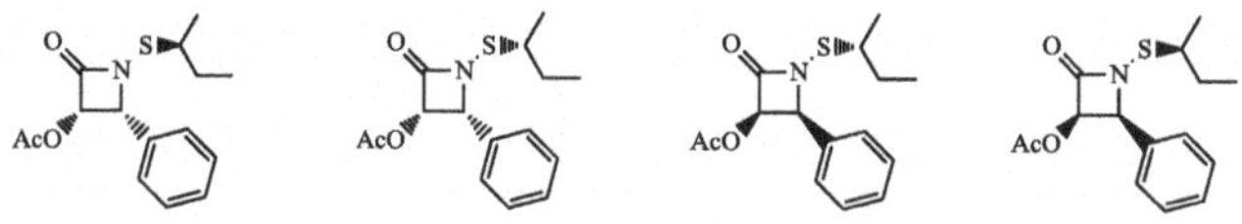

97 **98** **ent-97** **ent-98**

Fig. 25 Chiral *N*-*sec*-butylthio β-lactams **97**, **98**, *ent*-**97**, and *ent*-**98**

Curiously, diastereomeric β-lactams **97** and **98** exhibited the same in vitro anti-MRSA activities, as did their enantiomeric counterparts *ent*-**97** and *ent*-**98** to each other, but *ent*-**97** and *ent*-**98** were about 15% more active than **97** and **98**. Although these differences in bioactivity may seem rather insignificant, and probably are, they were at least discernible. This intrigued us because this indicates the absolute stereochemistry of the β-lactam ring itself only has an effect on bioactivity when the *N*-organothio group is also chiral (*N*-*sec*-butylthio), but not when the *N*-moiety is achiral (*N*-methylthio). We have no explanation for this, yet, but are certainly interested in exploring this further.

Overall, our preliminary screenings of all these analogues helped us find some general trends in the structural requirements of the C_3, C_4, and N substituents of *N*-alkylthio β-lactams. First, the nitrogen substituent must be either an organothio moiety in the sulfenyl or sulfinyl oxidation state, to induce its biological effect. Second, there appears to be a preference for balancing the lipophilic character of the C_3 and C_4 groups in order to attain optimal anti-MRSA activity, with C_3-alkoxy or acyloxy side chains and C_4 aryl moieties giving the best combinations so far. This structure–activity profile may relate to the compound's ability to penetrate through the bacterial cell membrane and cell wall to sites of action within the cytoplasm, rather than to any specific non-bonding interactions with a biological target such as an enzyme or cofactor. Polar substituents presumably make it more difficult for the molecule to enter the cytoplasm and, thus, reduce the antibacterial activity. On the other hand, compounds having too much lipophilicity and reduced water solubility are more likely to be trapped within the membrane or internal organelles, lowering the concentration in the cytoplasm.

7 Evaluating the Spectrum of Antibacterial Activity of *N*-Thiolated β-Lactams

During much of the early phase of our research on *N*-thiolated β-lactams, our focus was on structure–activity studies towards *Staphylococcus* bacteria. This led us to the general conclusion that ring substituents determine the in vitro anti-MRSA properties of the β-lactam, with the *N*-organothio moiety having the greatest role. To more fully assess the microbiological properties of these compounds, we carried out more detailed in vitro testing of selected compounds (**67–81**, **83**, **97**, and **98**)

Table 1 Susceptibility comparison of *N*-methylthio β-lactams **67–81**, **83**, and **97**, and **98**

		Zone of growth inhibition (mm)				
Bacteria genus (# species:strains)	Gram (±)	>30	21–30	15–20	<15	0
Bacteriodes sp. (1:1)	–			X	X	X
Bacillus spp. (7:7)	+		X	X	X	
Enterobacter sp. (1:1)	–					X
Enterococcus spp. (6:6)	+				X	X
Escherichia sp. (1:3)	–					X
Fusobacterium spp. (6:16)	–					X
Haemophilus sp. (1:3)	–				X	
Klebsiella sp. (1:1)	–					X
Lactococcus sp. (1:1)	+				X	X
Listeria sp. (1:2)	+				X	X
Micrococcus sp. (1:1)	+	X	X			
Mycobacterium sp. (1:1)	–					X
Neisserria sp. (1:2)	–			X	X	X
Peptostreptococcus sp. (1:1)	+					X
Porphyromonsa (1:2)	–					X
Proteus sp. (1:1)	–					X
Pseudomonas sp. (1:1)	–					X
Salmonella sp. (1:1)	–				X	X
Serratia sp. (1:1)	–					X
Sporobolomyces sp. (1:1)	–					X
Staphylococcus spp. (10:21)	+	X	X	X	X	X
Streptococcus spp. (2:2)	+			X	X	X
Vibrio sp. (1:2)	–				X	X
Susceptibilities		High		Moderate	Weak	None

against a wide range of Gram-positive and Gram-negative bacteria. These included 66 microbes from 23 genera, which are listed alphabetically in Table 1. For most of the bacteria examined, we included more than one species or strain, as indicated numerically in parentheses beside each genus. The table summarizes the relative susceptibilities each genera displays to these β-lactams, as measured by Kirby–Bauer agar diffusion assays. Microbes showing high susceptibility to the β-lactams are those having growth inhibition zones greater than 20 mm, denoted with an X for the zone size(s) seen, while others resulting in no growth inhibition zones are insensitive to the compounds.

Although in vitro activity varied considerably among the β-lactam derivatives and microbes, the data was fairly consistent in terms of which microbes were susceptible and which compounds were most active. In fact, most of the bacteria tested were not susceptible to the β-lactams, except for a notable few pathogenic species such as *Staphylococcus*, *Micrococcus*, and *Bacillus*. We also found more moderate activity against *Bacteroides*, *Streptococcus*, *Neisseria gonorrhoeae*, and weaker activity against *Salmonella typhimurium*, *Vibrio cholerae*, and *Mycobacterium tuberculosis*. None of the lactams were effective against *Enterobactor cloacae*, *Escherichia coli*, *Klebsiella pneumoniae*, *Listeria monocytogenes*, *Bacteroides fragalis*, *Serratia marcessens*, *Pseudomonas aeruginosa*, or *Proteus mirabilis*. This indicates that the microbes most affected by the β-lactams lie within a narrow window of just 9 genera among the 23 we evaluated and include both Gram-positive and Gram-negative bacteria. This finding alone is most unusual for any family or class of antibacterial agents and, certainly, not at all what one expects to find for a β-lactam antibiotic. The most highly susceptible microbes come from four distinct taxonomic orders as defined by their genetic, morphological, and metabolic traits, although few taxonomic relationships could be identified to account for this unusual spectrum of antimicrobial activity [18–21]. *N*-thiolated β-lactams affect mostly the "Bacilli," which consist of two orders: *Bacillales* and *Lactobacillales*, with *Bacillus*, *Staphylococcus*, and *Streptococcus* being the most highly susceptible members. However, we note that other "Bacilli," such as *Enterococcus*, *Lactococcus*, and *Listeria*, are rather insensitive to the lactams.

In screening specifically against staphylococci on agar test plates, we observed broad in vitro activity for most of the β-lactams against *S. aureus* (MSSA and MRSA), *S. epidermidis*, and *S. lugdunensis*; however, *S. lentus* and *S. simulans* were much less susceptible, with only some β-lactams showing decent activity. Since *Staphylococcus* and *Bacillus* are both in the *Bacillales* taxonomic order, we also tested the β-lactams against seven species of bacilli, including *Bacillus anthracis*, *B. globigii*, *B. thuringensis*, *B. megaterium*, *B. coagulans*, *B. subtilis*, and *B. cereus*. The most clinically important of these seven, *B. anthracis*, is a rod-shaped Gram-positive bacterium that causes anthrax infections. If inhaled, spores of *B. anthracis* rapidly migrate to lymph nodes of the lungs, where they begin to germinate and release toxins that disarm the immune response, causing bacteremia, toxemia, and often, death. The fact that these lactams show promising in vitro activity against *B. anthracis* is therefore rather significant. Structure–activity studies suggest that, in general, lipophilic acyloxy or alkoxy groups at C_3 of the β-lactam ring again provide the best activity against *Bacillus*. The propoxy and allyloxy side chains were also a little more active than the C_3-methoxy derivative. Spirocyclic groups at C_3 were, as for *S. aureus*, less active than the open chain variants. The most important determinant of anti-*Bacillus* activity, like we had seen earlier for MRSA, was the *N*-organothio moiety, with the *sec*-butylthio compound **88** once again having the best overall bioactivity. Compound **88** was significantly more active than *N*-methylthio β-lactam **30**, with zone sizes being more than double the diameter and equivalent to that of ciprofloxacin. The MIC values for lactam **4** against the avirulent Sterne and virulent Ames strains of *B. anthracis* were 0.5 μg/ml for each [24]. Anti-*Bacillus* activity requires the *N*-sulfenyl moiety, since the more highly oxidized *N*-sulfinyl, *N*-sulfonyl, and *N*-sulfonate derivatives were all ineffective. The similarity in the structure–activity patterns towards *B. anthracis* to that of MRSA suggests that the mode of action of these lactams in *Staphylococcus* and *Bacillus* (and probably other susceptible bacteria) is very likely identical.

8 Studies to Determine the Mode of Action of *N*-Thiolated β-Lactams as Antibacterials

The unique bacterial selectivities and structure–activity of these β-lactams suggest a highly unusual mode of antibacterial action. β-Lactam antibiotics are widely recognized for their cidal effects on bacteria, as a consequence of inhibiting the complete formation of the bacterial cell wall. For our *N*-thiolated β-lactams, we needed to determine whether they are bacteriostatic or bactericidal agents. We first examined bacterial cell growth by measuring the percentage of cell survival for methicillin-susceptible *S. aureus* (MSSA) and MRSA, respectively, in the absence or presence of β-lactam **3** over a 2 h time frame. In the absence of β-lactam **3**, both MSSA and MRSA grew logarithmically, but bacterial growth immediately ceased in the presence of β-lactam **3**. While bacterial growth was inhibited by β-lactam **3**, there was effectively no decrease in the number of viable cells at or above the MIC, which is indicative of a bacteriostatic effect.

This finding that the *N*-thiolated β-lactams are bacteriostatic compounds that act on only a narrow range of seemingly unrelated bacteria suggested to us that their mode of action and biological target are likely within the bacterial cell. In general terms, antibacterial agents that act on essential intracellular processes such as protein synthesis (chloramphenicol) or production of secondary metabolites (sulfa drugs) are often bacteriostatic agents, while those that act on the outer periphery of cells such as the bacterial membrane (polymyxins, bacitracin, cationic lipoproteins) or the cell wall (penicillin, vancomycin), or directly on DNA (metronidazole), are bactericidal. Although structure–activity profiling of bioactive compounds is not always all that informative in understanding the mechanism of action of a new antibiotic, particularly in the absence of clear trends, we did learn some valuable things from those prior studies. First, we determined that the *N*-organothio substituent is required for bioactivity, not the β-lactam ring or other ring substituents, and this enabled us to rule out various mechanisms such as the traditional one for β-lactams, the nucleophilic opening of the β-lactam ring. We were able to unambiguously, from the isolation of the *N*-protio lactam from *S. aureus* culture media treated with a *N*-thiolated β-lactam, and showing that this isolated substance engenders no bioactivity against the microbes. This finding told us that the β-lactam component is in effect just a carrier, not an essential unit, for the *N*-organothio side chain, which through some cellular event is picked up and sequestered by the bacterial cell. In fact, we followed this up by showing that *N*-alkylthio-2-oxazolidinones closely mirror these same trends, demonstrating that the four-membered β-lactam ring itself is not essential for antimicrobial activity [25]. This also coincides with our earlier conclusion that the other substituents on the β-lactam ring play a much less important function, with lipophilic ring substituents at C_3 and C_4 being preferable, while highly polar moieties work to decrease bioactivity. We believe that the side chain polarity influences the ability of the compound to passively cross the bacterial membrane to get to the target. This differs decidedly from classical water-soluble β-lactam antibiotics that act on β-lactam binding transpeptidases positioned on the outside of the cell membrane. For *N*-thiolated β-lactams, different levels of antibacterial activities that we find can be attributed to ring substituents that differ in their steric bulk or lipophilicity.

To examine the effects of these β-lactams on vital cellular processes in bacteria, we carried out radio-uptake experiments using β-lactam **3**. Typically, antibiotics that alter transcription or translation are bacteriostatic, suggesting the possibility that *N*-thiolated β-lactams could inhibit nucleic acid synthesis. We followed the uptake of radiolabeled ^{3}H-uridine into bacterial RNA, by culturing *S. aureus* with varying concentrations of β-lactam **3** versus rifampicin, a known RNA synthesis inhibitor. Indeed, we found that β-lactam **3** modulated ^{3}H-uridine incorporation to some extent, but certainly not sufficiently to identify transcription as the primary process targeted by the β-lactam. This was further delineated when we showed that β-lactam **3** does not change the rate of nucleotide incorporation into double-stranded DNA (DNA synthesis). Pulse-labeling studies with ^{3}H-thymidine allowed us to monitor DNA replication in living bacterial cells, in the presence versus the absence of β-lactam **3** and experimental controls. While the DNA synthesis inhibitor, ciprofloxacin, rapidly inhibited thymidine uptake, neither lactam **3** nor penicillin G did. To evaluate another possibility that β-lactam **3** could block protein transcription in bacteria, we measured the rate of uptake of radiolabeled ^{3}H-isoleucine in a culture of *S. aureus* treated with either β-lactam **3** or chloramphenicol (a translation inhibitor). Our results indicated that the lactam has a small, but delayed, response on protein synthesis in *S. aureus*, insufficient to be the primary cause of antimicrobial activity of these β-lactams. We then considered that the compounds may affect bacterial respiration by inhibiting cellular uptake and phosphorylation of glucose. Therefore, we examined the propensity for β-lactam **3** to cause metabolic starvation in *S. aureus* by measuring uptake of radiolabeled-glucose in the presence versus the absence of the β-lactam. We found that uptake was not inhibited and consequently were able to rule out metabolic respiration as a primary target of *N*-thiolated β-lactams. Finally, the last possibility we evaluated was the effect of β-lactam **3** on fatty acid biosynthesis in *S. aureus*. Here, we saw an immediate and powerful effect on the rate of uptake of radiolabeled ^{3}H-acetate from the media in *S. aureus* cells treated with β-lactam **3**. The blockage of acetate into the cell in the presence of the β-lactam, but not penicillin or DMSO, revealed decisively that the inhibition of fatty acid biosynthesis is a primary mode of action in *S. aureus*.

It is well documented that bacteria exposed to cell wall inhibitors (i.e., classical β-lactams) or to antibiotic agents that act on the cytoplasmic membrane (i.e., polymyxins) inflict severe morphological damage that can be revealed by scanning electron microscopy (SEM). To try to find such alterations in bacteria treated with our *N*-thiolated β-lactams, we inoculated cultures of *S. aureus* with β-lactam **3**, and after a period of exposure, we examined the cells visually by SEM. These samples were extracted from Kirby–Bauer agar diffusion plates from the outermost edges of the growth inhibition zones, where antibiotic concentration would be too low to completely stunt bacterial growth. *S. aureus* cells that had no exposure to any antibiotic or those surviving treatment with β-lactam **3** appeared morphologically identical, as grape-like clusters of uniformly formed, spherically shaped cells.

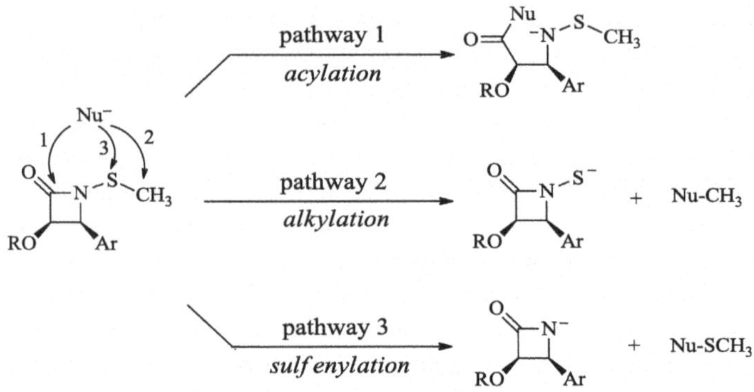

Scheme 9 Three alternative reaction pathways for *N*-thiolated β-lactams

However, *S. aureus* surviving treatment with penicillin G were severely damaged as a result of penicillin's effect on formation of the bacterial cell wall. The untreated and β-lactam **3**-treated *S. aureus* cells also retained the distinctive purple coloration upon Gram-staining that confirms the presence of a fully intact, resilient cell wall, while penicillin-treated *S. aureus* cells took on a light pink coloration revealing of a fragmented cell wall. These experiments thus provided us with additional evidence that *N*-thiolated β-lactams do not act on cell wall synthesis and are therefore different mechanistically from penicillin.

To ascertain the chemical pathway through which *N*-thiolated β-lactams execute their microbiological activity, we first considered three distinctly different alternatives by which a biological nucleophile could attack the molecule (Scheme 9). Thus, attack on the molecule by a biological nucleophile (Nu^-) directly on the β-lactam carbonyl results in acylation of the nucleophile with β-lactam ring opening (pathway 1), or on the carbon center of the *N*-alkylthio group to cause alkylation of the nucleophile (pathway 2), or at the sulfur center to transfer the sulfenyl chain onto the nucleophile by sulfenylation (pathway 3).

For all previously discovered β-lactams, acylation of the biological nucleophile (a serine transpeptidase) is the pathway that enables capture of the enzyme by formation of a hydrolytically stable acyl serine ester [26]. This acylation process requires that the β-lactam first be recognized by the murein transpeptidases, which requires the positioning of a carboxylic acid group on the β-lactam framework to form an important salt bridge within the enzyme active site. The lack of an acidic functionality on *N*-thiolated β-lactams suggests that transpeptidase binding, and subsequent ring opening via acylation, would be unlikely. Furthermore, we demonstrated that these compounds are highly resilient to acid or base hydrolysis, or to ring cleavage by β-lactamases. As a consequence, *N*-thiolated β-lactams do not act as β-lactamase inhibitors and possess no ability to slow down enzymatic hydrolysis of penicillin G. In terms of structure–activity trends, we also showed that steric crowding near the carbonyl center for our C_3-disubstituted analogues does not impede their bioactivity. Thus, this acylation pathway can be ruled out.

Fig. 26 Dethiolation of *N*-methylthio β-lactam **30** upon exposure to *S. aureus*

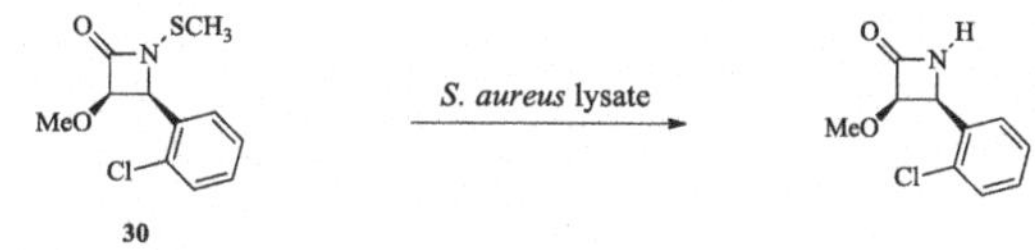

We then considered the second pathway that *N*-thiolated β-lactams could potentially act as biological alkylating agents. Again, we did not consider this to be a strong possibility, since our structure–activity studies would surely seem to rule this out. In effect, nucleophilic attack on the alkylthio carbon center would be predicted to be less favorable for bulky, branched alkylthio moieties on the lactam. Our data indicated just the opposite, such that the *N*-ethylthio, *N*-isopropylthio, and *N*-*sec*-butylthio lactams (**83**, **87**, and **88**, respectively) were increasingly *more bioactive* than the *N*-methylthio derivative **30**. We also substantiated this by examining the ability of β-lactam **3** to be an in vitro DNA alkylating agent. Antibiotics like leinamycin and mitomycin that are DNA alkylating agents are cidal. Nevertheless, we carried out experiments to determine whether *N*-thiolated β-lactam **3** could induce strand breakage of supercoiled DNA, by treating the plasmid pBR322 with varying concentrations of β-lactam **3** (5–100 μM) in sodium phosphate buffer for 24 h. Double-strand (fragmentation) and single-strand (linearization) breakage was then analyzed by agarose gel electrophoresis containing 1% ethidium bromide. It was immediately obvious that even at these artificially high concentrations, β-lactam **3** did not cause any fragmentation or relaxation of the plasmid superhelix by alkylation of the DNA. Since many well-known DNA alkylating agents require bioactivation by a cytoplasmic thiol, we also did experiments showing that β-lactam **3** had no effect on the plasmid even in the presence of glutathione, dithiothreitol, or β-mercaptoethanol. Our conclusion was that these β-lactams do not function antibacterially by alkylating bacterial components.

Finally, we consider the third pathway, sulfenylation of a biological nucleophile by transfer of the *N*-organothio moiety from the β-lactam. This pathway certainly seemed the most likely of the three, given that we had isolated the *N*-protio β-lactam from a culture of *S. aureus* treated with *N*-alkylthio β-lactam **30** (Fig. 26). Furthermore, Shah and Cama's early report that *N*-methylthio β-lactams are readily cleaved at the N–S bond by thiophiles, suggested that this could also be achieved in a bacterial cell by certain biological thiols [5]. The more ubiquitous organothiols in bacterial cells are those comprising the cytosolic redox buffer, which plays a vital role in preserving proteins in their proper oxidation state and sequestering cellular pollutants such as reactive oxygen species and alkylating agents. Most commonly, this thiol is glutathione, a tripeptide that is found in high concentrations in the cytoplasm of most bacteria and in human cells (Fig. 27).

To examine the sensitivity of β-lactams to glutathione, we ran Kirby–Bauer well diffusion experiments on β-lactam **30** by adding glutathione at varying concentrations directly to the diffusion well containing 20 μg of the lactam, prior to incubation with the bacteria. In this way, we were quickly able to verify that the

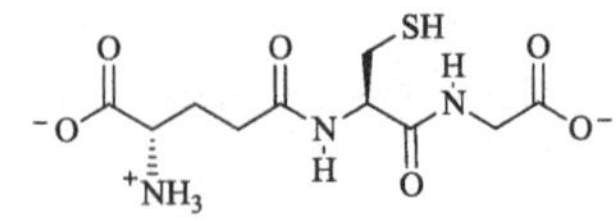

Fig. 27 Glutathione

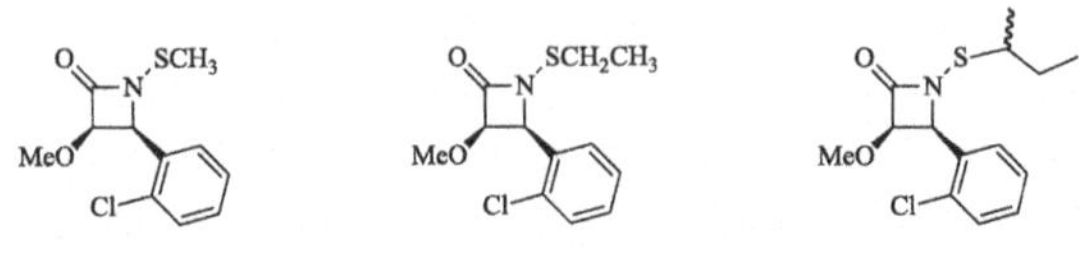

Fig. 28 *N*-methylthio β-lactam **30**, *N*-ethylthio β-lactam **83**, and *N*-*sec*-butylthio β-lactam **88**

in vitro bioactivity of these β-lactams against *S. aureus* can be attenuated by glutathione, and that this effect is enhanced as the amount of glutathione is increased. A second Kirby–Bauer well diffusion experiment was also carried out to assess whether glutathione could individually alter the anti-staphylococcal properties of *different N*-alkylthio β-lactam derivatives, in this case, *N*-methylthio β-lactam **30**, *N*-ethylthio β-lactam **83**, and *N*-*sec*-butylthio β-lactam **88** (Fig. 28). We prepared an agar plate such that 1 mg of glutathione was added (as an aqueous solution) into a well cut into the agar at the very center of the plate. We then added the three β-lactam derivatives (dissolved in DMSO) to separate wells on the outer perimeter of the plate, equidistant from the center well. The plate was then inoculated with *S. aureus* and incubated for 24 h to allow for bacterial growth. The shape and dimensions of the resulting growth inhibition zones around the lactam-containing wells were observed, and all appeared "indented" on the side facing the central well. This showed that the outward diffusion of the glutathione from the center well impeded the growth inhibitory effects of the β-lactams on the microbe, in all three cases. In follow-up to this, we then showed that these same three *N*-alkylthio β-lactams have different sensitivities to glutathione *under bacterial growth conditions*, by varying the amounts of glutathione added into the agar. As shown in Fig. 29, *N*-ethylthio lactam **83** and *N*-*sec*-butylthio lactam **88** both retained most of their antibacterial activity against *S. aureus* in the presence of 20 mg of added glutathione (left plate, Fig. 29a), while *N*-methylthio lactam **30** was completely inactivated and showed no growth inhibition zone. Of the three lactams, the *N*-methylthio derivative was most sensitive to glutathione, and its effect on bacterial growth was quickly neutralized. When we then increased the amount of glutathione mixed into the agar to 50 mg, both the *N*-methylthio and *N*-ethylthio β-lactams completely lost their activity, while the *N*-*sec*-butylthio compound was still strongly active (right plate, Fig. 29b). This enabled us to establish that the *N*-alkylthio β-lactams each differ in their stability to glutathione (and possibly other cellular thiols), with the order of stability being *N*-methylthio < *N*-ethylthio < *N*-*sec*-butylthio. This order of sensitivity parallels the in vitro activities of the three β-lactams towards *S. aureus*, and gave us the first insight that microbiological activities of the *N*-thiolated β-lactams can be highly dependent on differences in reactivity with thiophilic nucleophiles in or around the bacterial cell.

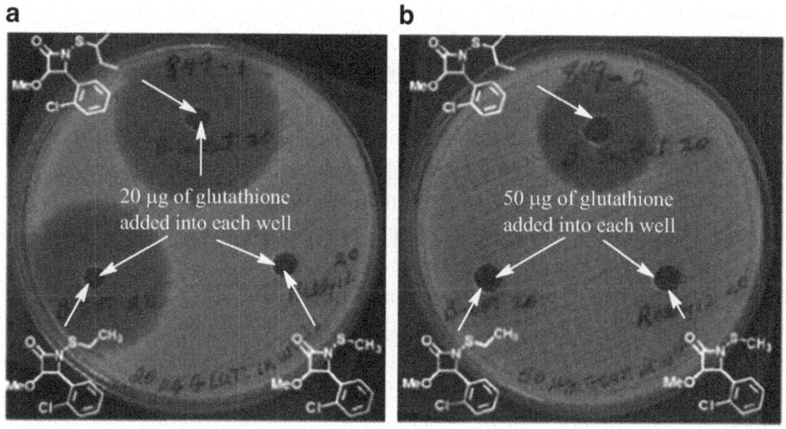

Fig. 29 Comparison of glutathione-sensitivities of *N*-methylthio β-lactam **30**, *N*-ethylthio β-lactam **83** and *N-sec*-butylthio β-lactam **88** in the presence of *S. aureus*

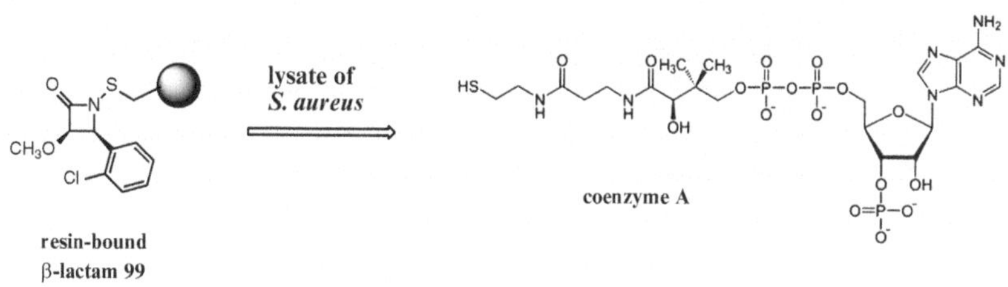

Fig. 30 Identification of coenzyme A from *S. aureus* lysate using a resin-bound *N*-thiolated β-lactam **99**

These agar plate experiments suggested to us that high levels of glutathione, such as that found in the cytoplasm of most microbial and human cells, protect against the *N*-thiolated β-lactams and could explain the differences in bioactivity these compounds have towards other microbes. For instance, we documented that *E. coli* and *P. aeruginosa* show no susceptibility to these lactams, and coincidentally, both have high cytosolic levels of glutathione. We made a plot of the cytosolic concentrations of glutathione in a selection of bacterial species, versus the antibacterial activities our lactams have against these microbes, and were surprised to find an inversely linear dependency. Although we know that high cytosolic concentrations of glutathione protect cells from the damaging cellular effects of the lactams, we still had little idea as to why the β-lactams can even possess antibacterial properties against microbes having low (or no) glutathione. What we still needed to do was to identify a cellular target in these bacteria that could be tied to their susceptibility to the lactams. To do this, we prepared a resin-bound β-lactam derivative **99** having a transferable *N*-alkylthio moiety that could be used to search for potential targets in the cytosol of *S. aureus* (Fig. 30) [27]. This resin-bound lactam was added directly into fresh lysate of cultured *S. aureus*, and the lysate was then extracted with ethyl acetate to pull out organic-soluble components. Among the various entities we found in the extract was the anticipated *N*-protio lactam that had been released from the resin polymer. The solid material we obtained from the lysate was treated with diisobutylaluminum hydride to cleave off the cellular component(s) that had been thiolated by the β-lactam **99**, and still bound to the polymer resin through a reducible disulfide linkage. HPLC analysis of the reduced product mixture revealed a cytosolic thiol that we quickly identified as coenzyme A (CoA). CoA is a low molecular weight thiol found predominantly in the cytoplasmic fluid of *S. aureus* and several other pathogenic bacteria, including *Bacillus* and *Micrococcus* species [28].

Serving as a cofactor in many enzymatic processes in both prokaryotic and eukaryotic cells, CoA also can function as an effective replacement for glutathione or other small organothiols as the cytosolic redox buffer that protects the cell from reactive electrophiles and oxidants. Bacteria generating the highest cytosolic concentration of CoA include *Staphylococcus*, *Bacillus,* and *Micrococcus*, which also have the lowest quantity of glutathione or small thiols. These are the microbes that are the most susceptible to the *N*-alkylthio β-lactams. We could not tell whether the sulfhydryl group of coenzyme A reacts directly with the *N*-thiolated β-lactams via sulfenylation. However, we found that at physiological pH in buffered media, CoA, glutathione, and cysteine do not cleave the *N*-organothio side chain off the lactams on their own. This was a bit surprising to us, given the earlier results we obtained in Kirby–Bauer testing that showed the attenuation effect of glutathione on the lactams. This has made us suspect that there may be another entity, perhaps a yet unidentified transferase protein that might assist in the transfer of the *N*-organothio group from the lactam onto the sulfhydryl of CoA. The determination that coenzyme A is a primary target of *N*-thiolated β-lactams in bacteria whose glutathione levels are low (or zero) has brought our focus to CoA and, specifically, CoA-dependent processes, as a potential target for the development of antibacterial drugs. Given the role of CoA as an essential cofactor in so many cellular processes, we also recognize the possibility that these "thiolating" β-lactams may inhibit or perturb multiple CoA-dependent pathways in bacteria.

9 Studies on the Effect of *N*-Thiolated β-Lactams on the Cytosolic Redox Buffer in *S. aureus*

All living cells have highly regulated thiol-based redox buffer systems designed to maintain proper stasis in the cell, as well as to intercept undesired chemicals that are detrimental to cellular components such as proteins, nucleic acids, and enzymatic cofactors. The redox buffer consists of an equilibrium mixture of a small sulfhydryl-containing molecule as the reducing agent and its disulfide counterpart as the oxidant. The concentration of free thiol in these cytosolic buffers is usually high, often millimolar. The elevated concentrations of CoA in *Staphylococcus* and *Bacillus* stem from the role CoA plays in the redox buffer of these particular microbes, in place of glutathione. The relative levels of "reduced CoA" (as the free thiol) versus "oxidized CoA" (CoA disulfide) are controlled by CoA disulfide reductase [29, 30] that contains a cysteine residue and a non-covalently bound flavin molecule [31]. The mechanism of biological reduction of CoA disulfide back to its reduced CoA thiol is through thiol-disulfide exchange with the active site cysteine, followed by flavin-mediated hydride transfer from NADPH to reduce the cysteine-CoA disulfide bond and regenerate the active site sulfhydryl. Coenzyme A disulfide reductase is substantially distinct in its mechanism and substrate (disulfide) selectivity compared to glutathione disulfide reductase (that regulates glutathione-based redox). The correlation between the antibacterial activity of the *N*-alkylthio β-lactams and the CoA-based redox in *S. aureus* (and other microbes) prompted us to study whether the lactams affect catalytic capabilities of CoA disulfide reductase. The reducibility of the N–S bond in these compounds provides the possibility that the active site cysteine of the reductase protein could become sulfenylated by the lactam. However, we determined that neither lactam **30** nor the CoA-alkyl disulfide formed by reaction of **30** with cytosolic CoA inhibited the reductase catalysis. Our belief is that sulfenylation of the active site cysteine of CoA disulfide reductase is most likely reversible due to NADPH-mediated reduction of the disulfide bond, thereby quickly regenerating the active sulfhydryl form of the enzyme and having no net effect on the CoA redox buffer.

10 Studies on the Effect of *N*-Thiolated β-Lactams on Fatty Acid Biosynthesis in *S. aureus*

Having ruled out that the CoA thiol-redox buffer or the CoA reductase enzyme controlling cell stasis is deleteriously altered by the *N*-alkylthio β-lactams, we then considered the possibility that the mixed CoA-alkyl disulfides formed by sulfenylation of CoA by the *N*-thiolated β-lactams could interact with CoA-dependent enzymes in bacterial fatty acid biosynthesis. The most obvious of these proteins, in our view, is β-ketoacyl-acyl carrier protein synthase III (FabH). Previous studies by our collaborator, Kevin Reynolds at Portland State University, showed that mixed CoA-alkyl disulfides (prepared synthetically) do in fact inhibit enzymatic activity of these proteins in a structure-dependent manner [32]. However, they had also found no evidence that these mixed disulfides had antibacterial capabilities when tested in vitro. This is likely due to the anionic charge on the CoA unit, which prevents the mixed alkyl-CoA disulfide from passing through the bacterial membrane to reach the enzyme in the cytosol. The role of the *N*-thiolated β-lactam, we postulate, is to pass through the cellular membrane as an uncharged, lipophilic molecule, and to then generate the anionically charged, active alkyl-CoA mixed disulfide in the cytoplasm by reaction with CoA sulfhydryl. Different *N*-alkylthio moieties on the β-lactam would lead to generation of different alkyl-CoA disulfides, which Reynolds has documented interact with their own

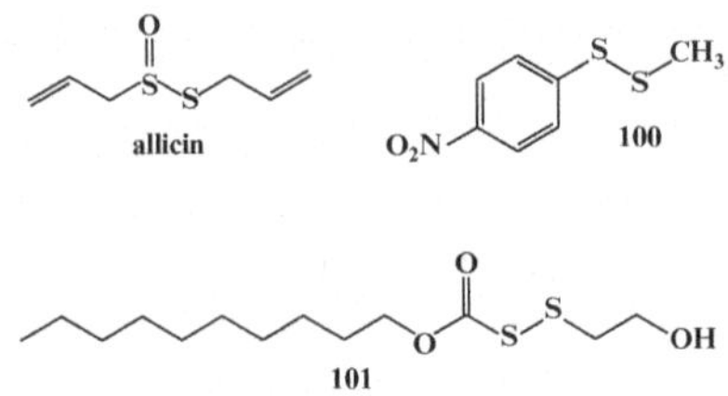

Fig. 31 Disulfide-based inhibitors of FabH

selectivities towards FabH proteins from different bacteria. We are interested in conducting further investigations into the effect of *N*-thiolated β-lactams on FabH and other related sulfhydryl enzymes.

11 Comparison of *N*-Thiolated β-Lactams to Other Known Fatty Acid Inhibitors

One can compare the properties of these *N*-thiolated β-lactams to those of the natural products, such as allicin, found in freshly crushed garlic cloves, which shows antifungal, antiviral, antiparisitic, and anticancer properties [33–43]. As for *N*-thiolated β-lactams, allicin inhibits fatty acid biosynthesis in bacteria, and shows the strongest bioactivity against microbes expressing low cytosolic concentrations of glutathione [44–46]. Both compounds partially inhibit protein and nucleic acid synthesis in bacteria [47]. Allicin is a specific inhibitor of acetyl-CoA synthetases through non-covalent reversible binding [48]. On the other hand, allicin reacts chemically with sulfhydryls of alcohol dehydrogenases, thioredoxin reductase, and RNA polymerase [49]. The similarities in the chemical reactivities and microbiological activities of allicin (and related compounds such as ajoene, thiolactomycin, and cerulenin) [50, 51] to those of *N*-thiolated β-lactams indicate that these compounds may all share common biochemical targets and mechanisms of action. Similarly, alkyl aryl disulfide **100** [52], acting as a generator of CoA-methyl disulfide in the cytosol, and disulfide **101** are FabH inhibitors displaying similarities in their mode of action to the *N*-thiolated β-lactams (Fig. 31).

12 Experiments to Determine Whether Extrogenous Fatty Acids in the Growth Media Attenuate the Antibacterial Properties of *N*-Thiolated β-Lactams

Recent studies have appeared in the literature indicating that exogenous fatty acids in growth media can override the inhibitory effects of antibacterial agents that block fatty acid biosynthesis in some bacteria. This finding of course concerned us, like many in the antibacterials area, given the interest in new antibiotics that act on fatty acid biosynthesis in pathogenic bacteria like MRSA, and their potential utility for clinical treatment of infections. Consequently, we set out to determine whether the antibacterial activity of *N*-thiolated β-lactams could be overcome by the addition of exogenous fatty acids during incubation. For this, we worked with Lindsey Shaw, a collaborator in the Biology Department at University of South Florida, to assess the MIC of the lactams for MRSA cultured in the presence of oleic acid or Tween 80. For these experiments, we chose to use a lethal clinical MRSA strain, USA-100, to ensure that the results in the laboratory had meaningful implications for clinical settings as well. The MRSA was cultured in growth media containing one of the fatty acids, Tween 80 or oleic acid, as well as in the absence of either, to compare the effects of the fatty acids on the antibacterial capabilities of the *N*-thiolated β-lactam. We also used bovine serum albumin, vancomycin (a cell wall synthesis inhibitor), and (a FAS inhibitor) as controls for these studies. Our data showed that the MIC (and thus the antibacterial activity) of vancomycin was unchanged by the supplementation with the fatty acids in the growth media, while the MIC for triclosan increased 100-fold and 500-fold in the presence of oleic acid and Tween 80, respectively, as we expected (based on the literature). In the case of *N*-*sec*-butylthio β-lactam **88**, the MIC showed a mere twofold increase in the presence of either fatty acid additive, relative to their absence in the media. We believe this shows definitively that FabH-mediated inhibition by *N*-thiolated β-lactams is only very weakly altered by exogenous fatty acids, and that the lactams may indeed act on a number of parallel cellular processes that affect growth or replication of bacteria. This matches our earlier findings and suppositions that *N*-thiolated β-lactams most likely have multiple modes of inhibitory effects that are CoA-dependent.

These observations reveal that *N*-thiolated β-lactams are unique not only among β-lactam antibacterials but also among agents that inhibit bacterial fatty acid biosynthesis. We contend that this uniqueness, coupled to their bacteriostatic properties, high selectivity for a few pathogenic bacteria including MRSA, and their ability to block more than one pathway make *N*-thiolated β-lactams good candidates for clinical development for the treatment of deadly bacterial infections.

13 Development of Nanoparticle-Bound *N*-Thiolated β-Lactams

Throughout much of these investigations, we were concerned about the limitations imposed by the low water solubility and chemical sensitivity *N*-thiolated β-lactams have with respect to their potential use as antibiotic agents. Thus far, we have had limited success in trying to enhance water solubility through the introduction of polar side chains at various locations in the molecule, given that in most instances these side chain alterations decreased antibacterial activity. About 5 years ago, we began experimenting with nanoparticles as a means to enhance both water solubility and protection towards chemical degradation of the *N*-thiolated β-lactams.

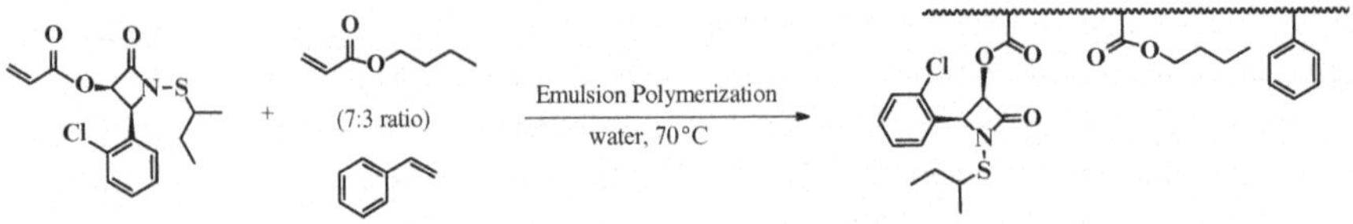

Scheme 10 Preparation of β-lactam-conjugated polyacrylate nanoparticles by emulsion polymerization

Fig. 32 Glycosylated bis-β-lactam **102**

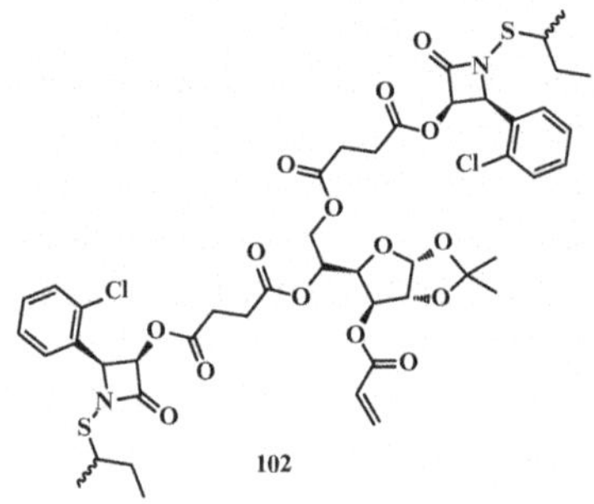

Using emulsion polymerization [52], acrylated monomers such as butyl acrylate and styrene can be polymerized by radical initiation in water, in the presence of an emulsified agent such as sodium dodecylsulfate. After a few hours with slight warming, the monomers are consumed and uniformly sized nanoparticles measuring around 50 nm are formed as a stable aqueous emulsion. What makes this procedure useful to us, in terms of our desired applications, is the β-lactam antibiotic can be introduced prior to polymerization, either as a free drug (to encapsulate it into the nanoparticle) or as an acrylated derivative (to attach it chemically within the matrix of the nanoparticle, as shown in Scheme 10). Microbiological testing indicated that the lactam-bound nanoparticles have strong antibacterial activity.

Glycosylated variants of these antibiotic-containing nanoparticles were also prepared and shown to have in vitro antibacterial activity as well [53]. For this, the acrylate monomer carrying the β-lactam is attached via a protected glycoside such as structure **102**, and this entire unit serves as a co-monomer for polymerization (Fig. 32). Dynamic light scattering indicated that the nanoparticles have an average diameter of 45 nm and are uniform in their size and shape. The bis-β-lactam acrylate **102** and the nanoparticle derived from it via emulsion polymerization have promising antibacterial activity against both MRSA and *B. anthracis*.

14 *N*-Thiolated β-Lactams as Antifungal Agents

Prior to our work, it was rare to find a β-lactam antibiotic that could inhibit the proliferation of both bacteria and fungi, given the bacteria-specific target that the classical β-lactams act upon. The mode of action of those β-lactam antibiotics ensured that the concentration needed to cause inhibitory effects in a fungal cell is in far excess to that required for bacteria. Certainly, the mode of action of the antifungal properties must be distinct to that in bacteria, given that fungal cells do not possess the transpeptidase enzymes that crosslink the bacterial cell wall. It is not coincidental that the β-lactam natural products are fungal metabolites made to protect fungi against bacterial invaders, and thus are specifically designed to act on bacterial enzymes, not fungal. Our results with bacteria, however, demonstrating that *N*-thiolated β-lactams act on cellular targets and process within the cytoplasm, not the cell wall, directed us to look for similar inhibitory effects on fungi [54]. The other significant finding is that *N*-thiolated β-lactams only function against microbes displaying low glutathione:coenzyme A levels in the cytoplasm. Among fungi, *Candida* happens to express high cytosolic concentrations of free thiol, including coenzyme A. We therefore carried out in vitro screenings of some of our *N*-thiolated β-lactams against *Candida*, and saw strong antifungal activity. We tested a dozen β-lactam analogues against seven *Candida* species, including *C. albicans*, *C. glabrata*, *C. tropicalis*, *C. parapsilosis*, *C. lusitaniae*, *C. kefyr*, *C. utilis*, and the resilient *C. krusei*. We saw some structural trends that closely paralleled those for the antibacterial activities. For instance, replacing the C_3 methoxy for an acetoxy or phenoxy, or more polar side chains, reduced bioactivity. For C_4 mono-halogenated aryl derivatives, those having fluorine attached to the aryl moiety had the weakest antifungal activities, but otherwise, all of the halogenated analogues were about equally active. It seemed preferable that the halogen be ortho or meta-substituted. Multiple halogens on the aryl ring increased activity as well. The most significant dependency in structure, however, at least for anti-*Candida* activity, is that of the *N*-organothio substituent. For antibacterial activity against MRSA, the *N*-*sec*-butylthio moiety gave the best results, but for *Candida*, there is a clear preference for the *N*-methylthio group. The most potent antifungal lactam in our collection turned out to *N*-methylthio β-lactam **17**. The MIC of compound **30** against *C. albicans* is 8 μg/ml, similar to that of the clinical antifungal drug clotrimazole. MICs of **30** against *Aspergillis niger* is 16 and 32 μg/ml for *Saccharomyces cervisiae*. As expected, for a fungistatic agent, the MIC value was increased as incubation time was increased from 24 to 48 h. Relative and absolute stereochemistry of the β-lactam was of no consequence, and the enantiomeric β-lactams (−)-**78** and (+)-**78** produced similar growth inhibition zone sizes to each other, against all of the *Candida* tested. This was the same outcome we observed for *S. aureus*.

Transmission microscopic images taken of *C. albicans* cells treated with lactam **30** showed appreciable deformities in the mitochondrial and intracellular membranes. The blockage of cell division by the β-lactam meant that the fungal cell could not maturate or replicate. Untreated cells did not display these effects. Trypan blue staining indicated that the fungal cells remained alive after treatment with lactam **30**, showing the fungistatic property of the β-lactams (identical to that of bacteria). Extraction of treated cells with an organic solvent once again allowed us to identify the *N*-protio β-lactam from dethiolation of *N*-methylthio β-lactam **30**.

This confirmed that the *N*-organothio substituent, and not the β-lactam ring, is required for antifungal activity.

15 *N*-Thiolated β-Lactams Also Display Anticancer Properties

Along with the investigations we did with bacteria and fungi, our laboratory collaborated with Q. Ping Dou's group (originally at the H. Lee Moffitt Cancer Research Center at University of South Florida, then at the Karmonos Cancer Research Center at Wayne State University) to study the effects of the β-lactams on human cancer cells. At first, the intention to probe for anticancer properties made little sense, since it is well known that β-lactams are highly selective for bacteria over human cells and should not be able to inhibit cancer cells. The only thing to suggest that the compounds might exhibit any anticancer activity at all was that they clearly operate through an unusual mode of action in microbial cells, certainly distinct to all previously known β-lactam compounds, and are highly lipophilic and susceptible to the effects of cellular thiols such as glutathione and coenzyme A that are also present in human cell lines. We published several papers from these investigations [55–58]. The first one published in 2002 reported that *N*-methylthio β-lactams induce apoptosis in human breast, prostate, leukemia, and head-and-neck cancer cell lines. The most potent β-lactam we found, compound **30**, induced DNA damage and the inhibition of DNA replication in Jurkat T cells. This caused the activation of p38 mitogen-activated protein kinase, as well as S-phase arrest leading to cellular apoptosis. We determined that p38 protein plays a key role in apoptotic induction, due to DNA damage, and induced caspase activation. In addition to caspase-8 activation, we noted cleavage of the Bcl-2 family protein Bid and release of cytochrome C from the mitochondria prior to caspase activation. These events suggest commitment by the cancer cell to undergo apoptosis in response to the damage caused by the uptake of the *N*-methylthio β-lactam **30**. Along with this, caspase-9 and caspase-3 became activated, and cleavage of PARP occurred. We saw a notable decrease in membrane permeability, which occurs late in cell apoptosis. All of these events were time dependent and β-lactam concentration dependent, as expected for caspase-induced apoptosis [33–37]. Our experiments, however, were not able to determine how duplex DNA is cleaved in the presence of the *N*-methylthio β-lactam. All we could ascertain was that DNA cleavage occurred in a time-dependent and concentration-dependent manner, prior to stagnation in the S-phase cell cycle. Activity was favored for the *N*-methylthio analogue over longer *N*-alkylthio variants. The *N*-protio β-lactam (lacking the *N*-methylthio moiety) did not induce the same effects, such as the induction of caspase-3 or PARP cleavage in the leukemia Jurkat T cells. Thus, as expected, the *N*-organothio group on the β-lactam is essential for the cancer cell apoptosis.

In a follow-up study, we focused more on whether the *N*-methylthio β-lactams could distinguish between cancer and normal cells in causing apoptosis. The β-lactam **30** selectively induces apoptosis in human leukemic Jurkat T cells and simian virus 40-transformed cells, but not in non-transformed, immortalized human natural killer (NK) cells or in parental normal fibroblast cells. All of the β-lactams having an *N*-methylthio side chain inhibited colony formation of human prostate cancer cells. *N*-methylthio β-lactam **24** having a nitrophenyl moiety at C_4 showed the strongest anticancer activity. Most recently, we examined anticancer activity for β-lactams having branched side chains off the C_3 center of the lactam, first for apoptosis induction and then for in vivo inhibition of breast tumor progression in mice. We determined that the same effects observed for the β-lactams in vitro also take place in vivo, in reducing the rate of growth of the xenograph and in inducing DNA cleavage and apoptosis. The mice treated with the β-lactam displayed a 50% decrease in tumor size with no indications of toxicity. This reduction of the tumor growth is directly due to apoptosis induced by the β-lactam. Although the details of the mechanism for this activity is not clear, the presumption is that the high glutathione levels maintained in normal mammalian cells may provide protection to apoptotic effects of the β-lactams that are otherwise found to occur, for unknown reasons, in transformed or tumor cells. The concentration of β-lactam required to see these anticancer effects in vitro and in vivo is 100–200 times that of the bacterial MIC of our most potent β-lactam analogues. At these very high lactam concentrations, and even several times higher, we found no cellular toxicity in normal human fibroblasts, indicating that the selectivities of the *N*-methylthio β-lactam for bacteria or fungi over human cells, and between human cancer versus healthy human cells, are each significant. Together, these studies indicate that *N*-thiolated β-lactams may have utility in cancer control or prevention and much more work towards assessing the value of these compounds for cancer treatment should be conducted.

16 Conclusions

Our investigations of *N*-thiolated β-lactams have led to some interesting findings about the chemistry and biological properties of these unique molecules. The structure-bioactivity and genera selectivity the compounds display are unprecedented and reveal a great deal about their potential development as clinical antibiotics. Through synthetic and computational modeling experiments, coupled to investigations into the cellular effects and mechanism of action of the lactams, we have learned about not only the compounds themselves but also the microbes they act upon. The bacteriostatic effects on select pathogenic microbes, unique SAR patterns, effects on cellular targets or processes, are distinctly different from how other β-lactam antibiotics behave. While we do not yet know all the intricacies of how these *N*-thiolated β-lactams execute their microbiological properties, we are able to see now that they act upon coenzyme A and lipid biosynthesis.

In addition to their antibacterial activity, however, they possess interesting bioactivities against eukaryotic cells, particularly yeast and cancer cells. The lactams show no obvious cytotoxicity or harmful effects in mice. It is of interest for us to consider at this point the direction that future research or the potential commercial development of *N*-thiolated β-lactams may take, particularly with regards to multi-drug resistant bacteria and cellular mechanisms for overcoming the most powerful, broad-spectrum antibiotics still in favor clinically today. Perhaps the continuing development of *N*-thiolated β-lactams, with their bacteriostatic properties and selective activity against some of the most deadly pathogenic bacteria such as MRSA and *B. anthracis*, and their antifungal and anticancer properties, may provide utility in the treatment of human diseases.

References

1. Morin RB, Gorman M (eds) (1982) Chemistry and biology of β-lactam antibiotics, vols 1–3. Academic, New York
2. Staudinger H (1907) Justus Liebigs Ann Chem 356:51
3. Fleming A (1929) Br J Exp Pathol 10:226
4. Crowfoot D, Bunn CW, Rogers-Low BW, Turner-Jones A (1949) In: Clarke HT, Johnson JR, Robinson R (eds) The chemistry of penicillin. Princeton University Press, Princeton, p 310
5. Shah NV, Cama LD (1987) Heterocycles 25:221
6. Turos E, Konaklieva MI, Ren R, Shi H, Gonzalez J, Dickey S, Lim DV (2000) Tetrahedron 56:5571
7. Ren X-F, Konaklieva MI, Shi H, Dickey S, Lim DV, Gonzalez J, Turos E (1998) J Org Chem 63:8898
8. Ren X-F, Konaklieva MI, Turos E (1995) J Org Chem 60:4980
9. Ren XF, Turos E, Lake CH, Churchill MR (1995) J Org Chem 60:6468
10. Ren XF, Turos E (1993) Tetrahedron Lett 34:1575
11. Imada A, Kitano K, Kintaka K, Muroi M, Asai M (1981) Nature 289:590
12. Breuer H, Cimarusti CM, Denzel T, Koster WH, Slusarchyk WA, Treuner UD (1981) J Antimicrob Chemother 8:21
13. Woulfe SR, Miller MJ (1984) Tetrahedron Lett 25:3293
14. Woulfe SR, Iwagami H, Miller MJ (1985) Tetrahedron Lett 26:3891
15. Woulfe SR, Miller MJ (1985) J Med Chem 28:1447
16. Woulfe SR, Miller MJ (1986) J Org Chem 51:3133
17. Boyd DB, Eigenbrot C, Indelicato JM, Miller MJ, Pasini CE, Woulfe SR (1987) J Med Chem 30:528
18. Turos E, Coates C, Shim J-Y, Wang Y, Roettgers JM, Long TE, Reddy GSK, Ortiz A, Culbreath M, Dickey S, Lim DV, Alonso E, Gonzalez J (2005) Bioorg Med Chem 13:6289
19. Long TE, Turos E, Konaklieva MI, Blum AL, Amry A, Baker EA, Suwandi LS, McCain MD, Rahman MF, Dickey S, Lim DV (2003) Bioorg Med Chem 11:1859
20. Coates C, Long TE, Turos E, Dickey S, Lim D (2003) Bioorg Med Chem 11:193
21. Turos E, Long TE, Konaklieva MI, Coates C, Shim J-Y, Dickey S, Lim DV, Cannons A (2002) Bioorg Med Chem Lett 12:2229
22. Kant J, Schwartz WS, Fairchild C, Qi G, Huang S, Long BH, Kadow JF, Langley DR, Farina V, Vyas D (1996) Tetrahedron Lett 37:6495
23. Heldreth B, Long TE, Jang S, Reddy SK, Turos E, Dickey S, Lim DV (2006) Bioorg Med Chem 14:3775
24. Turos E, Long TE, Heldreth B, Leslie JM, Guntireddygari SKR, Wang Y, Coates C, Konaklieva M, Dickey S, Lim DV, Alonso E, Gonzalez J (2006) Bioorg Med Chem Lett 16:2084

25. Mishra RK, Revell KD, Coates CM, Turos E, Dickey S, Lim DV (2006) Bioorg Med Chem Lett 16:2081
26. Ghuysen JM (1991) Annu Rev Microbiol 45:37
27. Revell KD, Heldreth B, Long TE, Jang S, Turos E (2007) Bioorg Med Chem 15:2453
28. Orford CD, Perry D, Adlard MW (1989) J Chromatogr 481:245
29. del Cardayre SB, Stock KP, Newton GL, Fahey RC, Davies JE (1998) J Biol Chem 273:5744
30. del Cardayre SB, Davies JE (1998) J Biol Chem 273:5752
31. Argyrou A, Blanchard JS (2004) Prog Nucleic Acid Res Mol Biol 78:89
32. Turos E, Revell KD, Ramaraju P, Gergeres D, Greenhalgh K, Young A, Dickey S, Lim D, Sathyanarayan N, Alhamadsheh MM, Reynolds K (2008) Bioorg Med Chem 16:6501
33. Koch HP, Lawson LD (1996) Garlic: the science and therapeutic application of *Allium sativum* L, vol 1. Williams and Wilkins, Baltimore, p 233
34. Ankri S, Mirelman D (1999) Microbes Infect 2:125
35. Block E (1985) Sci Am 252:94
36. Stall A, Seebeck E (1951) Adv Enzymol 11:377
37. Cavallito C, Bailey JH (1944) J Am Chem Soc 66:1944
38. Davis LE, Shen J, Royer RE (1994) Planta Med 60:546
39. Hughes BG, Lawson LD (1991) Phytother Res 5:154
40. Mirelman D, Monheit D, Varon SJ (1987) Infect Dis 156:243
41. Tsai Y, Cole LL, Dav Lockwood SJ, Simmons V, Wild GC (1985) Planta Med 5:460
42. Ankri S, Miron T, Rabinkov A, Wilchek M, Mirelman D (1997) Antimicrob Agents Chemother 10:2286
43. Mirelman D, Monheit D, Varon S (1987) J Infect Dis 156:243
44. Neuwirth Z, Sundstrom DC, Thompson NH (1988) Antimicrob Agents Chemother 32:1763
45. Ghannoum MA (1988) J Gen Microbiol 134:2917
46. Adetumbi M, Javor GT, Lau BH (1986) Antimicrob Agents Chemother 30:499
47. Feldberg RS, Chang SC, Kotik AN, Nadler M, Neuwirth Z, Sundstrom DC, Thompson NH (1988) Antimicrob Agents Chemother 32:1763
48. Focke M, Feld A, Lichtenthaler HK (1990) FEBS Lett 261:106
49. Ozolin ON, Uteshev TA, Kim IA, Deev AA, Kamzolova SG (1990) Mol Biol (Mosk) 24:1057
50. San-Blas G, Urbina JA, Marchan E, Contreras LM, Sorais F, San-Blas F (1997) Microbiology 143:1583
51. Urbina JA, Marchan E, Lazardi K, Visbal G, Apitz-Castro R, Gil F, Aguirre T, Piras MM, Piras R (1993) Biochem Pharm 45:2381
52. Turos E, Guntireddygari SKR, Greenhalgh K, Ramaraju P, Abeylath SC, Jang S, Dickey S, Lim DV (2007) Bioorg Med Chem Lett 17:3468
53. Abeylath SC, Turos E, Dickey S, Lim DV (2008) Bioorg Med Chem 16:2412
54. O'Driscoll M, Greenhalgh K, Young A, Turos E, Dickey S, Lim DV (2008) Bioorg Med Chem 16:7832
55. Smith DM, Kazi A, Smith L, Long TE, Heldreth B, Turos E, Dou QP (2002) Mol Pharmacol 61:1348
56. Kuhn D, Coates C, Daniel K, Chen D, Bhuiyan M, Kazi A, Turos E, Dou QP (2004) Front Biosci 9:2605
57. Chen D, Falsetti SC, Frezza M, Milacic V, Kazi A, Cui QC, Long TE, Turos E, Dou QP (2008) Cancer Lett 268:63
58. Frezza M, Garay J, Chen D, Cui C, Turos E, Dou QP (2008) Int J Mol Med 21:689

Top Heterocycl Chem (2013) 30: 183–222
DOI: 10.1007/7081_2012_88

Published online: 11 August 2012

Synthesis of β-Lactams and Their Chemical Manipulations Via Microwave-Induced Reactions

Indrani Banik and Bimal K. Banik

Abstract The β-lactam derivatives have many medicinal applications. Penicillins and cephalosporins antibiotics and a number of β-lactams have been discovered for the treatment of different medical disorders. As a result of this general trend of β-lactam use, the searches for clinically useful β-lactams will be pursued by many scientists. During the course of our studies on the synthesis of β-lactams and other heterocycles, we have found it convenient to conduct several types of synthetic steps under microwave irradiation. We have developed "microwave-induced organic reaction enhancement (MORE)" chemistry techniques for using nontraditional methods for rapid, safe, and environment-friendly reactions. These reactions are performed in unmodified domestic microwave ovens in a matter of minutes using very limited amounts of high boiling solvents (such as DMF and ethylene glycol) or no solvents if one of the reactants is a suitable liquid. It is not clear whether microwaves alter the transition state parameters of reactions. But, many laboratories (including our own) have reported that microwave-assisted reactions are much faster, comparatively free of by-products and sometimes susceptible to steric control.

Keywords β-Lactams · Chemical manipulation · Microwave reaction · Stereochemistry · Synthesis

---

Dedicated to Professor Paul Sale, Former Provost and Vice President of Academic Affairs, for his tremendous contribution to BKB's life and career.

I. Banik and B.K. Banik (✉)
Department of Chemistry, The University of Texas-Pan American, 1201 West University Drive, Edinburg, TX 78539, USA
e-mail: banik@utpa.edu

Contents

Abbreviations

Clay	Montmoriolonile
DMF	Dimethyl formamide
DMSO	Dimethyl sulfoxide
h	Hour (s)
IC_{50}	Cell growth inhibition at 50% concentration
MWI	Microwave Irradiation
NMR	Nuclear Magnetic Resonance
PAH	Polyaromatic hydrocarbon
SAR	Structure–activity relationship
TEA	Triethylamine

1 Introduction

β-Lactams have a variety of crucial medicinal applications. In addition to the well-known penicillins and cephalosporins antibiotics, many other β-lactam antibiotics have been discovered (for recent examples, see [1–4]). Potent and effective β-lactam antibiotics as well as more effective β-lactamase inhibitors are needed for the benefits of humankind. Therefore, chemists have been motivated to design new β-lactams [5]. These types of compounds have been used as starting materials for the preparation of various heterocyclic compounds of biological importance [6–10]. Substituted hydroxy β-lactams have been proved to be a versatile starting materials in the semisynthesis of paclitaxel (Taxol) and docetaxel (Taxotere) [11]. β-Lactams have been used for lowering cholesterol [12–15]. Human leukocyte elastase inhibitory mechanisms and the biological activity of these classes of compounds have been investigated [16] (and earlier references [17]). Recent remarkable developments include catalytic asymmetric (for catalytic asymmetric synthesis, see [18–22]; for polymer-supported synthesis, see [23–30]) novel synthesis of unique β-lactams. As a result of this general trend of β-lactam use in practical life, the targets for clinically effective β-lactams that are antibiotics and other medically important properties will continue to grow in the future (serine protease [31–34] and references cited therein; [35] and references cited therein). We have demonstrated numerous methods for the synthesis of β-lactams [36–44] and several other related biologically active compounds [45–58]. (Presented at the American Chemical Society (ACS) National Meeting, Orlando, FL, April 2002, MEDI-213. We sent this paper for presentation to the ACS on October 31, 2001 and it was accepted on December 16, 2001. The drug synthesis and chemistry branch of the NCI tested 24a in their 60 cell lines and sent us the report on August 8, 2001. This indicates that we are the pioneer in this field.) In continuation of our research in this area, we describe herein the synthesis of several β-lactams and products derived from them using microwave irradiation as one of the key steps. (Microwave activation has become very popular and useful technology in organic and medicinal chemistry. For some recent examples, see [59–62] and references cited therein. One of us was extensively involved in the microwave-assisted β-lactam synthesis. For example, see [63–70].)

2 Microwave-Assisted Eco-Friendly Synthetic Steps

During the course of our studies on the synthesis of β-lactams and other heterocycles, we have found it convenient to conduct several types of synthetic steps under microwave irradiation. (Microwave activation has become very popular and useful technology in organic and medicinal chemistry. For some recent examples, see [59–62] and references cited therein. One of us was extensively involved in the microwave-assisted β-lactam synthesis. For example, see [63–70].) "Microwave-induced organic reaction enhancement (MORE)" chemistry techniques have been used for nontraditional methods for rapid, safe, and environmentally benign reactions. These reactions are performed in unmodified domestic microwave ovens in a matter of minutes using very limited amounts of high boiling solvents (such as DMF, ethylene dichloride, and ethylene glycol) or no solvents if one of the reactants is a suitable liquid. Clearly, this method has received significant attention from scientists as can be proved by several thousands of published papers in various journals.

Domestic microwave ovens use radiation of 2,450 MHz frequency which is directed into the inner cavity of the oven. The level of the energy can be controlled by an on–off cycle that can be adjusted for various levels of energy. Therefore, selective microwave energy can be transferred to the reactants. Microwaves are nonionizing radiation with dipoles. Glass, ceramic materials, and many polymeric materials are transparent to microwaves. Great advantage is taken of this important property of microwaves to concisely reduce the amount of organic solvents needed as the reaction medium or microwave energy transfer agent. Upon irradiating a reaction mixture in an open glass vessel depending upon the scale of the experiment (a large beaker or Erlenmeyer flask with a loose cover), microwave irradiation is transferred to the reactants directly without the necessity for heating the glass vessel (extraction container) and setting up the convection currents. The energy input is controlled efficiently so that the solvent and/or the reaction mixture are not allowed to boil and the flask is removed from the oven easily. This process limits the amount of vaporization significantly and, therefore, no reflux condenser is essential. Stirrers are not necessary if the reaction mixture is placed as a thin layer in the glass vessel used so that microwaves can penetrate into the entire reaction mixture efficiently. The usual reaction time is a few minutes even on a few 100 g scale. However, an optimization of the reaction condition is required since it has been found that the time may differ significantly depending upon the size and nature of the microwave oven on–off cycle, solvent, and heat sink. A variation of one of these parameters may cause acceleration or retardation of reaction rate.

It is not clear whether microwaves alter the transition state parameters of reactions. Some evidence support that it is the radiation that is responsible for accelerated rate of the process. But, many laboratories including our own have reported that microwave-assisted reactions are much faster, high yielding comparatively free of by-products, and sometimes susceptible to stereochemistry control of the products. A major advantage of MORE chemistry method is the minimum amount of solvents used. A slurry at room temperature is sufficient and allows enough reactants to go into solution at the comparatively high temperatures reached very rapidly in a microwave oven in 1 or 2 min. This process facilitates the reaction dramatically. Reduction in the use of solvents as reaction media and clean reaction with better stereochemistry control reduce pollution at the source and ensure high levels of efficiency from a practical point of view. Another advantage of MORE chemistry method is the lowered energy consumption compared to conventional heating under reflux. The use of high boiling polar solvent adds significant value. The reaction mixture under irradiation is not allowed to come to the boiling point so there is minimum vaporization. Also, microwave-assisted reactions on several 100 g scale require only 10–15 min of irradiation in a domestic microwave oven of 800–1,000 W power. Commercial microwave ovens are available for kilogram scale preparative reactions. The availability of such equipment certainly adds more use of MORE chemistry techniques for process development research. There is no doubt that microwave-enhanced chemical synthesis has played an important role for the manufacture of specialty pharmaceuticals such as peptides, β-lactam, heterocycles, steroids, alkaloids, and their analogs.

3 Catalytic Transfer Hydrogenation in β-Lactam Chemistry

A few laboratories have employed catalytic transfer hydrogenation (CTH) in their research [71–86] (for a recent review, see [87–94]). This is a safe and efficient operation in which a catalyst and hydrogen gas are replaced with a catalyst and a hydrogen donor such as cyclohexene, hydrazine, formic acid, sodium formate, ammonium formate, cyclohexadiene, and phosphinic acid, sodium hypophophite

Scheme 1

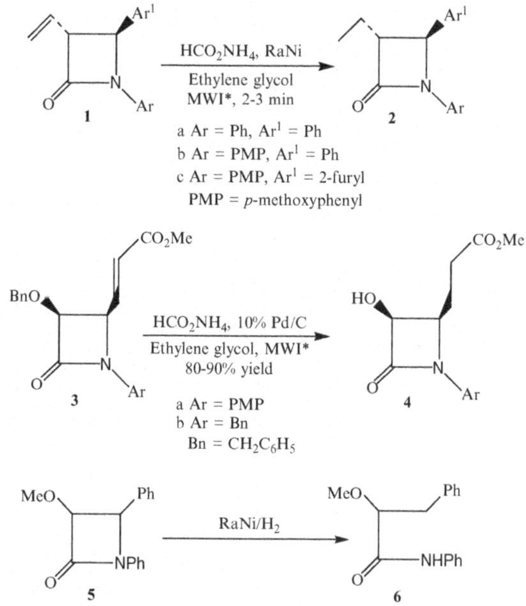

[71–86] (for a recent review, see [87–94]). This type of hydrogenation is usually conducted in flasks fitted with a magnetic stirrer and a reflux condenser under anhydrous conditions. Ethyl alcohol is a common solvent for CTH. However, CTH can be conducted very rapidly and in quantitative yield inside an unmodified domestic microwave oven in ethylene glycol. For example, during the reduction of **1** and **2** (Scheme 1), the vinyl group at C-3 was converted to the ethyl group [92, 93]. No cleavage of the β-lactam ring was observed.

The β-lactam **3** underwent selective hydrogenolysis of the *O*-benzyl group and reduction of the unsaturated ester group to a saturated ester side chain to yield the β-lactam **4**, which retained the *N*-benzyl group intact when 10% Pd/C was used as the catalyst following this method (Scheme 1). This reaction demonstrates the selectivity of hydrogenation reactions.

Ojima et al. have cleaved N–C_4 bonds in 4-phenyl-2-azetidinones to prepare phenylalanine derivatives in high yield (for example, see [8, 95, 96]). This was performed using conventional catalytic hydrogenation method (ambient pressure of hydrogen at 50°C in methanol with Pd/C as catalyst). In the presence of a large excess of Raney nickel catalyst and hydrogen, 3-methoxy-1,4-diphenyl-2-azetidinone underwent β-lactam cleavage to provide a small amount of the anilide of α-methoxy-β-phenylpropionic acid (Scheme 1). However, cleavage of the β-lactams ring did not occur under milder conditions. However, further investigation on this reaction was very successful in obtaining exciting results.

Microwave-assisted catalytic transfer hydrogenolysis was extensively investigated. This was performed at 120–130°C using 10% Pd/C as the catalyst. In every case, extremely fast scission of 4-phenyl-2-azetidinones was observed in the presence of ammonium formate as the hydrogen donor (Scheme 2).

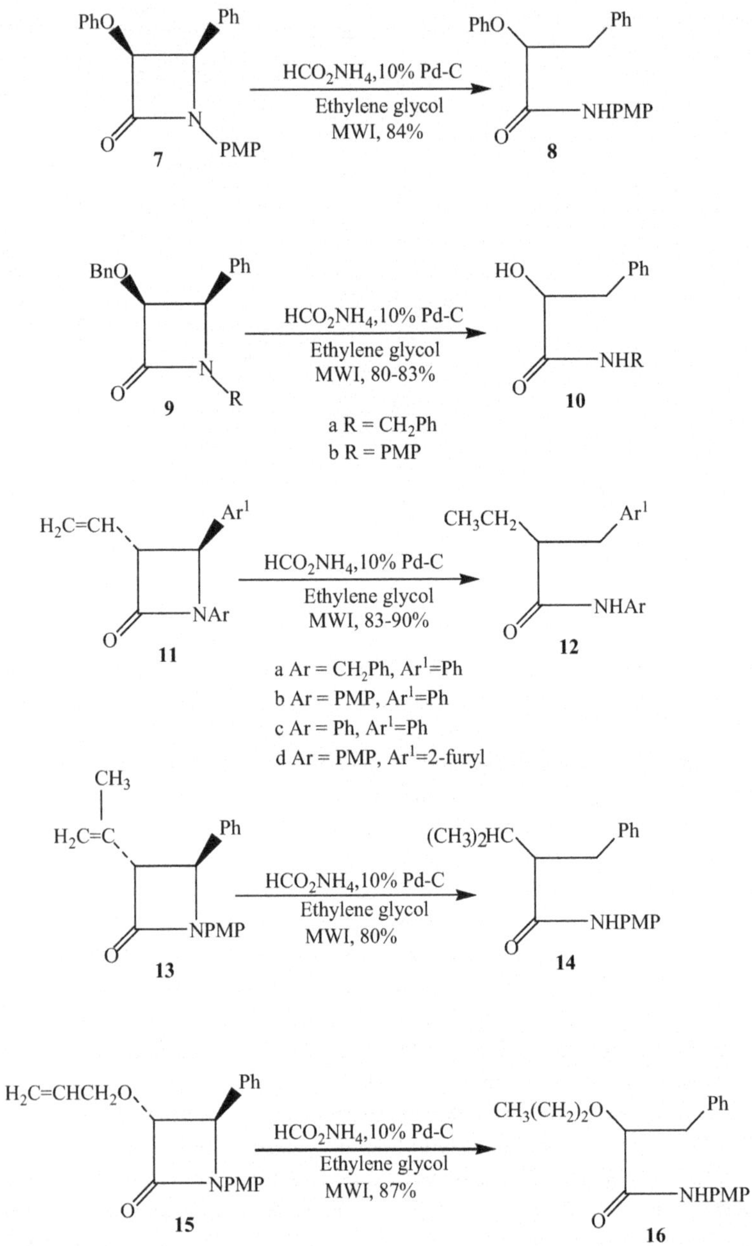
PhO
Ph
HCO2NH4,10% Pd-C
Ethylene glycol
MWI, 84%
N
PMP
O
7
PhO
Ph
NHPMP
O
8
BnO
Ph
HCO2NH4,10% Pd-C
Ethylene glycol
MWI, 80-83%
N
R
O
9
HO
Ph
NHR
O
10
a R = CH2Ph
b R = PMP
H2C=CH
Ar1
HCO2NH4,10% Pd-C
Ethylene glycol
MWI, 83-90%
NAr
O
11
CH3CH2
Ar1
NHAr
O
12
a Ar = CH2Ph, Ar1=Ph
b Ar = PMP, Ar1=Ph
c Ar = Ph, Ar1=Ph
d Ar = PMP, Ar1=2-furyl
CH3
H2C=C
Ph
HCO2NH4,10% Pd-C
Ethylene glycol
MWI, 80%
NPMP
O
13
(CH3)2HC
Ph
NHPMP
O
14
H2C=CHCH2O
Ph
HCO2NH4,10% Pd-C
Ethylene glycol
MWI, 87%
NPMP
O
15
CH3(CH2)2O
Ph
NHPMP
O
16

Scheme 2

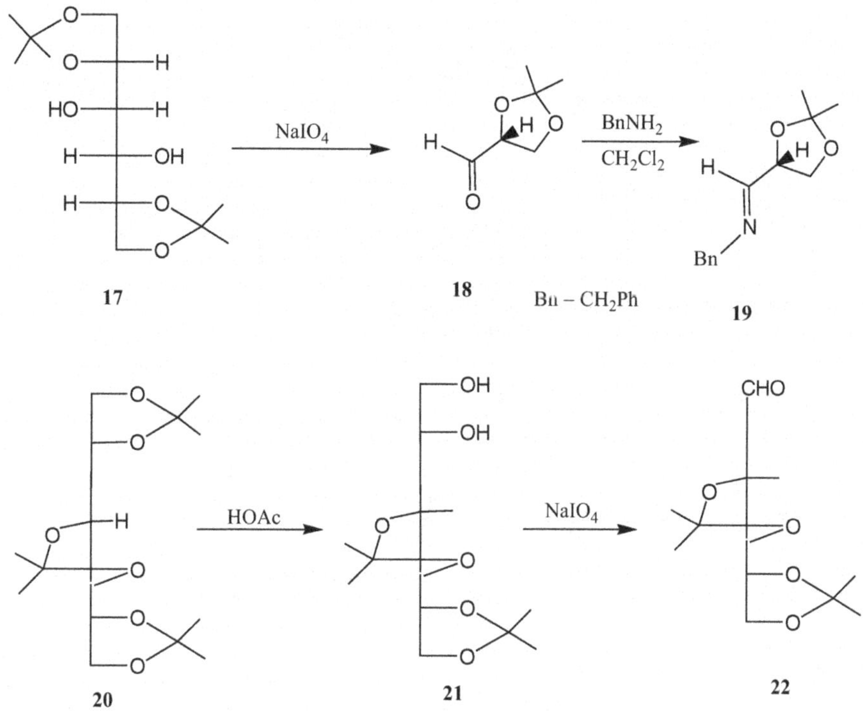

Scheme 3

The *N*-benzyl group of the β-lactam **9** under this condition was not hydrogenolyzed. But, the O–Bn group at C-3 was converted to an alcoholic OH group; alkene β-lactams (**11**, **13**, and **15**) were reduced to alkyl β-lactams (Scheme 2). The reduction product was obtained in high yield and very quickly. Interestingly under these conditions Ra–Ni did not cause cleavage of the β-lactam ring in **1** (Scheme 1).

4 Enantiospecific Synthesis of (−)-2*S*,3*S*-2-Amino-3,4-Dihydroxybutyric Acid

β-Lactams can be used for the synthesis of optically pure polyhydroxy amino acid. Microwave-induced reactions were used in preparation of these types of compounds (for example, see [67–69, 97–105]). An aldehyde from a sugar derivative permitted the synthesis of α-hydroxy β-lactams of predictable absolute configuration. These 3-hydroxy-2-azetidinones were then used to generate polyhydroxy amino acids (Scheme 3).

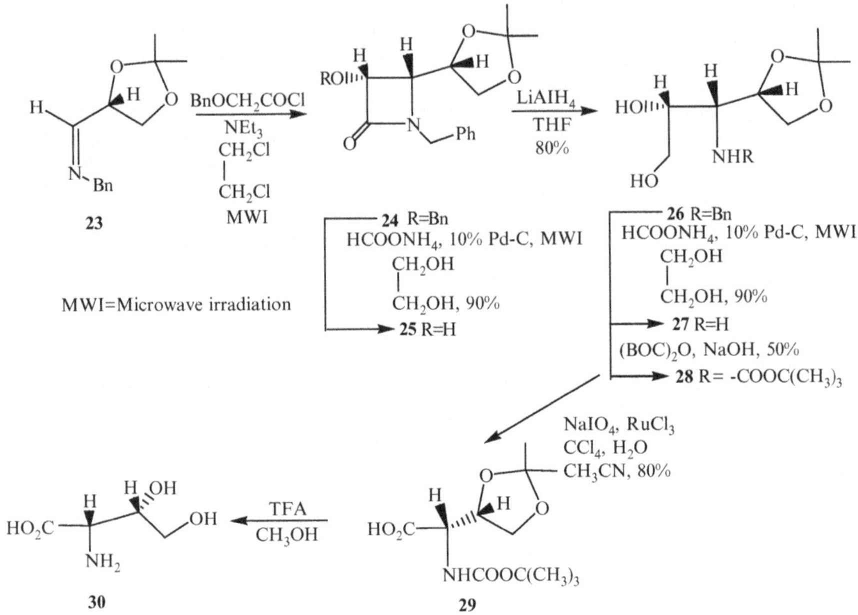

Scheme 4

Commercially available D-mannitol was selected as one of the starting compounds. D-mannitol was utilized in two different ways for the preparation of polyhydroxy amino acids (for example, **17** and **20**). An unusual amino acid, also known as L-γ-hydroxythreonine, received significant attention. Klenk and Diebold reported that it is one of the oxidative products of sphingosine derivative. Consequently, several syntheses of this compound have been published [6–8, 10, 98].

The starting compound for this synthesis (Scheme 4) was D-glyceraldehyde acetonide **18** which was allowed to react with benzyl amine to form an imine **23**. Reaction of **23** with benzyloxyacetyl chloride and triethylamine using microwave irradiation produced a single *cis*-β-lactam **24** with known absolute configuration.

Microwave-assisted CTH of **24** was conducted at approximately 130°C with ammonium formate as the hydrogen donor and 10% Pd/C as the catalyst [71–86] (for a recent review, see [87–94]). Selective hydrogenolysis to **25** was observed: the benzyloxy group was converted to a hydroxyl group but the *N*-benzyl group was unaffected. Lithium aluminum hydride-induced reaction of **25** cleaved the β-lactam ring and led to the vicinal diol **26**. Oxidation of **26** failed to yield the desired product. It was found that protection of the amino group with electron-withdrawing group was necessary before conducting the oxidation step. Microwave-assisted CTH of **26** was successful to remove the *N*-benzyl group to give the primary amino compound **27**. The amino compound was converted to *t*-butoxycarbonyl derivative **28** using standard reaction conditions.

At the beginning of the process it was decided to prepare an L-amino acid. Therefore, the diol (acetonide) was not disturbed and an acid was generated by

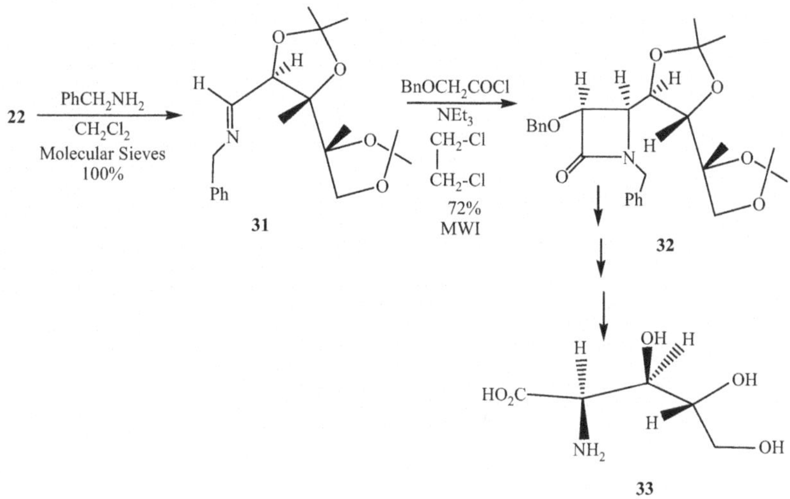

Scheme 5

ruthenium tetroxide oxidation of the unprotected diol **28**. The amino acid derivative **29** obtained in this manner was then reacted with trifluoroacetic acid to remove the ketal and Boc groups. The levorotatory (natural) form of γ-hydroxy threonine, the desired amino acid **30**, was obtained through the above sequence of reactions.

This polyhydroxyamino acid and its mirror image **33** have been the target of synthesis in several laboratories because of their biological activities [106]. (For some examples of bismuth nitrate-catalyzed reactions from our laboratory, see [107–112].) The imine **31** obtained by the condensation of **22** with benzyl amine was allowed to react with benzyloxyacetyl chloride and triethylamine, under microwave irradiation and a single optically active β-lactam **32** was the product. The absolute configuration of the chiral center next to the aldehyde group in **22** was helpful to predict the absolute configuration of this *cis*-β-lactam as shown in **32**. Optically active (−)-polyoxamic acid was then prepared through a series of steps. As discussed before, many of the synthetic steps for the preparation of **33** were conducted in domestic microwave oven. Successful synthesis of **30** and **33** dictates that many other optically active polyhydroxy amino acids can be prepared starting from β-lactams with predictable absolute configuration using microwave irradiation as one of the key steps. For example, deprotection of the ketal group in **24** and oxidation would provide an acid. Reduction of the 4-carboxy-β-lactam by mild conditions should afford an amino acid derivative. The above sequence of reaction suggests a variety of polyhydroxyamino acid could be synthesized by appropriate selection of the reaction pathways. Reduction of β-lactam ring followed by deprotection of the ketal group would lead to a different amino acid if deprotection of the ketal remains the first choice rather than reductive cleavage of the β-lactam ring [6] (Scheme 5).

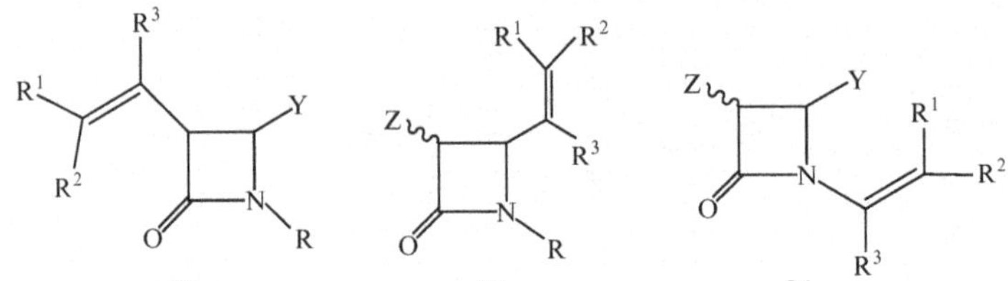

34 35 36

Scheme 6

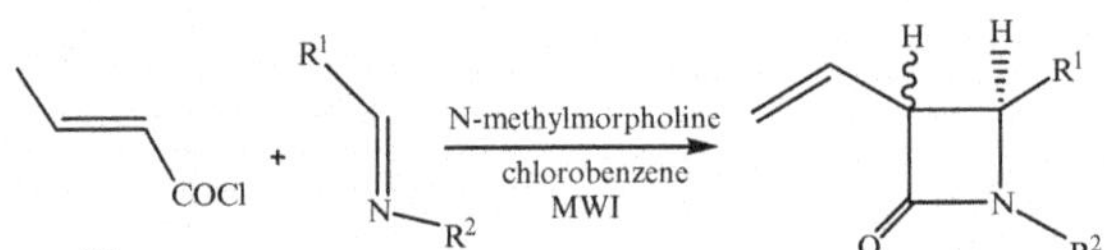

37 38 39

Scheme 7

Table 1 Preparation of vinyl-β-lactams under microwave condition

Entry	R^1	R^2	Stereo-chemistry	Yield (%)
a	Phenyl	Phenyl	*Trans*	65
b	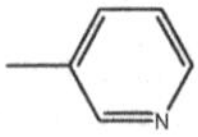	*p*-Anisyl	*Trans*	52
c	–CO–Ph	*p*-Anisyl	*Cis:trans* (70:30)	80
d	–CO–Ph	–CH(CH_3)Ph	*Cis*	55
e	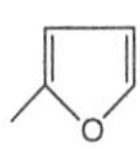	CH(CO_2PNB)CH(CH_3)OH	*Cis*	15
f	*p*-Anisyl	*p*-Anisyl	*Trans*	55
g	Phenyl	*p*-Anisyl	*Trans*	50
h	*Trans*-cinnamyl	CH(CO_2PNB)CH(CH_3)OH	*Cis:trans* (90:10)	60

5 Steric Course of Cyclization of α-Vinyl β-Lactams

Vinyl β-lactams (**34–36**) are important starting compounds for different types of heterocycles (Scheme 6). These compounds can be synthesized using microwave irradiation. A direct synthesis of α-vinyl-β-lactams **39** and **42** by the reaction of α, β-unsaturated acid chloride (for example, *trans*-crotonyl chloride **37**) with imine **38** in the presence of triethylamine in refluxing dichloromethane or benzene was developed [7]. Exclusive formation of *trans*-β-lactams **39** was observed from *trans* crotonyl chloride **37** and the diaryl Schiff base **38** (Scheme 7, entry a, b, f, and g; Table 1).

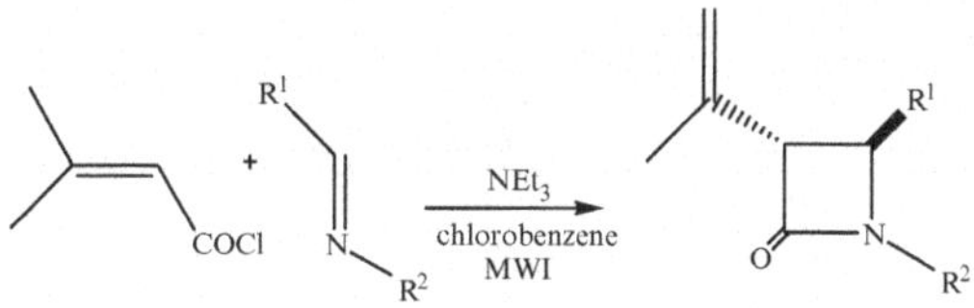

40 41 42

Scheme 8

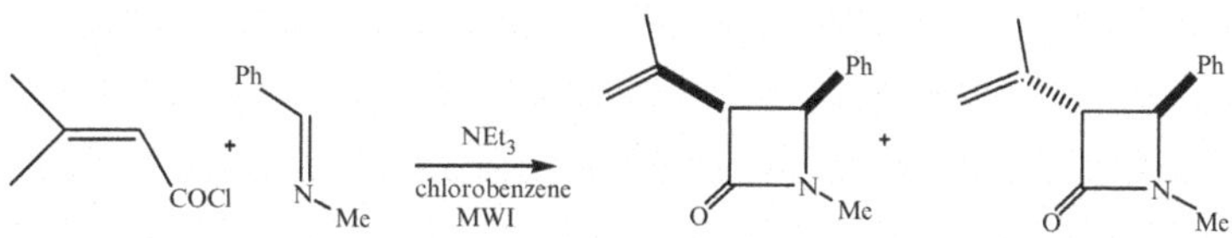

43 44 45 46

Scheme 9

Zamboni and Just used this method for preparing numerous *trans*-α-vinyl-β-lactams as potential intermediates for β-lactam antibiotics [43] (Scheme 8). Thus, the reaction of β-dimethylacryloyl chloride **40** and the imine **41** to obtain the *trans*-β-lactam **42** (Scheme 8) was reported. It was shown that crotonyl chloride **37** and dimethylacryloyl chloride **40** worked in a similar way during the formation of α-vinyl-β-lactams **39** and **42**, respectively. The imines **38** (R^1 = Ph or furfuryl or styryl) produced from aniline or *p*-anisidine lead only to *trans*-β-lactams. But, the imine from cinnamaldehyde and aniline produced a mixture of *cis*- and *trans*-β-lactams. At room temperature the *cis* isomer was formed in predominant form but at higher temperature the *trans* product was formed in major proportion. Imine **44** prepared from aliphatic amine produced mostly the *cis*-α-vinyl-β-lactam **45** (Scheme 9).

The main features of MORE chemistry for the synthesis of α-vinyl β-lactams were: (a) the use of open glass systems (this avoided the danger of explosions because of rapid rise of pressure and temperature as in some sealed system reactions); (b) control of the microwave irradiation input into the reaction mixture, such that the temperature remained below the boiling point of the mixture; (c) no solvents required if one of the reactants is a polar liquid; (d) minimal amounts of polar solvents required since it was observed that a slurry is adequate for successful reaction at higher temperature under microwave irradiation; (e) beakers, conical flasks, and glass trays were used as reaction vessels; (f) reactants were energized directly, no stirrers were essential; (g) most organic reactions on a large scale required only a short time of irradiation instead of hours as in conventional set ups; (h) rapid increase of temperature favored a few reaction pathways and thus led to selectivity. Selective formations of products with control of stereochemistry were important.

Imines from aromatic aldehydes and substituted anilines produced only *trans*-α-vinyl-β-lactams **39** (Schemes 7 and 8). But, the imine **44** from benzaldehyde and

Table 2 Preparation of hydroxy-β-lactams by microwave irradiation

Length of irradiation	*Trans:cis*	Temperature of reaction mixtures (°C)	Amount of water as heat sink (mL)
1 min	40:60	78	200
2 min	60:40	92	200
3 min	70:30	95	200
4 min	70:30	105	200
5 min	80:20	108	200
10 min	80:20	118	200
30 s	25:75	55	800
1 min	30:70	62	800
10 min	80:20	115	–
30 min	65:35	RT-115	–
6 h	40:60	40	–

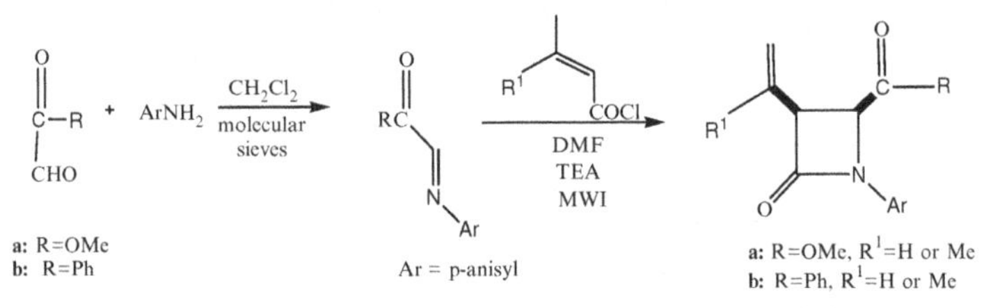

Scheme 10

methylamine produced a mixture of *cis*- and *trans*-β-lactams **45** and **46** in a ratio depending upon the temperature of the reaction mixture (Scheme 9). Higher temperature favored the formation of the *trans*-β-lactam, which is thermodynamically more stable (Table 1). However, this process was not a base-catalyzed isomerization. Unchanged *cis*-β-lactam **45** was recovered after irradiation in the presence of *N*-methylmorpholine. At the beginning of the procedure as depicted in (Scheme 9), *cis*-β-lactam was produced in higher amount. After 10 min of irradiation, *trans*-β-lactam was formed as the predominant isomer (Table 2). It was necessary to keep water in the microwave oven to control the temperature of the reaction. Water absorbed a portion of microwave energy thereby prevented reactants not to boil. Interestingly, *cis*-β-lactams **49a**, **b** were obtained exclusively from acid chlorides **37** or **40** and imine **48** (obtained from glyoxalic ester **47a** or phenyl glyoxal **47b**)-irrespective of the temperature of the reaction (Scheme 10). The reason for the formation of *cis*-β-lactams from glyoxalic ester and phenyl glyoxal under microwave irradiation could not be explained based on the other results as described above. However, the role of an electron-withdrawing group could be important in the formation of *cis*-β-lactams.

Isomerization of β-lactam formation could be avoided using a novel concept. For example, the problem of *cis*/*trans* isomerism disappeared when an imine

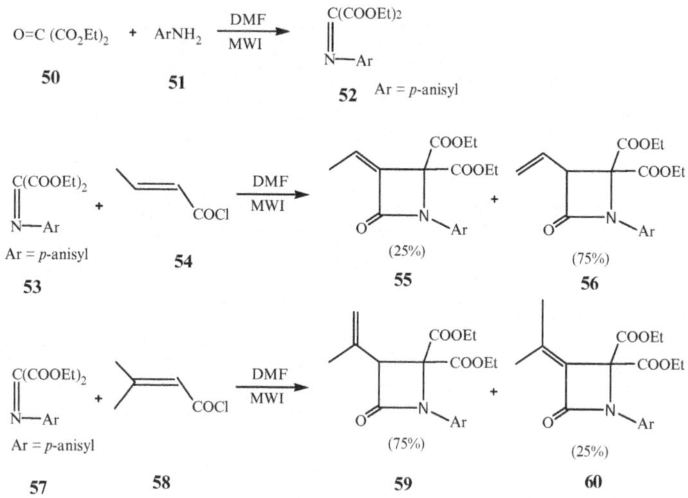

Scheme 11

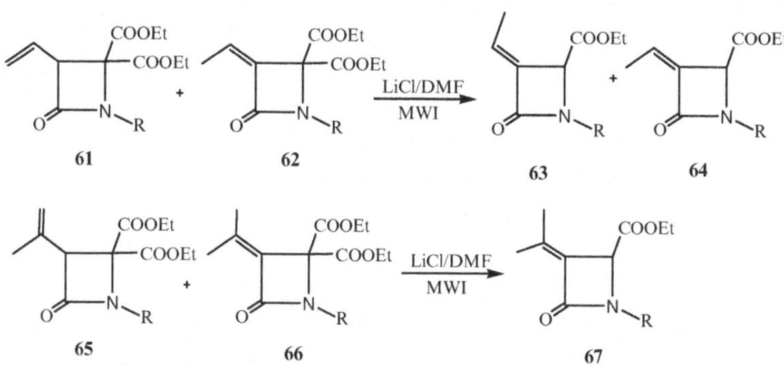

Scheme 12

52 derived from a ketomalonate **50** was used for the preparation of the vinyl-β-lactam **56**. However, an isomer **55** was also produced in which the double bond in the side chain is conjugated with the β-lactam carbonyl group (Scheme 11) (for example, see [67]). A similar reaction between **57** and **58** produced **59** and **60**. This reaction was extended with other substrates: using different groups at nitrogen of the β-lactam ring. Dealkoxycarbonylation of β-lactams **61** and **62** was conducted under the influence of lithium chloride in DMF.

Using microwave irradiation the products were found to be a mixture of E and Z isomers **63** and **64** (1:1 ratio). Conjugation of the alkene group was possible during microwave irradiation. Similar reactions of the mixtures of **65** and **66** afforded a single product **67** in good yield (Scheme 12). The mixture of unsaturated

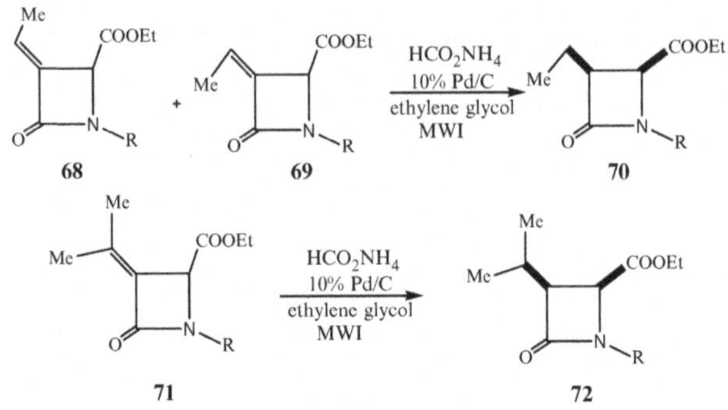

Scheme 13

compounds **63** and **64** was then reduced to the saturated compounds **70** and **72**, respectively, by microwave-assisted CTH method.

Ethylene glycol was used as the solvent, ammonium formate as the hydrogen donor, and 10% Pd/C as the catalyst. The stereochemistry of the products **70** and **72** from this study was found to be exclusively *cis* (Scheme 13). Preparation of these types of alkyl-substituted β-lactam was very difficult through direct cycloaddition of saturated acid chloride with imine.

6 Glycosylation of Alcohols and Acid-Induced Cleavage of the β-Lactam Glycosides

We reported the use of bismuth nitrate pentahydrate in several organic transformations [106]. (For some examples of bismuth nitrate-catalyzed reactions from our laboratory, see [107–112].) For example, nitration of aromatic hydrocarbons, phenolic compounds, and the aromatic group in β-lactams were realized in excellent yield. These were exciting because phenolic compounds and β-lactams could be functionalized using bismuth nitrate even at high temperature. No oxidation of phenol and cleavage of β-lactam rings were observed. A facile Michael reaction of indoles and carbamates with unsaturated ketones was also performed with bismuth nitrate [111]. These reactions, in principle, required the presence of mineral acids, or Lewis acids. Alternatively, a coordination of electron pairs of electronegative elements to the vacant d-orbital of bismuth might also facilitate these reactions. The mechanism of these bismuth nitrate reactions was not explored. However, in some examples, involvement of catalytic amounts of nitric acid was speculated. In some instances coordination of bismuth salts with electronegative species was proved. There were unique advantages of using bismuth salts; the nontoxicity and mildness were the main important features.

Ferrier rearrangement is a very useful reaction in organic synthesis. Since Ferrier rearrangement is an acid-catalyzed process, we envisioned that bismuth nitrate

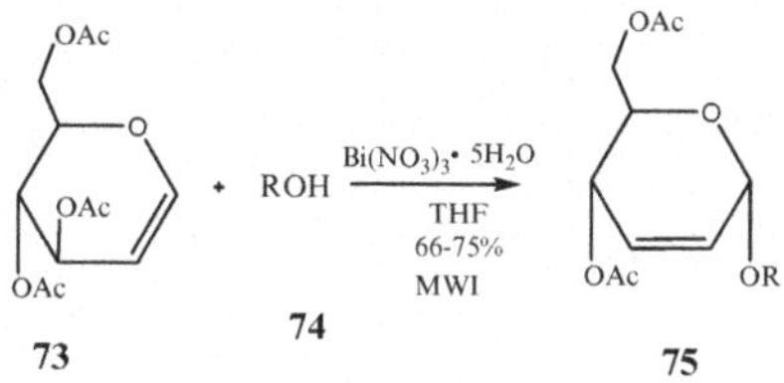

Scheme 14

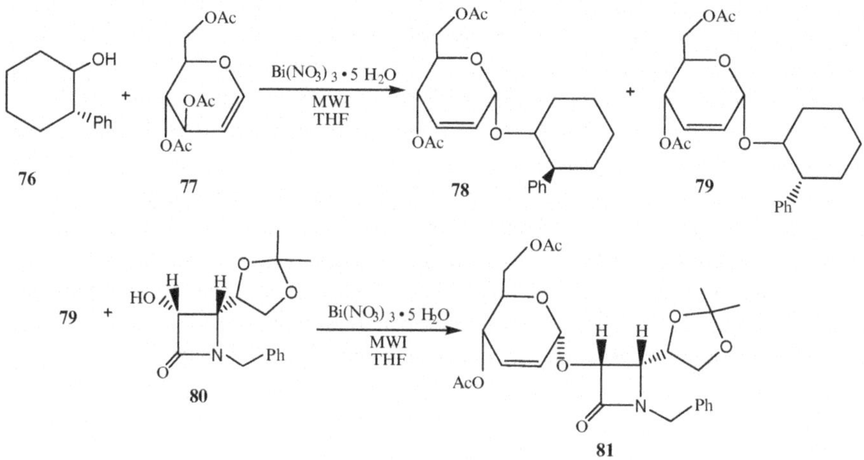

Scheme 15

might be a good catalyst in the glycosylation of alcohols in the presence of glycal. Interestingly, to our surprise bismuth nitrate proved to be effective in catalyzing a Ferrier type of rearrangement [100, 113].

Reaction of methanol, ethanol and isopropanol, and other alcohols with 3,4,5-tri-*O*-acetyl D-glucal in the presence of bismuth nitrate (10–30 mol%) produced a single glycoside in each case in 68–75% yield [43]. To control the reaction temperature further, it was necessary to use a beaker containing water (depending upon the scale of the experiment) next to the reaction vessel during microwave exposure. Moreover, the "on–off" cycle present in the microwave oven was also used to allow solvent not to be evaporated. The anomeric stereochemistry of these glycosides was determined as α-from the coupling constant of the anomeric hydrogen (1–2 Hz) of a reduced isomer prepared by a hydrogenation experiment (Schemes 14 and 15). (One of us was extensively involved in the microwave-assisted β-lactam synthesis. For example, see [70].) The β-glycoside should have a higher coupling constant because of axial–axial interactions.

We reported the use of iodine in organic synthesis [53, 114–116]. In the course of our continuing studies on antibiotics and anticancer agents, we synthesized and studied medicinal activities of several β-lactams. In an extension of iodine-catalyzed

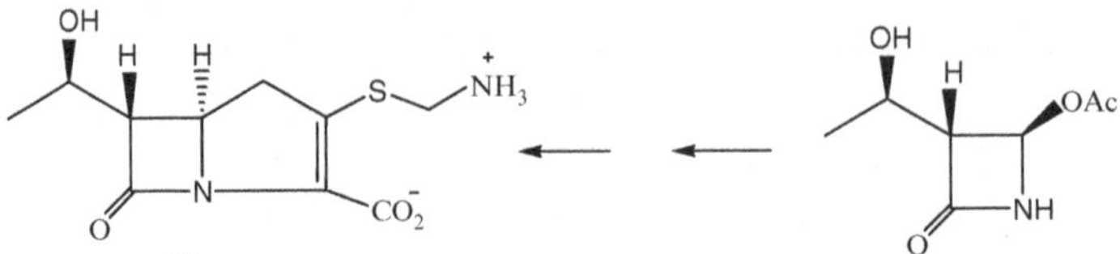

82

83

Scheme 16

glycosylation of alcohol, we studied chiral resolution of a racemic alcohol that is present in thienamycin side chain for obtaining both enanatiomers of this important antibiotic [117–121]. The strategy was to prepare two separable diastereomeric *O*-glycosides by the Ferrier rearrangement as one of the important steps [122].

In 1979, the clinical use of the β-lactam antibiotic thienamycin (**82**, Scheme 16) was started. PS-5 and carptimycin also belonged to this family of antibiotics. Various synthetic methods were published for the preparation of thienamycin and related antibiotics. Most of these methods required optically pure 3-(1-hydroxyethyl)-4-acetoxy-2-azetidinone (**83**) as the starting materials. Since enantiomers differ in their biological properties, availability of both enantiomeric forms of this compound is highly desirable [123, 124]. One way to prepare both enantiomers was to perform an easy chiral resolution of a readily available racemic alcohol. Our attempt for the synthesis of thienamycin provided a few glycosides via iodine-catalyzed reactions.

Microwave-induced hydrogenation using ethylene glycol as the reaction medium and ammonium formate as the hydrogen donor in the presence of 10% Pd–C produced the saturated products **85** and **87** very easily. The reaction was completed within 1 min (Scheme 17). Similar hydrogenation of the β-lactam containing a reducible group and an aromatic ring at C_4 afforded a propionamide derivative as a result of the N_1–C_4 bond fission as discussed before (for example, see [93–95]). It appeared that the sugar group in the glycosides **84** and **86** exerted steric hindrance to the hydrogenolysis of the N_1–C_4 bond. However, conducting the hydrogenation reaction for a long period afforded many other uncharacterized materials. In the complex reaction mixtures several compounds were noted, but they were not identified thoroughly; these compounds were formed because of the cleavage of the N_1–C_4 bond, reduction of the olefinic group and allylic deacetoxylation of the carbohydrate group. The ^{1}H NMR spectra of the 2,3-dideoxy carbohydrate derivatives **85** and **87** showed only small couplings (1–2 Hz) for the anomeric hydrogen indicating axial stereochemistry of the glycoside bonds. A higher coupling constant (8–10 Hz) was expected with a β-glycoside product. A similar reaction with benzyl-protected glycal failed to produce any glycoside. In conformity with the observation presented earlier, the importance of leaving group properties was also found in the present study. The glycal that was derived from galactal with identical protective group proved to be ineffective.

The two diastereomeric **88** and **90** were separated by column chromatography over silica gel. Prolonged exposure of **88** and **90** to silica gel caused decomposition. Then the sugar group of **88** and **90** was removed by mild aqueous hydrochloric acid

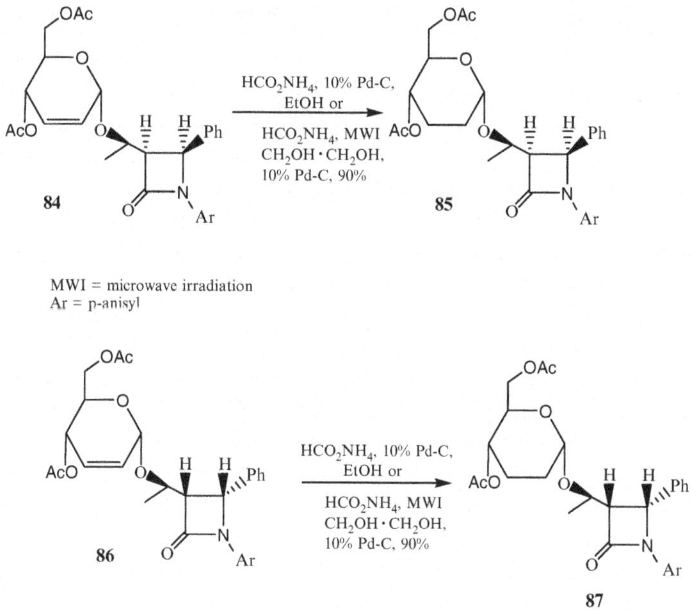

Scheme 17

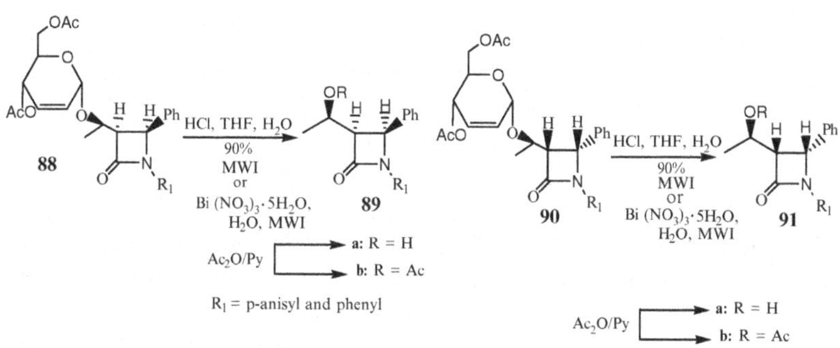

Scheme 18

treatment. The cleavage reaction was also performed using bismuth nitrate under microwave irradiation to afford alcohol derivatives. β-Lactams **89a** and **91a** were converted into their respective acetates **81b** and **91b**. Followed by usual method β-lactam **89b** was found to be enantiomerically pure as revealed by ^{1}H NMR spectroscopy using an optically active shift reagent [125]. This NMR study further confirmed that the compounds **89b** and **91b** are mirror images to each other (Scheme 18). The structure of **89b** and **91b** was also confirmed by chemical conversion to alkenes (not shown). Specific alkenes were formed from **89** and **91a**. This method established the stereochemistry of the products as described in (Scheme 18).

7 Stereoselectivity in the β-Lactam Formation Reaction

The stereochemistry (*cis* and *trans*) of the β-lactams **93** and **94** formed by the reaction of acid chloride and imine in the microwave oven with irradiation for a short period of time (1–2 min) at low power setting was identical as obtained by the traditional method. Our studies showed that it is possible to control the stereochemistry of β-lactam formation using high level irradiation in a domestic microwave oven following cycloaddition reaction. For a study of the effect of microwave irradiation on the formation of hydroxy-β-lactam compounds, we selected several imines **92** and the high boiling tertiary amine, *N*-methylmorpholine in place of lower boiling triethylamine. Chlorobenzene (b.p. 132°C) was chosen as the reaction medium in place of toluene which absorbs microwave energy only poorly because of low dipole movement. Polar solvents were superior for microwave-induced reactions. The approximate temperature of the reaction (110–120°C) mixture was determined by using a thermometer after the microwave irradiation was stopped. A beaker of water was kept next to the reaction vessel in the microwave oven as a "heat sink" for control of the amount of microwave energy entering into the reaction mixtures of small size (1–5 g). For large-scale (100 g) reaction, the proportion of solvent was reduced. Microwave-induced reactions were very effective for 50 g scale for this reaction.

Observations on the preparation of α-benzyloxy-β-lactams from benzyloxyacetyl chloride **91a**, **92a**, and *N*-methylmorpholine showed the formation of varying amounts of *cis*- and *trans*-β-lactams **93a** and **94a**. These results were interesting because this reaction should produce only the *cis* isomer. Based on these results a systematic study was undertaken. It was demonstrated that *cis*-β-lactam **94a** formation was favored by lower power of microwave irradiation (power levels 4 and 5). At about 112°C reaction temperature, there was more of the *trans* product **93a** than the *cis* isomer **94a**. When the reaction of the same acid chloride **91a** with imine **92e** derived from optically active D-glyceraldehyde acetonide was conducted under MORE chemistry conditions, only the *cis*-β-lactam **94e** was obtained both at low and high power settings. Different way to obtain *trans*-isomer of **94e** was attempted. However, the product remained *cis*. This is in clear contrast with the observation made on **92a**, **92b**, **92c**, and **92d**. Various attempts in the presence of a base were used to obtain *trans*-β-lactam with **92e**. However, all attempts failed: only a single *cis*-β-lactam was obtained.

We investigated the β-lactam forming reaction when the imine was derived from an aromatic aldehyde and an aryl alkyl amine as in **92b** and **92c**. When the cycloaddition was conducted in the traditional manner at 0°C by the "normal addition" method, only the *cis*-β-lactam was formed. However, if the "inverse addition" method was used, 30% *cis* and 70% *trans*-β-lactams were obtained under the same conditions in excellent yield. Therefore, the order of addition of reagents had a tremendous role on the stereochemistry of the β-lactam formation reaction through cycloaddition method. This suggests that different mechanistic routes are possible to obtain different β-lactams by changing the order of addition of reagents.

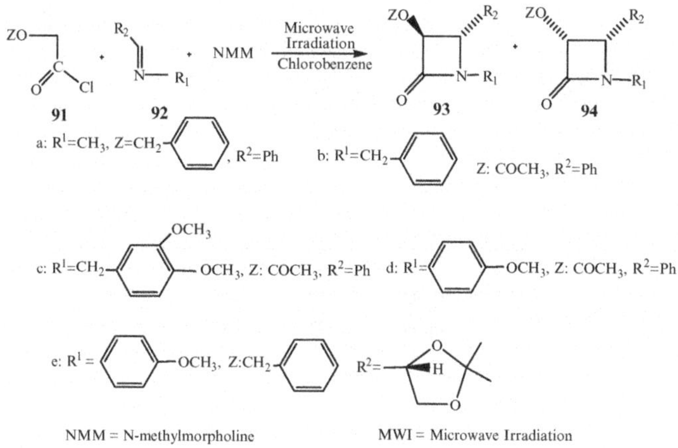

Scheme 19

Interestingly, when this reaction was conducted in a microwave oven using chlorobenzene (preheated to about 110°C) as the reaction medium, the ratio of *trans*- to *cis*-β-lactam was 90:10 irrespective of the sequence of addition (i.e., "normal addition" or "inverse addition"). It is difficult to explain these results. However, we hypothesized that fast reactions and additions were unable to differentiate the sequence of addition of the reagents. When the *cis*-β-lactam **94c** was heated with *N*-methylmorpholine in chlorobenzene solution for 30 min, there was no isomerization to the thermodynamically more stable *trans*-β-lactam **93c**. This study confirmed that the formation of *cis*- and *trans*-β-lactams followed multiple mechanisms. The transition state structures were altered by minor variation of the conditions of the experiments. Microwave irradiation seems to be responsible for the alteration of the transition state structures.

β-Lactam ring was used as the starting compound for the preparation of anticancer drug [R]. A compound of particular interest was the *cis*-β-lactam **94d** (and its *trans* isomer **93d**), which was used as intermediates for the side chain of Taxol and its analogs such as Taxotere. Coupling of **94d** with baccatin and chemical manipulation of the structures provided Taxol and Taxotere. When this cycloaddition reaction was conducted under microwave irradiation for 5 min, the ratio of the *trans* to *cis* isomer was 95:5. It was reported the preparation of the racemic form of the *trans*-β-lactam **93d** by the slow addition of triethylamine to a refluxing solution of the imine and the acid chloride in toluene solution [122]. This *trans*-β-lactam **93d** is an intermediate for 2′-epi-taxol (Scheme 19). Therefore, microwave-induced reactions can produce isomeric β-lactams of medicinal significance in a very simple manner and within a very short time.

From the results obtained above, it is concluded that the cycloaddition reaction involves multiple pathways some of which are highly accelerated and significant by rapid microwave irradiation. Higher temperature also facilitates *trans*-β-lactam

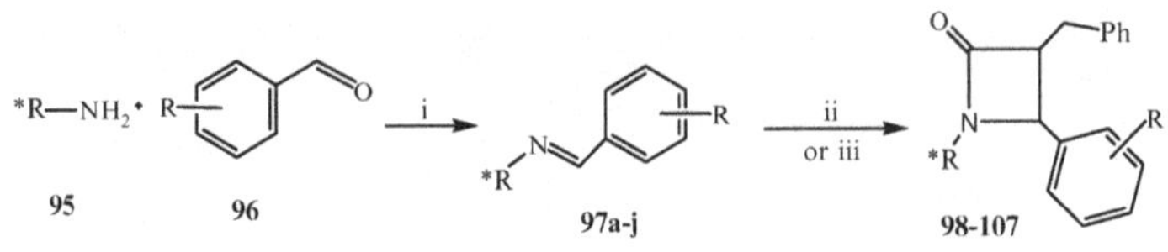

R*-NH$_2$ R	4-Me$_2$N	3,4-(MeO)$_2$	2,4,6-(MeO)$_3$	2-OH-3-OMe/ 2-(OCOR)-3-OMe
(+)-3-aminomethylpinane	**97a/98**	**97b/99**		
(S)-1-phenylethylamine	**97c/99**	**97e/102**	**97g/104**	**97i/106**
(R)-1-phenylethylamine	**97d/101**	**97f/103**	**97h/105**	**97j/107**

Scheme 20

formation. However, microwave irradiation is much more effective in the formation of specific *trans*-β-lactams.

By irradiating mixtures of an amine **95** and aldehyde **96** in a microwave oven (Scheme 20), the imine intermediate **97** was prepared quantitatively. Surprisingly, azeotropic distillation with Dean–Stark trap as well as in the presence of water trapping agents led to significantly low conversions and partial decomposition. This was because the microwave-induced reaction was very fast which allowed using these relatively unstable products without chromatography purification. The imine **97** was converted into the target azetidinones **98–107** via Staudinger [2 + 2] ketene-imine cycloaddition. This was done using triethylamine as a base and 3-phenylpropionyl chloride as a ketene component. The cycloaddition reaction was performed in twofold excess of ketene precursor due to the presence of a free hydroxyl group. However, in the case of azetidinones **106** and **107** formation, reactions were initially performed in refluxing toluene as the solvent. This was in an attempt to achieve β-lactam ring formation with *trans*-selectivity [122]. β-Lactams **98–107** were isolated in good to high yields since the transformations proceeded efficiently. The products were obtained diastereoselectively as pairs of isomers with the desired *trans*-C-3 and C-4-configuration, i.e., (3*R*,4*S*) and (3*S*,4*R*). The isomeric ratios were calculated on the basis of the relative integral intensities of the proton NMR spectra. High performance flash chromatography (HPFC) on silica gel separated the diastereoisomers. X-ray analysis of selected samples of the isomers with lower R_f-values determined the absolute configurations of the phenylethylamine azetidinones **99–107**. The isomers are solid products in most of the cases, while the less polar ones are always viscous oils. Crystals were grown by slow diffusion of hexane into chloroform solutions of the enantiomeric compounds. By using phenylethylamine auxiliary low to medium asymmetric induction in the ring construction was obtained. However, equimolar mixtures of *trans*-isomers were formed from aminomethylpinane derivatives. Comparison between 4-dimethylamino and methoxy derivatives showed that all phenylethylamine derivatives are formed efficiently. Aldehyde group moderately influenced the substituents selectivity. Commensurable superiority of the less polar isomer was observed in the formation of dimethylamino and 3,4-dimethoxy substituted β-lactams. However, a better selectivity was achieved with bulky electron donating substituent in *ortho*-position. The best result was confirmed with three methoxy groups.

Tetrachlorophthalolyl glycine **110** was prepared in 94% yield in 90 s in an unmodified domestic microwave oven. Protection of glycine **109** on a molecular scale of commercially available tetrachlorophthalic anhydride **108** was performed.

Tetrachlorophthaloyl (TCP)-protected glycine was found to be isolated in more than 90% yield after 8 min of irradiation followed by crystallization of the crude material from methanol. Treatment of acid **110** with thionyl chloride for 4 h provided the acid chloride **111** in very good yield [126].

The reaction of **111** with various imines in dichloromethane in the presence of triethylamine using microwave irradiation provided the TCP-protected α-amino-β-lactams **113a** and **113b** in moderate yields with varying *cis*/*trans* ratio. Acid chloride **111** reacted with imine **112** in chlorobenzene in the presence of *N*-methylmorpholine under microwave irradiation for 3–5 min and provided TCP-protected β-lactams **113a** and **113b** in excellent yield. *Trans* selectivity was observed (more than 90%) when the reaction was performed under microwave irradiation. Under conventional conditions, however, the product was a mixture of *cis*/*trans* isomers in varying proportions. Temperature of the reaction was found to affect the stereoselectivity. The imine solution in chlorobenzene with *N*-methylmorpholine was irradiated in a microwave oven extensively for 1 min in order to reach a temperature of about 110°C. The acid chloride was quickly added and the mixture of the reactants was then further irradiated for 3 min at a low power level. After cooling the reaction mixture, *n*-hexane was added, and the product was filtered off and washed with water in order to remove the inorganic salt. The protected group then was removed by reacting **113** with ethylenediamine to afford **114**. The resulting amine **114** was converted to amide **115**. Removal of the protective group was also performed in microwave oven. This reaction was selective since no cleavage of the β-lactam ring was observed (Scheme 21).

Interestingly, in the case of 4-styryl-2-azetidinone only the *cis*-β-lactam was the product obtained by this method. This was the product even under a high level of microwave irradiation. In another method, the imines **117** (**a–h**) were prepared by a general method. This was performed by refluxing 2-amino thiazole with different aromatic aldehydes. On cyclocondensation of imines with phenyl acetyl chloride and in the presence of triethylamine, azetidinones as shown in Scheme 22 were afforded in 60–65% yields. The ^{1}H NMR showed a *cis* configuration of the C-3 and C-4 protons of the 2-azetidinone ring. Each proton at C-3 and C-4 showed as a doublet. Polarization of the molecules under microwave irradiation caused rapid reaction to occur. Importantly, the yields of **118** (**a–h**) were high (85–95%) under microwave irradiation conditions. Using conventional heating however, the yields were comparatively low (60–65%).

The negative charges of the oxygen and nitrogen atoms of 1,3-thiazolyl group as well as the partial pi positive charges of sulfur and supplementary arm 2-OH contributed in favor of an antibacterial activity. Additionally it was established

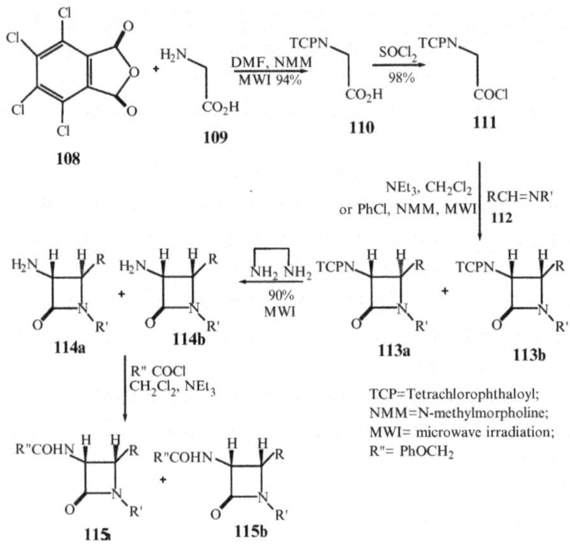

Scheme 21

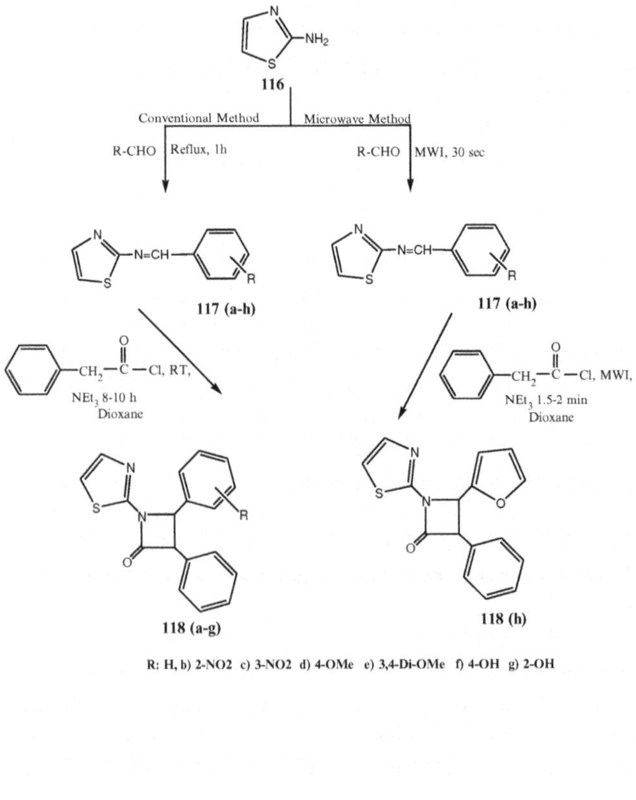

Scheme 22

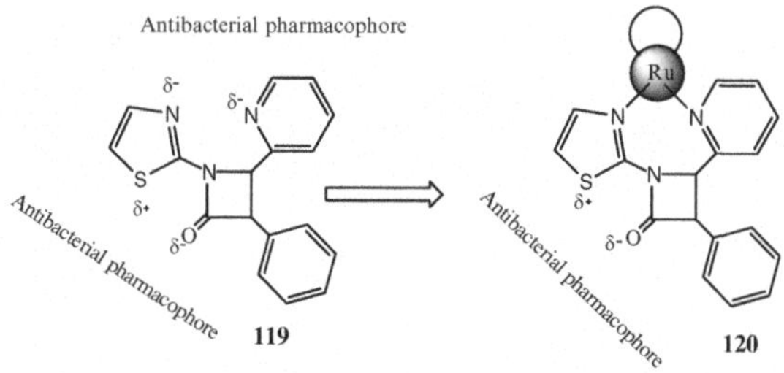

119 120

Scheme 23

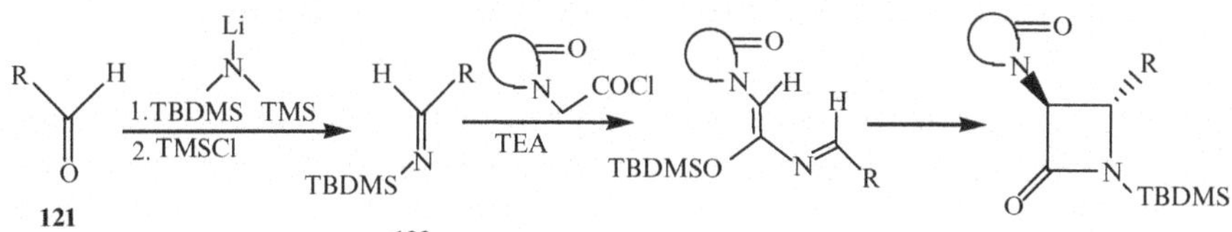

121 122 123 124

Scheme 24

that the activity is improved with increase in negative charge of one heteroatom of the pharmacophore fragment of the compounds (**119** and **120**) (Scheme 23) [127].

A systematic study to address the use of microwave techniques in the synthesis of β-lactam rings from 1,3-azadienes in the absence of solvents, following a two-step Staudinger reaction was performed. The cyclization was conducted on a scale from milligrams to 50 g in 1–20 min using inexpensive equipment. The adequate choice of irradiation time and control of the energy source from the microwave oven into the reaction mixture was the key to solvent-free organic reactions. This may in turn be responsible for the decomposition of the starting azadiene. A number of experiments were performed to identify the best conditions for obtaining high cyclization yield. The starting azadiene was prepared from a silylimine. The silylimine was in turn obtained from an aldehyde and an acyl chloride. The *N*-trialkylsilyl-substituted azadienes **123** was prepared with the hydrolytically stable TBDMS-group. According to the procedure (Scheme 24) the required *N-tert*-butyldimethylsilyl imine **123** was prepared from lithium *N,N*-(*tert*-butyldimethyl) trimethylsilyl amide **122** and an aldehyde **121**. Azadienes **123** was used as such or purified by flash chromatography on a short silica gel-column. This was depending on the purity of the crude reaction mixture as evaluated by ^{1}H NMR. Azadienes resulting from *tert*-butyldimethylsilylimines could be stored at low temperature to prepare β-lactam **124** [128].

A new procedure for the synthesis of β-aminoesters and β-lactams starting from imines and silyl ketene acetals was developed. Reaction of benzylidene aniline **125a** (R = Ph) with the silyl ketene acetal **126** over montmorillonite K10 afforded the β-aminoester [129] (Scheme 25).

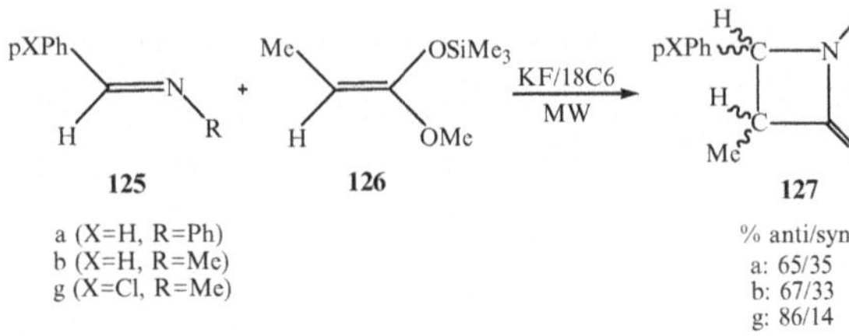

Scheme 25

8 Formation of β-Lactams with Polyaromatic Imines

The Staudinger reaction has been used extensively for the synthesis of polyaromatic β-lactams. This reaction requires an imine, base, and acid chloride, (or equivalent). The stereochemistry of resulting β-lactams varies, including *cis*, *trans*, and a *cis–trans* mixture, depending on the substituents present in the imine and acid chloride and conditions of the reactions. In general, the reaction of acyloxy-, alkoxy-, and nitrogen-containing acid chloride with diaryl imines produces *cis*-β-lactam under Staudinger reaction conditions. However, the reaction of polyaromatic imines with acetoxy, phenoxy, and phthalimido acid chloride in the presence of triethylamine at −78°C to room temperature produced *trans*-β-lactams (**128–141**) (Scheme 26). Isomeric polyaromatic imines (**142–144**) in which the aromatic moieties were interchanged produced *cis*-β-lactams (**145–147**) in good yield (Scheme 27). Although formation of β-lactams using the Staudinger reaction was discovered more than 90 years ago, surprisingly, there was no report in the literature regarding the use of tetracyclic or pentacyclic aromatic systems in β-lactam formation reaction. Interestingly, the formation of *trans*-β-lactams with polyaromatic imines was not described in the literature [57, 130]. Some previous studies were performed toward the formation of *trans*-β-lactams. However, the conditions in those experiments were different from those in the present study. For example, synthesis of some *trans*-β-lactams was performed using high-power microwave irradiation and changing the order of the addition of the reagents. In addition, *trans*-β-lactams were obtained in low yield as the only isolated products using cyclic imines, but they were not derived from an aromatic amine.

The above concept has been extended to similar polyaromatic imines. As a model study, reaction of diarylimine **149** with *O*-acetylmandelic chloride **148** in the presence of triethyl amine was performed at 0°C-room temperature and a single β-lactam **150** was obtained in 70% yield. The acetate and the C-4 hydrogen in **150** resonate at δ 1.6 and 5.7, respectively. Microwave-induced method, however, produced a mixture of two β-lactams **150** and **151** in a ratio of 1:1 (70% yield). High temperature reaction also produced an identical mixture of products **150** and **151** in 70% yield. The acetate and the C-4 hydrogen in β-lactam **151** resonate at δ 2.2 and 5.67, respectively (Scheme 28).

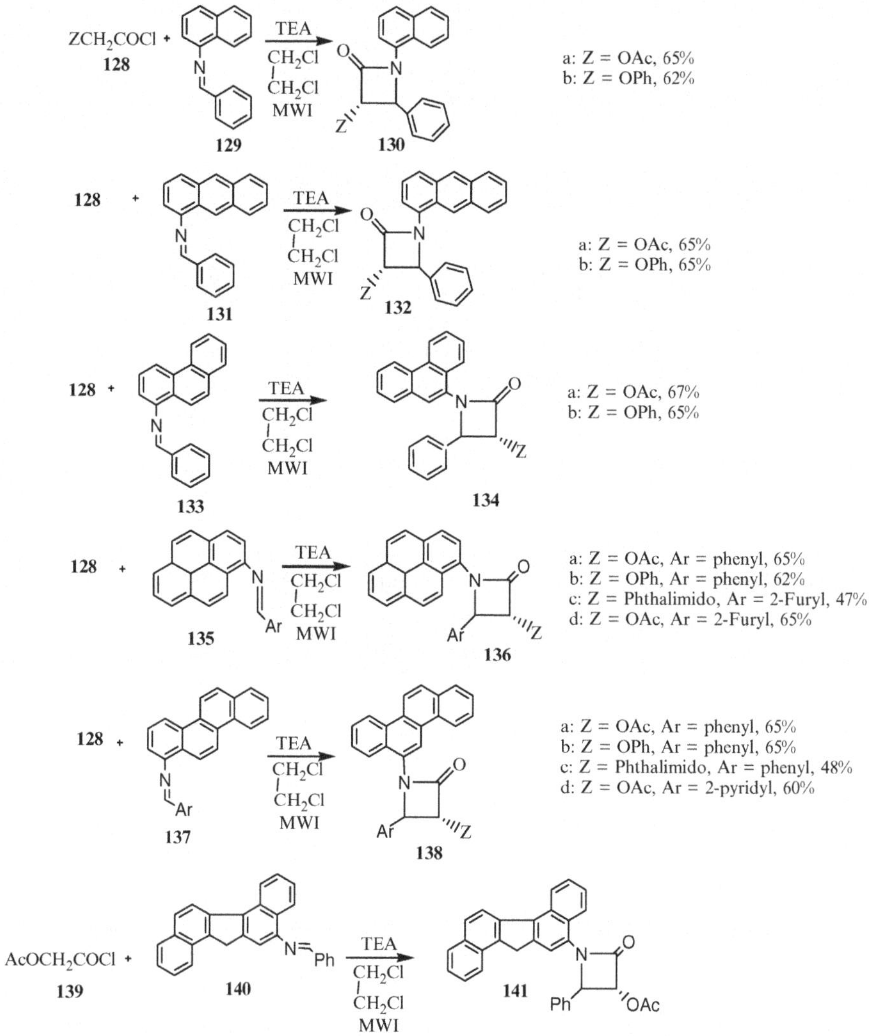

Scheme 26

Cycloaddition of imine **152** derived from a multicyclic aromatic amine at 0°C room temperature afforded two products **153** and **154** in a ratio of 7:3. Microwave irradiation, however, produced a mixture of two products **153** and **154** in a ratio of 2:8 (Scheme 29). The acetate and C-4 hydrogen in **153** resonate at δ 2.17 and 5.95, respectively. The acetate and C-4 hydrogen in **154** resonate at δ 2.33 and 6.46, respectively. Similar observation was obtained with imine **155**.

When irradiated in a microwave oven using chlorobenzene and triethylamine, β-lactam **153–155** did not isomerize. The β-lactams also did not produce to isomeric β-lactams when they were refluxed in ethylenedichloride and triethylamine. These

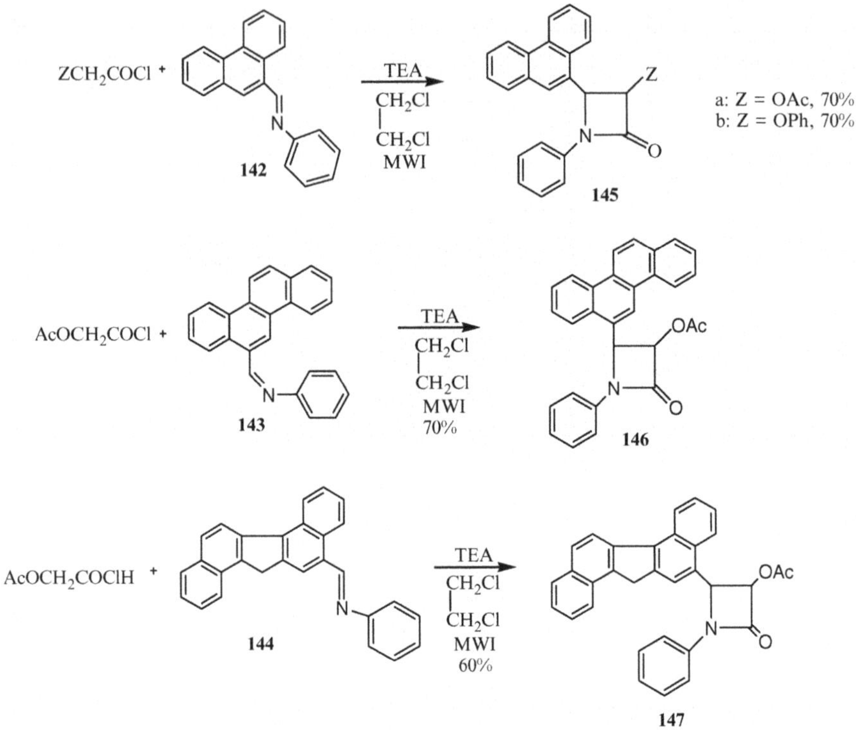

Scheme 27

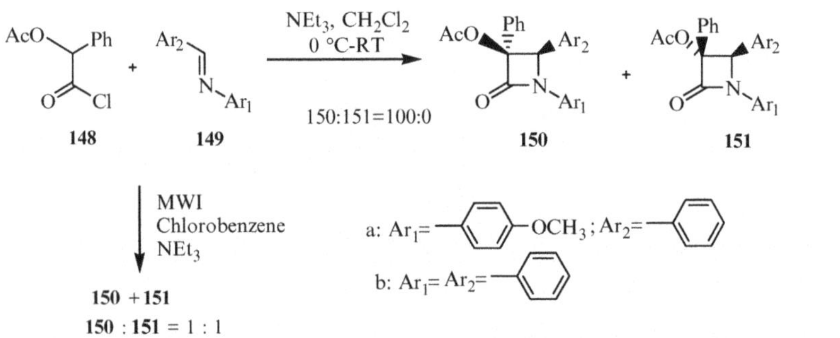

Scheme 28

experiments established that there were no isomerization of the β-lactams during reaction at a high temperature and/or under microwave irradiation.

This results in a *trans*-β-lactam in which C-4 phenyl and C-3 acetoxy groups are *trans* to each other in the predominant product. The classical condition using imine **149** followed the normal cycloaddition path as reported in the literature and this

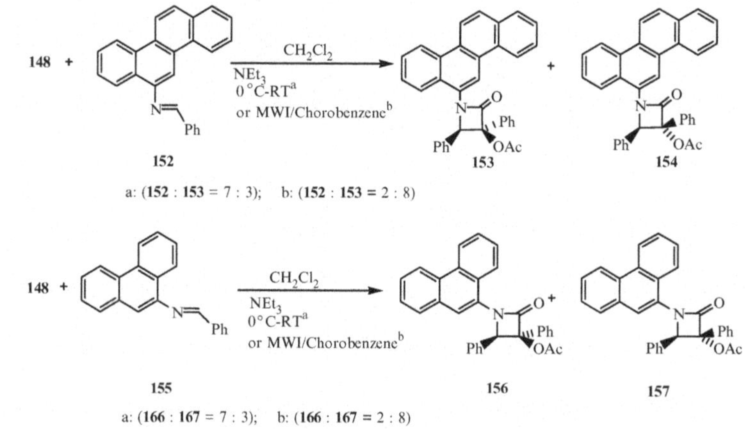

Scheme 29

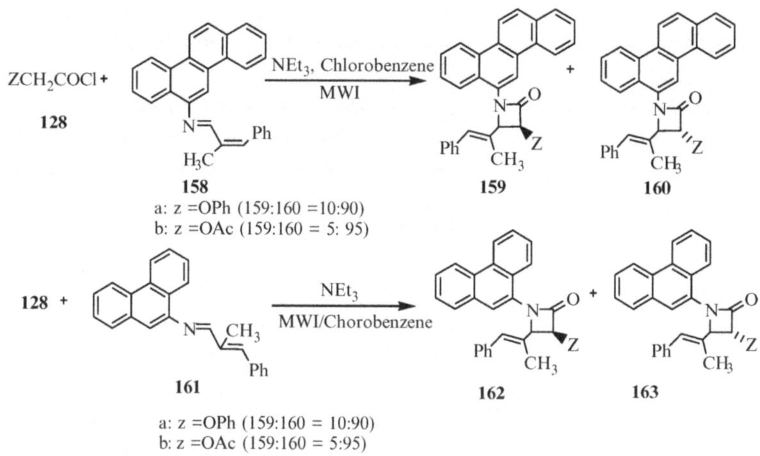

Scheme 30

gave *cis*-type of compounds (C-3 acetoxy and C-4*H* are *trans* to each other). However, drastic energy through microwave radiation altered the structure of the intermediate presumably through a rotation of the bond and these result in the formation of different stereoisomers in major proportion. A similar approach with different imines **158** and **161** was performed and the corresponding β-lactams were obtained (Scheme 30). Acetoxyacetyl chloride provided *trans*-β-lactam predominantly even with conjugated system.

Reaction of ferrocene aldehyde with 6-aminochrysene in toluene produced imine **164** after 72 h reflux using a Dean–Stark water separator. The imine **164** was then reacted with acid chloride **128** at 0°C room temperature and a single

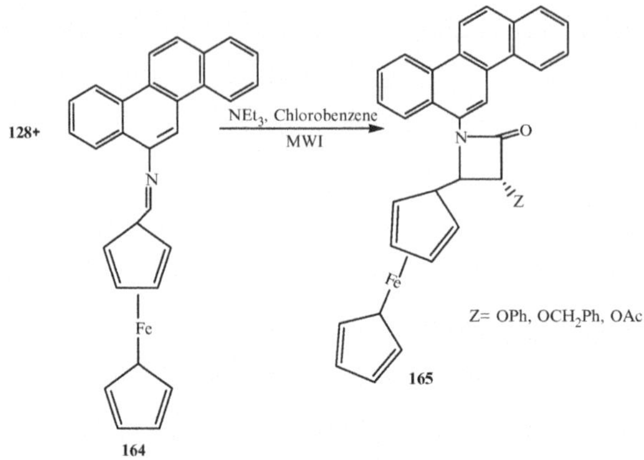

Scheme 31

trans-β-lactam **165** was obtained in 70% yield (Scheme 31). This result suggests that ferrocene group exerts a similar behavior like aromatic aldehydes in the Staudinger reaction.

The formation of a *trans* isomer as observed in the present study can be rationalized through isomerization of the enolates (Scheme 32, **166–176**).

The electron-withdrawing polyaromatic system at the nitrogen stabilizes the iminium ion [131]. This process allows rotation of the bond (**169–170**) and results in the formation of *trans*-β-lactam **171**. This observation is similar to that described by Just et al. in which formation of a *trans*-isomer having electron-withdrawing nitro-substituted imines was performed [132]. In contrast, the exclusive formation of a *cis*-β-lactam having a polyaromatic group and cinnamyl at C-4 prompted us to develop a hypothesis regarding a mechanism previously described by Doyle et al. [133]. In this context, extended conjugation of the cinnamyl and polyaromatic system stabilizes the acyliminium ion **172**. Furthermore, the presence of the cinnamyl or polyaromatic system at C-4 outweighs the contribution of the *N*-polyaromatic system, resulting in *cis*-β-lactam formation. Subsequent proton abstraction from complex **172** produced *cis*-β-lactam **175–176** (90° bond rotation and closure) or **174** (anion inversion and closure). This hypothesis was further strengthened by the possible formation of donor–acceptor complex **176** as suggested by Bose et al. [134]. This complex formation effectively stabilized the transition states of the reaction.

9 Mechanism of β-Lactam Formation Reactions

The mechanism of β-lactam formation has been investigated extensively, and the rationale for the observed diastereoselectivity in certain cases remains unknown. It has been shown that the stereoselectivity depends on a number of factors: the

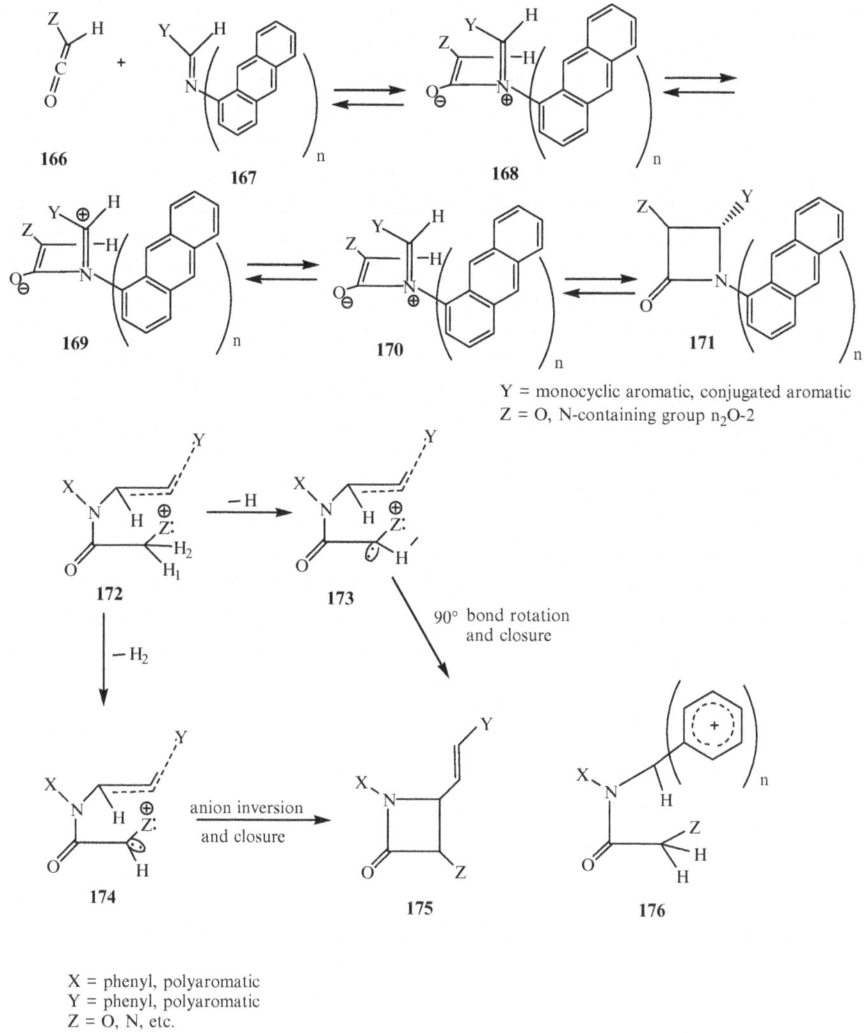

Scheme 32

structure of the imine, acid chloride (equivalent), sequence of reagent addition, solvent, amounts of solvent, time, temperature, and bases. In a large number of observations, *cis*-β-lactam was found to be the exclusive or major product when acid chloride (equivalent) was added dropwise at low to room temperature to the solution of imines and tertiary base. On the other hand, a *trans*-β-lactam is the major or exclusive product when a tertiary base is slowly added to the imine and acid chloride (equivalent) solution at room to high temperature. Based on the large amount of data in the literature, some predictions concerning their stereoselectivity have been made. For instance, Georg and Ravikumar have established some general rules regarding stereoselectivity in the formation of β-lactam rings [134]. Also, computer-assisted theoretical calculations have been advanced to explain the stereochemical outcome. Cossio and coworkers and Sordo and coworkers have explained the stereochemical results on the basis of torquoelectronic effects (for example, see [135–143]). Low-temperature infrared spectroscopy has been used by Lynch et al. to identify the reactive intermediates [144]. In general, two mechanisms have been proposed to explain the product distribution in the β-lactam formation reaction. One of these, the ketene mechanism, was observed in a low-temperature infrared spectroscopy study while the other, the acylation of imine mechanism was also believed to be involved. Both mechanisms were supported by numerous lines of evidence in several studies. In particular, it has been hypothesized that cycloaddition of the imine occurs from the least hindered side of the hetene, a process that generates zwitterionic intermediates; conrotatory cyclization of these intermediates can then provide *cis*- and *trans*-β-lactams. In addition, the latter mechanism proposes acylation of the imine by the acid chloride to form *N*-acyliminium chloride, which produces zwitterionic intermediates.

Our current thoughts regarding the likely mechanisms by which this cycloaddition with polyaromatic imines proceeded are described below. Infrared spectra had shown a strong band at 2,200 cm^{-1} when acid chloride and imine were reacted in the presence of triethylamine within 30 min after the start of the reaction. This band mostly disappeared after 24 h or reaction. Therefore, a ketene was involved in this reaction as an intermediate.

However, Cossio et al. described that the SN_2 intramolecular mechanism favored the preferential or exclusive formation of *trans*-β-lactams, particularly when the reactions were allowed to take place in the absence of a tertiary base in the initial stages of the reaction. In contrast, triethylamine was used as one of the reactants at the beginning of our experiment, yet the product was a *trans*-β-lactam in many cases as described here. The use of diisopropylethyl amine as the base did not improve the yield of the products. However, the stereochemistry of the products with polyaromatic imines remained identical. The stereoselectivity of *trans*-β-lactam can be explained on the basis of steric effects. It is postulated the isomerization of C=N bond prior to ring closure as a result of severe steric interactions. Undoubtedly, this mechanism explained our *trans*-selectivity, but did not explain *cis* selectivity with similar types of compounds. Formation of a mixture of *cis*- and *trans*-isomers with naphthalenyl and anthracenyl imines could not be explained using the mechanisms described above. If the electron withdrawal properties or the steric crowding of the *N*-polyaromatic system were solely responsible for the *trans*-β-lactam formation, then an identical stereochemical distribution would have been observed in the isomeric naphthalenyl and anthracenyl compounds. Comparison of these results and examination of the *N*-polyaromatic systems with which *trans* isomers were the only products revealed a structural similarity. These imines have a peri hydrogen very close to the C=N bond, whereas this peri hydrogen is relatively distant from the same bond in similar structures. But, it has been established that microwave irradiation can alter the stereochemical distribution when naphthalenyl and anthracenyl imines were used irrespective of

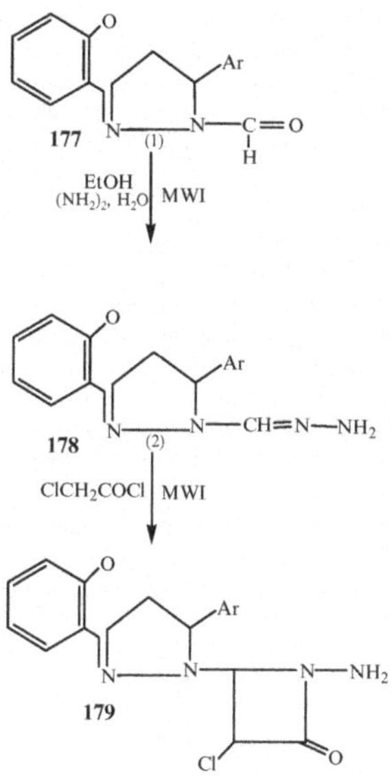

Scheme 33

the absence of the peri hydrogen. In contrast, several other imines with extended conjugation never produced *trans*-β-lactams even after performing the reactions under forcing conditions. In this respect, it appeared that stabilization of the positive charge by the extended conjugation is the major contributor in dictating the stereochemistry of the final β-lactams. These extensive results also indicate that it is the nature of the C-4 group that controls the isomer distribution in this type of reaction. The products as described in Scheme 26 were obtained based on hypothesis-driven work on conformationally restricted compounds (for example, see [135–142]).

10 Synthesis of Diverse β-Lactams (Scheme 33)

Polycyclic novel β-lactams were prepared starting from cyclic imine through microwave irradiation as shown in (Scheme 38) [143]. It has been found that clay can help to form imine and Staudinger reaction can be performed with success in a one-pot operation (Scheme 34) [144].

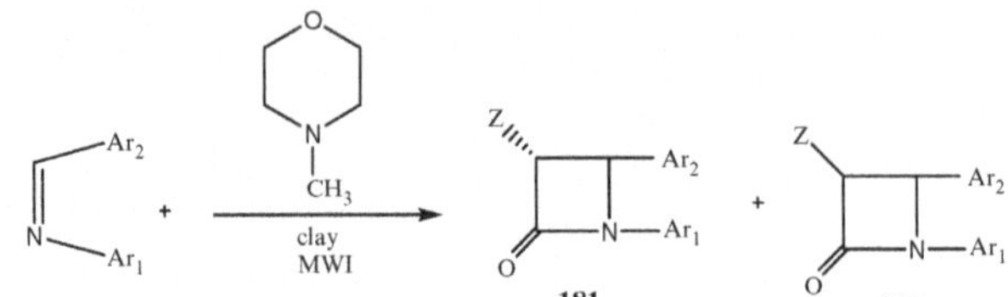

180 181 182

a: Ar1= Ar2=phenyl
b: Ar1= *p*-mehoxyphenyl; Ar2= phenyl
c: Ar1= phenyl, Ar2=*p*-methoxyphenyl
d: Ar1= Ar2=*p*-methoxyphenyl
e: Ar1= phenyl, Ar2= benzyl
f: Ar1= *p*-methoxyphenyl, Ar2= cinnamyl

Scheme 34

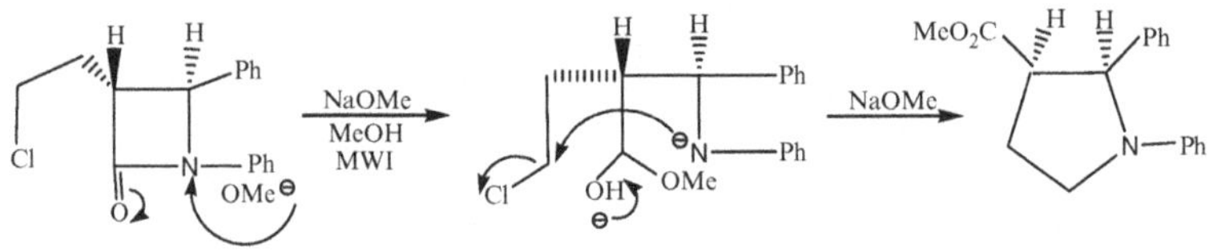

183 184 185

Scheme 35

11 Rearrangement of the β-Lactam Ring

Pyrimidines and oxazines were prepared from β-lactam rearrangement reaction using domestic microwave oven. For example, *trans*-β-lactam **183** on treatment with sodium methoxide in methanol produced pyrrolidine **185** due to the cleavage and subsequent rearrangement (Scheme 35) [145]. Oxazines were derived under reductive conditions with indium metal-induced reactions (Schemes 36 and 37). The nitro group can be reduced to amino group. On nucleophilic attack by the amino group to the β-lactam carbonyl group, the ring was cleaved (Scheme 37) [146].

12 Preparation of Taxol and Taxotere Side Chains

The formation of a single glycoside from optically active α-hydroxy β-lactams studied raised the possibility of synthesizing both enantiomeric forms of racemic β-lactams. On this basis, glycosylation of racemic *cis*-1-(*p*-anisyl)-3-hydroxy-4-phenyl-2-azetidinone **194** was investigated in detail. On treatment with glucal triacetate **193** and iodine in tetrahydrofuran solution, **194** led in 60% yield to a mixture of two diastereomeric compounds **196** and **195** in the ratio of 55:45 (Scheme 38).

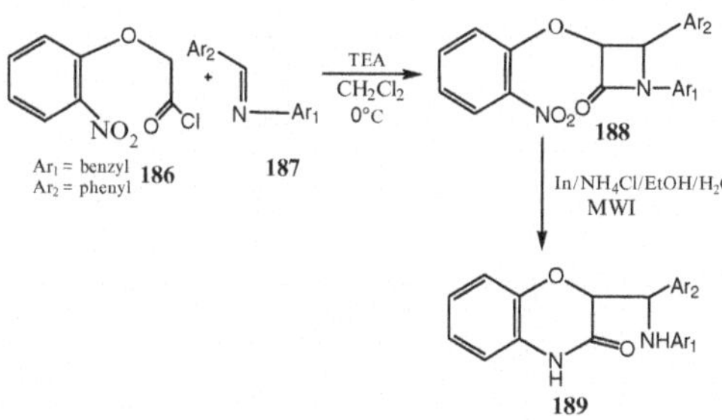

Scheme 36

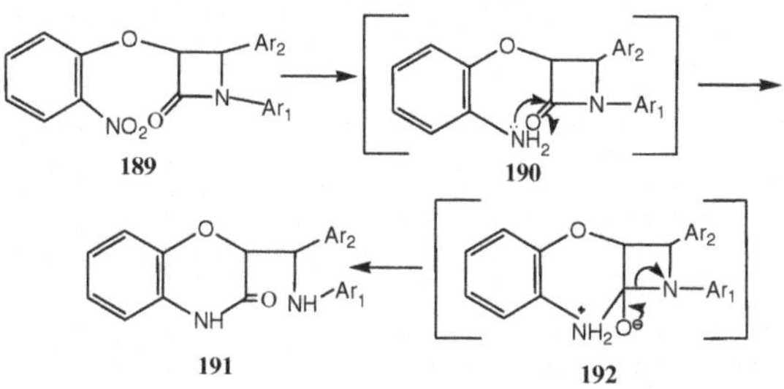

Scheme 37

Catalyst transfer hydrogenation of **196** in ethanol produced a 2,3,4-trideoxyglucoside **197**. The CTH reaction using a domestic microwave oven in ethylene glycol produced an identical product within 2 min. The small couplings for the anomeric proton indicated that an α-glycosidic linkage had been formed (Scheme 38). This reaction was performed by placing the reaction mixture in a large beaker or an Erlenmeyer flask with a loose cover for a top and a beaker of water by its side. Irradiation for 2 min under low energy setting heated the reaction mixture to about 80°C.

The diastereomers **196** and **195** were separated through column chromatography and treated individually with aqueous hydrochloric acid solution or bismuth nitrate-catalyzed reaction to remove the sugar moiety. As expected, the hydroxy β-lactams **197** and **199** were obtained, and these were then converted to their corresponding acetates **198** and **200** in 90% yield (Scheme 39). Although aqueous hydrochloride acid worked well in the cleavage of the hemiacetal bond in **195** and **195**, we discovered that an aqueous solution of bismuth nitrate produced products **197** and **199** in comparable yields.

2-Fluoro-3-formyl substituted quinolines gave the corresponding hydrazones upon reduction with ocranoid acid hydrazide in ethanol. β-Lactams are afforded through conventional heating and by reaction with chloroacetyl chloride and triethylamine in a microwave oven [147].

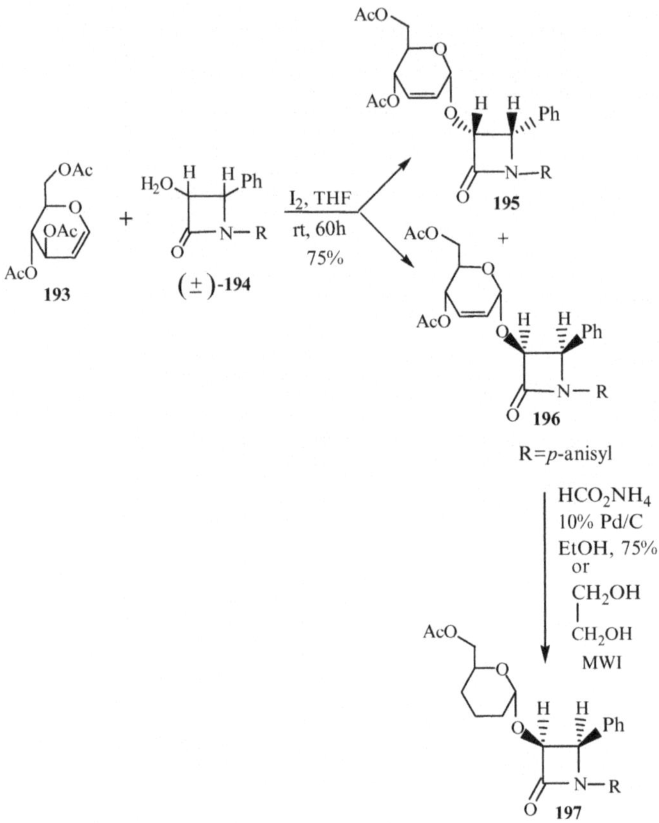

Scheme 38

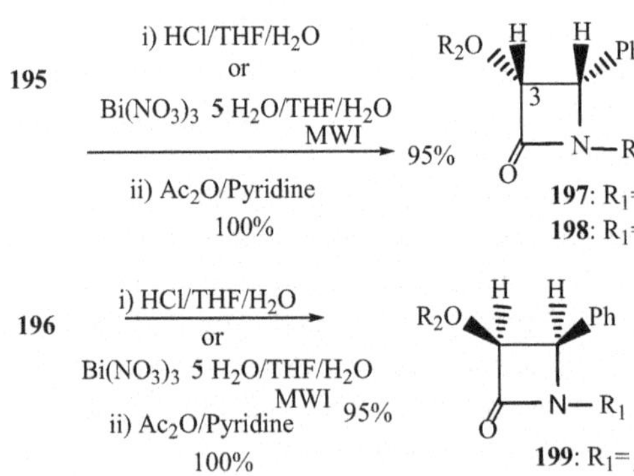

197: R_1= *p*-anisyl, R_2= H
198: R_1= *p*-anisyl, R_2= Ac

199: R_1= *p*-anisyl, R_2= H
200: R_1= *p*-anisyl, R_2= Ac

Scheme 39

13 Conclusion

Domestic and automated microwave-induced reactions are explored systematically with different other substrates. Most of the instances, there is a tendency to form *trans*-β-lactams. The synthesis of *trans*-isomer was investigated using related methods [148–152]. The mechanism of *trans*-β-lactam formation was also explained in some elegant studies [153–156]. Higher activation energy transformations that are difficult and impossible to complete with conventional heating (oil bath, steam bath, and mantle) can be performed very easily with microwave irradiation because of the facile energy transfer process. Heat is applied externally and it passes through the walls of the reaction vessel and solvent during thermally assisted chemical reactions. Therefore, some of the reactions under thermal conditions may not prove efficient. However, microwave-induced reactions have a number of advantages because of microwave-coupling, microwave heating, microwave irradiation, and molecular heating. Microwave coupling is the direct transfer of microwave energy to a substrate that results in instantaneous heating. Microwave heating is the direct energy transfer process to the reaction mixtures; this raises kinetic excitation and is characterized by rapid energy transfer. Microwave irradiation is a form of nonionizing radiation that transfers energy by interacting with polar molecules. In addition, the direct energy transfer from microwave to the molecules will greatly enhance the speed of the chemical reaction (solventless reaction). Therefore, microwave-induced reactions will continue to attract chemists and industrialists for the synthesis of novel molecules with controlled stereochemistry, higher yield and because of the green nature of the process.

Acknowledgments We gratefully acknowledge the funding support from Kleberg Foundation of Texas and NCI to BKB.

References

1. Southgate R, Branch C, Coulton S, Hunt E (1993) Springer 2:621
2. Kidwai M, Sapra P, Bhushan KR (1999) Curr Med Chem 6:195–215
3. Bose AK, Manhas MS, Banik BK, Srirajan V (2000) b-Lactams: cyclic amides of distinction. In: Greenberg A, Breneman CM, Liebman JF (eds) The amide linkage: selected structural aspects in chemistry, biochemistry, and material science. Wiley-Interscience, New York, pp 157–214 (Chap. 7)
4. Banik BK (ed) (2004) β-Lactams: synthesis, stereochemistry, synthons and biological evaluation; current medicinal chemistry, vol 11. Bentham Science, San Francisco
5. Buynak JD (2004) Curr Med Chem 11:1951–1964
6. Manhas MS, Banik BK, Mathur A, Vincent JE, Bose AK (2000) Tetrahedron 56:5587–5601
7. Bose AK, Mathur C, Wagle DR, Manhas MS (2000) Tetrahedron 56:5603–5619
8. Ojima I (1995) Acc Chem Res 28:383–389
9. Banik BK, Manhas MS, Bose AK (1994) J Org Chem 58:4714–4716
10. Banik BK, Manhas MS, Bose AK (1993) J Org Chem 58:307–309
11. Suffness M (1995) Taxol science and applications. CRC, Boca Raton

12. Clader JW, Burnett DA, Caplen MA, Domalski MS, Dugar S, Vaccaro W, Sher R, Browne ME, Zhao H, Burrier RE, Salisbury B, Davis HR (1996) J Med Chem 39:3684–3693
13. Burnett DA, Caplen MA, Darris HR Jr, Burrier RE, Clader JW (1994) J Med Chem 37:1734–1736
14. Burnett DA (2004) Curr Med Chem 11:1873–1887
15. Clader JW (2004) J Med Chem 47:1–9
16. Finke PE, Shah SK, Fletcher DS, Ashe BM, Brause KA, Chandler GO, Dellea PS, Hand KM, Maycock AL, Osinga DG, Underwood DJ, Weston H, Davies P, Doherty JB (1995) J Med Chem 38:2449–2462
17. Magriotis PA (2001) Angew Chem 113:4507–4509
18. Taggi AE, Hafez AM, Wack H, Young B, Drury WJ III, Lectka T (2000) J Am Chem Soc 122:7831–7832
19. Banik BK et al (2005) Bioorg Med Chem 13:3611–3622
20. Hodous BL, Fu GC (2002) J Am Chem Soc 124:1578–1579
21. Cordova A, Watanabe S, Tanaka F, Notz W, Barbas CF Jr (2002) J Am Chem Soc 124:1866–1867
22. France S, Weatherwax A, Taggi AE, Lectka T (2004) Acc Chem Res 37:592–600
23. Ruhland B, Bhandari A, Gordon EM, Gallop MA (1996) J Am Chem Soc 118:253–254
24. Furman B, Thurmer R, Kaluza Z, Lysek R, Voelter W, Chmielewski M (1999) Angew Chem Int Ed 38:1121–1123
25. Gordon K, Bolger M, Khan N, Balasubramanian SA (2000) Tetrahedron Lett 41:8621–8625
26. Schunk S, Enders D (2000) Org Lett 2:907–910
27. Annunziata R, Benaglia M, Cinquini M, Cozzi F (2000) Chem Eur J 6:133–138
28. Gordon KH, Balasubramanian S (2001) Org Lett 3:53–56
29. Schunk S, Enders D (2002) J Org Chem 67:8034–8042
30. Delpiccolo CML, Mata EG (2002) Tetrahed Asym 13:905–910
31. Mascaretti OA, Boschetti CE, Danelon GO, Mata EG, Roveri OA (1995) Curr Med Chem 1:441–470
32. Edwards PD, Bernstein PR (1994) Med Res Rev 14:127–194
33. Sandanayaka VP, Prashad AS, Yang Y, Williamson T, Lin YI, Mansour TS (2003) J Med Chem 46:2569–2571
34. Buynak JD, Rao AS, Fod GP, Carver C, Carver C, Adam G, Geng B, Bachmann B, Shobassy S, Lackey S (1997) J Med Chem 40:3423–3433
35. Bonneau PR, Hasani F, Plouffe C, Malenfant E, Laplante SR, Guse I, Ogilvie WW, Plante R, Davidson WC, Hopkins JL, Morelock MM, Cordingley MG, Deziel R (1999) J Am Chem Soc 121:2965–2973
36. Ghatak A, Becker FF, Banik BK (2000) Heterocycles 53:2769–2772
37. Banik BK, Ghatak A, Becker FF (2000) J Chem Soc Perkin Trans 1:2179–2181
38. Banik BK, Becker FF (2000) Tetrahedron Lett 41:6551–6554
39. Ng S, Banik I, Okawa A, Becker FF, Banik BK (2001) J Chem Res 3:118–119
40. Dasgupta SK, Banik BK (2002) Tetrahedron Lett 43:9445–9447
41. Banik BK, Samajdar S, Banik I (2003) Tetrahedron Lett 44:1699–1701
42. Banik BK, Banik I, Hackfeld L (2003) Heterocycles 59:505–508
43. Banik BK, Adler D, Nguyen P, Srivastava N (2003) Heterocycles 61:101–104
44. Banik BK, Banik I, Samajdar S, Wilson M (2003) Heterocycles 63:283–296
45. Becker FF, Banik BK (1998) Bioorg Med Chem Lett 8:2877–2880
46. Banik BK, Venkatraman MS, Mukhopadhyay C, Becker FF (1998) Tetrahedron Lett 39:7247–7250
47. Becker FF, Mukhopadhyay C, Hackfeld L, Banik I, Banik BK (2000) Bioorg Med Chem 8:2693–2699
48. Banik BK, Ghatak A, Mukhopadhyay C, Becker FF (2000) J Chem Res 108–109
49. Mukhopadhyay C, Becker FF, Banik BK (2001) J Chem Res 1:28–31
50. Banik BK, Becker FF (2001) Bioorg Med Chem 9:593–605

51. Banik BK, Becker FF (2001) Curr Med Chem 8:1513–1533
52. Banik BK (ed) (2001) Current approaches to the development of new chemotherapeutic agents, current medicinal chemistry, vol 8. Bentham Science, San Francisco
53. Banik BK, Samajdar S, Banik I (2004) J Org Chem 69:213–216
54. Becker FF, Banik BK (2000) US Patent 6,015,811
55. Becker FF, Banik BK (2001) US Patent 6,184,224
56. Becker FF, Banik BK (2002) US Patent 6,362,200
57. Banik I, Becker FF, Banik BK (2003) J Med Chem 46:12–15
58. Banik BK, Becker FF, Banik I (2004) Bioorg Med Chem 2:2523–2528
59. Perreux L, Loupy A (2001) Tetrahedron 57:9199–9223
60. Lidstrom P, Tierney J, Wathey B, Westman J (2001) Tetrahedron 57:9225–9287
61. Bose AK, Manhas MS, Ganguly SN, Sharma AH, Banik BK (2002) Synthesis 1578–1591
62. Bose AK, Banik BK, Lavlinskaia N, Jayaraman M, Manhas MS (1997) Chemtech 27:18–24
63. Bose AK, Manhas MS, Ghosh M, Shah M, Raju VS, Bari SS, Newaz SN, Banik BK, Barakat KJ, Chaudhury AG (1991) J Org Chem 56:6968–6970
64. Banik BK, Manhas MS, Kaluza Z, Barakat KJ, Bose AK (1992) Tetrahedron Lett 33:3603–3606
65. Banik BK, Manhas MS, Newaz SN, Bose AK (1993) Bioorg Med Chem Lett 3:2363–2368
66. Bose AK, Banik BK, Barakat KJ, Manhas MS (1993) Synlett 575–577
67. Bose AK, Manhas MS, Banik BK, Robb EW (1994) Res Chem Intermed 20:1–20
68. Banik BK, Manhas MS, Robb EW, Bose AK (1997) Heterocycles 44:405–415
69. Banik BK, Jayaraman M, Srirajan V, Manhas MS, Bose AK (1997) J Ind Chem Soc 74:951
70. Banik BK, Barakat KJ, Wagle DR, Manhas MS, Bose AK (1999) J Org Chem 64:5746–5753
71. Rajagopal S, Anwer MK, Spatola AF (1994) In: Basava C, Anantharamaiah GM (eds) Peptides: design, synthesis and biological activity. Birkhauser, Boston, p 11 (Chap. 2)
72. Rao HSP, Reedy KS (1994) Tetrahedron Lett 35:171–174
73. Ram S, Ehrenkaufer RE (1988) Synthesis 91
74. Viswanatha V, Hruby V (1980) J Org Chem 45:2010
75. Jackson AE, Johnstone RAW (1997) Synthesis 685
76. Brieger G, Nestrick T (1974) J Chem Rev 74:567
77. Braude EA, Listead RP (1954) J Chem Soc 3544–3548
78. Anwer MK, Khan SA, Sivanandaiah KM (1978) Synthesis 751
79. Furst A, Berlo RC, Hooton S (1965) Chem Rev 65:51–68
80. Gray BD, Jeffs PW (1987) J Chem Soc Chem Commun 1329–1330
81. Elamin B, Anantharamaiah GM, Royer GP, Means GE (1979) J Org Chem 44:3442–3444
82. Anwer MK, Spatola AF (1980) Synthesis 929
83. Adger BM, O'Farrell C, Lewis NJ, Mitchell MB (1987) Synthesis 53
84. Carpino L, Tunga A (1986) J Org Chem 11:1930
85. Ram S, Ehrenkaufer RE (1986) Synthesis 133
86. Overman LE, Sugai S (1985) Helv Chim Acta 68:745
87. Ranu BC, Sarkar A, Guchhait SK, Ghosh KJ (1998) Ind Chem Soc 75:690–694
88. Bajwa JS (1992) Tetrahedron Lett 33:2299
89. Felix AM, Heimer EP, Lambros TJ, Tzougraki C, Meienhofer J (1978) J Org Chem 43:4194–4196
90. Entwistle ID, Gilkerson T, Johnstone RAW, Telford RP (1978) Tetrahedron 34:313
91. Entwistle ID, Johnstone RAW, Telford RP (1977) J Chem Res 117
92. Marques CA, Selva M, Tundo P (1993) J Chem Soc Perkin Trans 1:529
93. Bose AK, Banik BK, Barakat KJ, Manhas MS (1993) Synlett 575
94. Bose AK, Manhas MS, Ghosh M, Shah M, Raju VS, Bari SS, Newaz SN, Banik BK, Chaudhary AG, Barakat KJ (1991) J Org Chem 56:6968–6970
95. Ojima I, Suga S, Abe R (1980) Chem Lett 9:853–856
96. Newaz SN, Manhas MS (1993) Synlett 897
97. Banik BK, Newaz SN, Manhas MS, Bose AK (1993) Synlett 897

98. Banik BK, Manhas MS, Bose AK (1994) J Org Chem 59:4714–4716
99. Banik BK, Subbaraju GV, Manhas MS, Bose AK (1996) Tetrahedron Lett 37:1363–1366
100. Banik BK, Manhas MS, Bose AK (1997) Tetrahedron Lett 38:5077–5080
101. Banik BK, Zegrocka O, Manhas MS, Bose AK (1997) Heterocycles 46:173–176
102. Bose AK, Manhas MS, Ghosh M, Raju VS, Tabei K, Urbanczyk-Lipkowsa Z (1990) Heterocycles 30:741–744
103. Bose AK, Manhas MS, Ghosh M, Shah M, Raju VS, Bari SS, Newaz SN, Banik BK, Chaudhury AG, Barakat KJ (1991) J Org Chem 56:6968–6970
104. Bari SS, Bose AK, Chaudhury AG, Manhas MS, Raju VS, Robb EW (1992) J Chem Educ 69:938–939
105. Banik BK, Newaz SN, Manhas MS, Bose AK (1993) Bioorg Med Chem Lett 3:2363–2368
106. Samajdar S, Becker FF, Banik BK (2000) Tetrahedron Lett 41:8017–8020
107. Samajdar S, Becker FF, Banik BK (2001) Arkivoc 8:27–33
108. Samajdar S, Basu MK, Becker FF, Banik BK (2002) Synth Commun 32:1917–1921
109. Samajdar S, Becker FF, Banik BK (2001) Synth Commun 31:2691–2695
110. Srivastava N, Dasgupta S, Banik BK (2003) Tetrahedron Lett 44:119–194
111. Srivastava N, Banik BK (2003) J Org Chem 68:2109–2114
112. Banik BK, Banik I, Renteria M, Dasgupta S (2005) Tetrahedron Lett 46:2643–2646
113. Banik BK, Manhas MS, Bose AK (1994) J Org Chem 56:4714–4718
114. Samajdar S, Basu MK, Becker FF, Banik BK (2001) Tetrahedron Lett 42:4425–4427
115. Basu MK, Samajdar S, Becker FF, Banik BK (2001) Synlett 519
116. Banik BK, Fernandez M, Alvarez C (2005) Tetrahedron Lett 46:2479–2482
117. Salzmann TN, Ratcliffe RW, Christensen BG (1980) J Am Chem Soc 102:6161–6163
118. Hanessian S, Desilets D, Bennani YL (1990) J Org Chem 55:3098–3103
119. Nagahara T, Kametani T (1987) Heterocycles 25:729–806
120. Kametani T, Fukumoto K, Ihara M (1982) Heterocycles 17:463–506
121. DiNinno F, Beattie TR, Christensen BG (1977) J Org Chem 42:2960–2965
122. Todorov AR, Kurteva VB, Bontechev RP, Vassilev NG (2009) Tetrahedron 65:10339–10347
123. Ariens EJ (1993) Stereochemistry and biological activity of drugs. Blackwell Scientific Publishers, Oxford
124. Banik BK (ed) (2004) β-Lactams: synthesis, stereochemistry, synthons and biological evaluation. Curr Med Chem, vol 11. Bentham Science, San Francisco
125. Chiral praseodymium *trans*-3-(heptafluropropylhydroxymethylene) camphorate was used as shift reagent in the NMR study
126. Bose AK, Jayraman M, Okawa A, Bari SS, Robb EW, Manhas MS (1996) Tetrahedron Lett 37:6989–6992
127. Ali P, Meshram J, Tiwari V (2010) Int J Chem Tech Res 2:956–964
128. Matelli G, Spunta G, Panunzio M (1998) Tetrahedron Lett 39:6257–6260
129. Texier-Boullet F, Latouche R, Hamelin J (1993) Tetrahedron Lett 34:2123–2126
130. Banik BK, Becker FF, Banik I (2004) Bioorg Med Chem 12:2523–2528
131. Georg GI, Ravikumar VT (1992) In: Georg GI (ed) In the organic chemistry of β-lactams. VCH, New York
132. Just G, Ugolini A, Zamboni R (1979) Synth Commun 9:117–121
133. Doyle TW, Belleau B, Luh BY, Ferrari CF, Cunningham MP (1977) Can J Chem 55:468–507
134. Bose AK, Chiang YH, Manhas MS (1972) Tetrahedron Lett 13:4091–4094
135. Lin TH, Rogers TS, Hill DL, Simpson-Herren L, Farnell DR, Kochhar DM, Alam M, Brouillette W, Muccio DD (1996) Toxicol Appl Pharmacol 139:310–316
136. Garratt PJ, Vonhoff S, Rowe SJ, Sugden D (1994) Bioorg Med Chem Lett 4:1559–1564
137. Mathe-Allainmat M, Gaudy F, Sicic S, Dangy-Caye A-L, Shen S, Bremont B, Benatalah Z, Langlois M, Renard P, Delagrange P (1996) J Med Chem 39:3089–3095
138. Anderson A, Boyd AC, Clark JK, Fielding L, Gemmell DK, Hamilton NM, Maidment MS, May V, McGuire R, McPhail P, Sansbury FH, Sundaram H, Taylor R (2000) J Med Chem 43:4118–4125

139. Barboni L, Lambertucci C, Appendino G, Vander Velde DG, Himes RH, Bombardelli E, Wang M, Synder JP (2001) J Med Chem 44:1576–1587
140. Hernandez AI, Balzarini J, Karlsson A, Camarasa MJ, Perez MJ (2002) J Med Chem 45:4254–4263
141. Selvakumar N, Srinivas D, Khera MK, Kumar MS, Mamidi RNVS, Sarnaik H, Charavaryamath C, Rao BS, Raheem MA, Das J, Iqbal J, Rajagopalan R (2002) J Med Chem 45:3953–3962
142. Ullrich T, Krich S, Binder D, Mereiter K, Anderson DJ, Meyer MD, Pyerin M (2002) J Med Chem 45:4047–4054
143. Lopez R, Sordo TL, Sordo JA, Gonzalez J (1993) J Org Chem 58:7036–7037
144. Ramos K, Banik BK (2011) Heterocyclic Lett 1:27–30
145. Bandyopadhyay D, Yanez M, Banik BK (2011) Heterocyclic Lett 1:65–67
146. Banik I, Samajdar S, Banik BK (2011) Heterocyclic Lett 1:69–72
147. Kidwai M, Sapra P, Bhushan KR, Saxen RK, Guptha R, Singh M (2000) Monatshefte Chem 131:85–90
148. Alcaide B, Vicente-Rodriguez A (1999) Tetrahedron Lett 40:2005–2006
149. Afonso A, Rosenblum SB, Puar MS, McPhail AT (1998) Tetrahedron Lett 39:7431–7434
150. Bolli MH, Marfurt J, Grisostomi C, Boss C, Binkert C, Hess P, Treiber A, Thorin E, Morrison K, Buchmann S, Bur D, Ramuz H, Clozel M, Fischli W, Weller T (2004) J Med Chem 47:2776–2795
151. Endo M, Droghini R (1993) Bioorg Med Chem Lett 3:2483–2486
152. Bose AK, Banik BK, Manhas MS (1995) Tetrahedron Lett 36:213–216
153. Arrieta A, Lecea B, Cossio FP (1998) J Org Chem 63:5869–5876
154. Cossio FP, Arrieta A, Lecea B, Ugalde JM (1994) J Am Chem Soc 116:2085–2093
155. Cossio FP, Ugalde JM, Lopez X, Lecea B, Palomo C (1993) J Am Chem Soc 115:995–1004
156. Hegedus LS, Montgomery J, Narukawa Y, Snustad DC (1991) J Am Chem Soc 113:5784–5791

Index

The manufacturer's authorised representative in the EU is Springer Nature Customer Service Centre GmbH, Europaplatz 3, 69115 Heidelberg, Germany. If you have any concerns regarding our products, please contact ProductSafety@springernature.com

Printed and bound by CPI Group (UK) Ltd, Croydon, CR0 4YY
25/09/2026
02222894-0008